Practical Recording Techniques, Fourth Edition

Practical Recording Techniques

Fourth Edition

Bruce Bartlett
Jenny Bartlett

ELSEVIER

Amsterdam • Boston • Heidelberg • London
New York • Oxford • Paris • San Diego
San Francisco • Singapore • Sydney • Tokyo

Focal Press is an imprint of Elsevier

Acquisitions Editor: Emma Baxter and Catharine Steers
Project Manager: Paul Gottehrer
Assistant Editor: Becky Golden-Harrell
Marketing Manager: Lucy Lomas-Walker
Cover Design: Eric DeCicco

Focal Press is an imprint of Elsevier
30 Corporate Drive, Suite 400, Burlington, MA 01803, USA
Linacre House, Jordan Hill, Oxford OX2 8DP, UK

Recognizing the importance of preserving what has been written, Elsevier prints its books on
acid-free paper whenever possible.

Library of Congress Cataloging-in-Publication Data
Application submitted

British Library Cataloguing-in-Publication Data
A catalogue record for this book is available from the British Library.

ISBN 13: 978-0-240-80685-3
ISBN 10: 0-240-80685-9

For information on all Focal Press publications
visit our website at www.books.elsevier.com

05 06 07 08 09 10   10 9 8 7 6 5 4 3 2 1

Printed in the United States of America

Working together to grow
libraries in developing countries

www.elsevier.com | www.bookaid.org | www.sabre.org

ELSEVIER   BOOK AID International   Sabre Foundation

To family, friends, and music.

TABLE OF CONTENTS

4 Equipping Your Studio   41

5 Monitoring 73

6 Microphones 87

7 Microphone Technique Basics   109

8 Microphone Techniques   131

9   Digital Recording   171

10   Effects and Signal Processors   201

13 Computer Recording   283

14 Judging Sound Quality   317

18 On-Location Recording of Classical Music   429

19 Surround Sound: Techniques and Media   439

D Further Education   521

E Impedance   527

PREFACE

Recording is a highly skilled craft combining art and science. It requires technical knowledge as well as musical understanding and critical listening ability. By learning these skills, you can capture a musical performance and reproduce it with quality sound for the enjoyment and inspiration of others.

Your recordings will become carefully tailored creations of which you can be proud. They will be a legacy that can bring pleasure to many people for years to come.

This book is intended as a hands-on, practical guide for beginning and intermediate recording engineers, producers, musicians—anyone who wants to make better recordings by understanding recording equipment and techniques. I hope to prepare the reader for work in a home studio, a small professional studio, or an on-location recording session.

Practical Recording Techniques offers up-to-date information on the latest recording technology, such as hard-disk and memory recorders, computer recording, keyboard and digital workstations, SMPTE and MIDI, surround sound, and audio for the Internet. But it also guides the beginner through the basics, showing how to make quality recordings with the new breed of inexpensive home-studio equipment.

The first chapter answers the question, "Why do we record?" Next, the book overviews the recording-and-reproduction chain to instill a system concept. The basics of sound and signals are explained so that you'll know what you're controlling when you adjust the controls on a piece of recording equipment. Then advice is given on equipping a home studio, from low budget to advanced.

Studio setup is covered next, including suggestions for improving your studio acoustics, choosing monitor speakers, and preventing hum.

Each piece of recording equipment is explained in detail, as well as the control-room techniques you'll use during actual sessions. Two chapters are devoted to the technology of digital recording and MIDI sequencing. A major chapter on computer recording covers the latest ways of creating and recording music. Two sections on remote recording cover techniques for both popular and classical music.

A special chapter explains how to judge recordings and improve them. The engineer must know not only how to use the equipment, but also how to tell good sound from bad.

The latest developments in recording are surround sound and audio for the Internet. Both these topics are covered in detail in their own chapters.

Finally five appendices explain the decibel, suggest how to optimize your computer for digital audio, introduce SMPTE time code, suggest further education, and explain impedance.

The CD included with this book demonstrates various topics explained in the book. Throughout the text, references to specific CD tracks guide the reader to relevant audio demonstrations.

Based on my work as a professional recording engineer, the book is full of tips and shortcuts for making great-sounding recordings, whether in a professional studio, project studio, on-location, or at home. You'll find many topics not covered in similar texts:

Loop-based recording

Hum prevention tips

The latest monitoring methods

Examples of mic models by type

Microphone selection guide

Tonal effects of microphone placement

Glossary of sound-quality descriptions

The latest types of digital recorders

Up-to-date coverage of computer recording

Optimizing your computer for digital audio

Documenting the recording session

Audio-for-video techniques

On-location recording

Troubleshooting bad sound; guidelines for good sound

Audio on the Web

Surround sound and DVD

ACKNOWLEDGMENTS

Thank you to Nick Batzdorf of Recording magazine for giving me permission to draw from my "Take One" series. For my education, thank you to the College of Wooster, Crown International, Shure Brothers Inc., Astatic Corporation, and all the studios I've worked for.

Thank you to Beth Howard, Emma Baxter, Becky Golden-Harrow, Paul Gottehrer, and Joanne Tracy at Focal Press/Elsevier for their fine work and support.

My deepest thanks to Jenny Bartlett for her many helpful suggestions as a layperson consultant and editor. She made sure the book could be understood by beginners.

A note of appreciation goes to the Pat Metheny Group and Samuel "Adagio for Strings" Barber, among many others, whose music inspired the chapter "Music: Why You Record."

Finally, to the musicians I've recorded and played with, a special thanks for teaching me indirectly about recording.

MUSIC: WHY YOU RECORD

As you learn about recording techniques for music, it's wise to remember that music is a wonderful reason for recording.

Music can be exalting, exciting, soothing, sensuous, and fulfilling. It's marvelous that recordings can preserve it. As a recording engineer or recording musician, it's to your advantage to better understand what music is all about.

Music starts as musical ideas or feelings in the mind and heart of its composer. Musical instruments are used to translate these ideas and feelings into sound waves. Somehow, the emotion contained in the music—the message—is coded in the vibrations of air molecules. Those sounds are converted to electricity and stored magnetically or optically. The composer's message manages to survive the trip through the mixing console and recorders; the signal is transferred to disc or computer files. Finally, the original sound waves are reproduced in the listening room, and miraculously the original emotion is reproduced in the listener as well.

Of course, not everyone reacts to a piece of music the same way, so the listener may not perceive the composer's intent. Still, it's amazing that anything as intangible as a thought or feeling can be conveyed by tiny magnetic patterns on a hard disk or by pits on a compact disc.

The point of music lies in what it's doing now, in the present. In other words, the meaning of "Doo wop she bop" is "Doo wop she bop." The meaning of an Am7 chord followed by a Fmaj7 chord is the experience of Am7 followed by Fmaj7.

Increasing Your Involvement in Music

Sometimes, to get involved in music, you must relax enough to lie back and listen. You have to feel unhurried, to be content to sit between your stereo speakers or wear headphones, and listen with undivided attention. Actively analyze or feel what the musicians are playing.

Music affects people much more when they are already feeling the emotion expressed in the song. For example, hearing a fast Irish reel when you're in a party mood, or hearing a piece by Debussy when you're feeling sensuous, is more moving because your feelings resonate with those in the music. When you're falling in love, any music that is meaningful to you is enhanced.

If you identify strongly with a particular song, that tells you something about yourself and your current mood. And the songs that other people identify with tell you something about them. You can understand individuals better by listening to their favorite music.

Different Ways of Listening

There are so many levels on which to listen to music—so many ways to focus attention. Try this. Play one of your favorite records several times while listening for these different aspects:

- Overall mood and rhythm
- Lyrics
- Vocal technique
- Bass line
- Drum fills
- Sound quality
- Technical proficiency of musicians
- Musical arrangement or structure
- Reaction of one musician to another musician's playing
- Surprises versus predictable patterns

By listening to a piece of music from several perspectives, you'll get much more out of it than if you just hear it as background. There's a lot going on in any song that usually goes unnoticed. Sometimes when you play an old familiar record and listen to the lyrics for the first time, the whole meaning of the song changes for you.

2

Most people react to music on the basic level of mood and rhythmic motivation. But as a recording enthusiast, you hear much more detail because your focus demands sustained critical listening. The same is true of trained musicians focusing on the musical aspects of a performance.

It's all there for anyone to hear, but you must train yourself to hear selectively and to focus attention on a particular level of the multidimensional musical event. For example, instead of just feeling excited while listening to an impressive lead-guitar solo, listen to what the guitarist is actually playing. You may hear some amazing things.

Here's one secret of really involving yourself in recorded music: Imagine yourself playing it! For example, if you're a bass player, listen to the bass line in a particular record, and imagine that you're playing the bass line. You'll hear the part as never before. Or respond to the music visually—see it as you do in the movie *Fantasia*.

Follow the melody line and see its shape. Hear where it reaches up, strains, and then relaxes. Hear how one note leads into the next. How does the musical expression change from moment to moment?

There are times when you can almost touch music: some music has a prickly texture (many transients, emphasized high frequencies); some music is soft and sinuous (sine-wave synthesizer notes, soaring vocal harmonies); and some music is airy and spacious (lots of reverberation).

Why Record?

Recording is a real service. Without it, people would be exposed to much less music. They would be limited to the occasional live concert or to their own live music, played once and forever gone.

With recordings, you can preserve a performance for thousands of listeners. You can hear an enormous variety of musical expressions whenever you want. Unlike a live concert, a record can be played over and over for analysis. Tapes or discs are also a way to achieve a sort of immortality. The Beatles may be gone, but their music lives on.

Records can even reveal your evolving consciousness as you grow and change. A tape or disc stays the same physically, but you hear it differently over the years as your perception changes. Recordings are a constant against which you can measure change in yourself.

Be proud that you are contributing to the recording art—it is done in the service of music.

2

THE RECORDING CHAIN

Welcome to the brave new world of 21st-century recording! The digital technologies of the past few years have given us possibilities undreamed of just 15 years ago. This book will show you an overview of current recording technology, help you choose the equipment that best suits your needs, and guide you in using it to create great recordings. And it will explain the technical jargon in plain English.

Thanks to the shift from analog to digital technology, the excitement and satisfaction of recording are accessible to more people than ever before. It used to take a whole roomful—or truckful—of expensive equipment to produce a good recording. But the new generation of smaller, cheaper gear means you may be able to tuck your studio into a corner of your bedroom or the back seat of your Toyota. As a result, many more people are involved in the process of recording—as musicians recording their own albums, or as engineers offering services to others.

As a recording engineer, you are a key player. Your skills help artists realize their visions in sound. Your miking techniques capture the vibrancy of the performance, whether it's the shimmering overtones of a string quartet or the sonic assault of an electric blues band. Your "post" work in the studio—adding effects, tweaking levels, etc.—will take the raw material of the performance and shape and blend it into a polished musical statement. Mastering the technology by becoming fluent with the audio tools at hand, you will produce exciting recordings that will delight your clients and give you a real sense of pride and achievement.

Be sure to practice what you learn in this book. There's no substitute for hands-on experience. You might offer to record a band's rehearsal for free while you experiment and master the gear. Be patient, let yourself make mistakes, and above all, listen to how the sound changes when you move a mic or tweak a knob.

Types of Recording

Let's get started. Currently there are six main ways to record music:

1. Live stereo recording: Record with a stereo microphone or two microphones into a recorder.

2. Live-mix recording: Pick up the musicians with several mics plugged into a mixer. Adjust the mic levels and record the mix into a recorder.

3. Separate multitrack recorder and mixer: Record with several mics into a mixer, which is connected to a multitrack recorder. Each track on the recorder contains the sound of a different instrument. After the recording is done, you mix or combine the tracks to stereo or surround.

4. Stand-alone Digital Audio Workstation (DAW, recorder-mixer): This is a multitrack recorder and a mixer combined in one portable chassis. The multitrack recording is done on a hard drive or MiniDisc.

5. Computer DAW: This system includes a computer, recording software, and an audio interface that gets audio into and out of your computer. You record on the computer's hard drive.

6. MIDI sequencing: A musician performs on a MIDI controller, such as a piano-style keyboard or drum pads. The controller puts out a MIDI signal, a series of numbers that indicates which keys were pressed and when they were pressed. The MIDI signal is recorded into computer memory by a sequencer. When you play back the sequence, it plays the tone generators in a synthesizer or plays samples: digital recordings of musical notes. Some recording software includes a sequencer application.

Let's look at each type of recording in more detail.

Live Stereo Recording

This method is most commonly used to record an orchestra, symphonic band, pipe organ, small ensemble, quartet, or soloist. The microphones pick up the overall sound of the instruments and the concert-hall acoustics. You might use this minimalist technique to record a folk group, acoustic jazz group, or classical-music ensemble in a good-sounding room.

Figure 2.1 shows the stages of this method—the links in the recording chain. Let's look at each stage from left to right (beginning to end).

1. The musical instruments or voices make sound waves.
2. The sound waves travel through the air and bounce or reflect off the walls, ceiling, and floor of the concert hall. These reflections add a pleasing sense of spaciousness.
3. The sound waves from the instruments and the room reach the microphones, which convert the sound into electrical signals.
4. The sound quality is greatly affected by mic technique. Microphone choice and placement are critical in this method because you have no way to adjust the sound later.
5. The signals from the microphones go to a 2-track recorder. It may be a hard-drive recorder, CD-R burner, DVD-R burner, Flash memory recorder, or computer hard drive. The signal changes to a pattern stored on a medium, such as magnetic patterns on a hard disk. During playback, the patterns on the medium are converted back into a signal.

As the medium moves during recording, signals are stored along a track—a path or channel on the medium containing a recorded signal.

Figure 2.1 The recording chain for live stereo recording.

One or more tracks can be recorded on a single medium. For example, a 2-track hard-disk recording stores two tracks on hard disk, such as the two different audio signals required for stereo recording.

6. To hear the signal you're recording, you need a monitor system: headphones or a stereo power amplifier and loudspeakers. You use the monitors to judge how well your mic technique is working.

The speakers or headphones convert the signal back into sound. This sound resembles that of the original instruments. Also, the acoustics of the listening room affect the sound reaching the listener.

Live-Mix Recording

Now let's look at a more complex way to record (Figure 2.2). This one is seldom used except for live broadcasts or recordings of PA mixes.

1. You use several microphones. Each one is placed close to each instrument or singer. As a result, each mic picks up very little room acoustics. This gives a close, clear sound that's desirable in recorded pop music or narration. For more clarity, you might add some sound-absorbent material on the floor, walls, and ceiling.

2. All the mics plug into a mixer, which blends all the microphone signals into one signal, stereo or mono. The mixer also has a volume control for each microphone. While listening to the mixer's signal, you adjust the volume of each instrument to make a pleasing loudness balance.

For example, if the guitar is too quiet relative to the voice, simply turn up the volume control for the guitar microphone until it blends well

Figure 2.2 The recording chain for a live-mix recording.

with the voice. That's a lot easier than grouping the musicians around a single microphone and moving the musicians until you hear a good balance.

Many mixers let you control other aspects of sound besides volume. You can control tone quality (bass and treble), stereo position (left, right, or center), and effects (such as artificial reverberation, which sounds like room acoustics). You monitor the mixer's signal with headphones or speakers.

3. When the mix sounds okay, you record the mixer's output signal with a 2-track recorder.

Separate Multitrack Recorder and Mixer

One problem with the previous setup is that you have to mix while the musicians are playing. If you make a mistake while mixing—say, one instrument is too quiet—the musicians have to play the song again until you get the balance right. And if you're recording a live gig, there's only one chance to perfect the mix.

The solution is to use a multitrack recorder, which records four or more tracks. It's as if several 2-track recorders were locked together. You record the signal of each microphone on its own track, then mix these recorded signals after the performance is done. You can either record a different instrument on each track or record different groups of instruments on each track. Figure 2.3 shows the stages in this method.

1. Place microphones near the instruments.

2. Plug the mics into a mixing console: a big, sophisticated mixer. During multitrack recording, the mixing console amplifies the weak

Figure 2.3 The recording chain for multitrack recording.

microphone signals up to the level needed by the recorder. This console is also used to send each microphone signal to the desired track.

3. Record the amplified mic signals on the multitrack recorder.

You can record more instruments later on unused tracks—a process called overdubbing. Wearing headphones, the performer listens to the recorded tracks and plays or sings along with them. You record the performance on an unused track.

After the recording is done, you will play all the tracks through the mixing console to mix them with a pleasing balance (Figure 2.4). Here are the steps:

1. Play back the multitrack recording of the song several times, adjusting the track volumes and tone controls until the mix is just the way you want it. You can add effects to enhance the sound quality. Some examples are echo, reverberation, and compression (explained in Chapter 10). Effects are made by signal processors that connect to your mixer, or by software applications that are part of a recording program.

2. Record or export your final mix on a 2-track stereo recorder (hard-disk recorder, memory recorder, CD-R, DVD-R, or computer hard drive).

Three types of a multitrack recorder are a Modular Digital Multitrack (MDM), a hard-disk recorder, and a Flash memory recorder. MDM recording, which is done on a videocassette, can be slow: you must fast-forward or rewind to the part you want to hear. In contrast, a hard drive

Figure 2.4 The recording chain for a multitrack mixdown.

and Flash memory have random access: you can instantly go to any part of the recorded program.

Stand-Alone DAW (Recorder-Mixer)

A recorder-mixer (Figure 2.5) combines a multitrack recorder and mixer in a single chassis. It's relatively easy to use. Other names for a recorder-mixer are "stand-alone Digital Audio Workstation," "digital multi-tracker," or "portable studio." The recorder is a hard drive, a MiniDisc recorder, or a Flash memory card.

Most recorder-mixers have built-in effects, or they can be used with outboard effects units. Figure 2.6 shows a small home studio setup using a recorder-mixer and outboard effects.

Computer DAW

This low-cost system includes a computer, recording software, and an audio interface that gets audio into and out of your computer (Figure 2.7). Four types of audio interface are

- A sound card that plugs into a slot in the computer.
- An I/O interface (sometimes called a breakout box): a chassis with input/output connectors, wired to your computer via a USB or FireWire port.
- A controller or control surface: a device resembling a mixer that controls the virtual controls you see on screen. Some controllers have input and output connectors built in.

Figure 2.5 A recorder-mixer.

Figure 2.6 A small home studio with a recorder-mixer and outboard effects.

Figure 2.7 Computer with a choice of audio interface and recording/editing software.

- A FirePort, a device by Alesis that accepts an Alesis-formatted hard drive with audio recordings on it, and converts the audio files to a signal sent via a FireWire connection to your computer.

Using the recording software, you perform these operations:

1. Record music on the computer's hard drive.
2. Edit the tracks to fix mistakes or to copy/move song sections.
3. Mix the tracks with a mouse or controller by adjusting virtual controls that appear on your computer screen.

You might also assemble a song from samples or from loops. Samples are recordings of single notes of various instruments. Loops are repeating musical patterns.

MIDI Sequencing

Like a player piano, MIDI sequencing records your performance gestures rather than audio. Figure 2.8 shows the process.

1. Play music on a MIDI controller, such as a piano-style keyboard, drum machine pads, or breath controller.
2. As you play the controller, it sends a MIDI signal from its MIDI OUT connector. This signal is a string of numbers—computer code—that tells which keys you pressed, when you pressed them, how fast you pressed them, and so on. In other words, the MIDI signal represents your performance gestures. It's not an audio signal.
3. The MIDI signal goes to a synthesizer or sound module, which might be part of a computer sound card. In these devices are tone

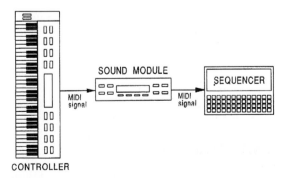

Figure 2.8 MIDI sequencing system.

generators that create musical sounds, such as a bass, piano, or drums. The MIDI signal plays the tone generators. You hear them with speakers or headphones.

The MIDI signal could also drive a drum machine or sampler. They contain samples, which are recordings in computer memory of single notes played by real musical instruments. The MIDI signal plays the samples—a process called wavetable synthesis. Samples are actually small wav files or aiff files.

4. The MIDI signal of the notes you play also goes to a sequencer—a device that records the MIDI signal. Sequencers come in three forms: a circuit built into a stand-alone synthesizer, a stand-alone sequencer device, or a computer sequencer program that records MIDI sequences on a hard drive.

5. When you play the sequencer recording, it activates the sound generators or samples to play the same notes you played. You can edit the sequencer recording; for example, fix wrong notes, change the instrument sounds without having to redo the performance, or change the tempo without changing the pitch.

MIDI/digital-audio software lets you record MIDI sequences and digital audio on hard disk. First record a few tracks of MIDI sequences onto hard disk, then add audio tracks (lead vocal, sax solo, or whatever). All these elements will stay synchronized.

Pros and Cons of Each Method

Live stereo recording is simple, cheap, and fast. But it usually sounds too muddy with rock music, and you must adjust balances by moving musicians. It can work well with classical music, and sometimes with folk or acoustic jazz music.

Live-mix recording is fairly simple and quick. However, loud instruments might sound distant in the recording because their sound "leaks" into distant mics. And if the mix or performance has mistakes, the band has to re-record the entire song. Also, the live sound of the band can make it hard to hear the monitored sound clearly.

Multitrack recording has many advantages. You can punch-in—fix a musical mistake by recording a new, correct part over the mistake. You can overdub—record one instrument at a time. This reduces leakage and gives a tighter sound. Also, you can postpone mixing decisions until after the performance. Then you can monitor the mix in quiet surroundings. This method is more complex and expensive than live-mix recording.

If you use a separate multitrack recorder and mixer, each component can be used independently. For example, you might do an on-location live recording with just the recorder, or do a PA job with just the mixer. Or, if you already have a mixer, all you need to buy is a recorder. This system is a little difficult to set up because you need to connect cables between the mixer and recorder and between the mixer and outboard effects units.

A stand-alone DAW (recorder-mixer) is easy to use because it is a single portable chassis that includes most of your studio equipment: recorder, mixer, effects, and often a CD burner. It doesn't require cables except for mics, instruments, and monitor speakers. High-end units let you edit the music. They also have automated mixing—memory chips in the mixer remember your mixdown settings—and reset the mixer accordingly the next time you play back the recording.

A computer DAW is inexpensive, powerful, and flexible. It lets you do sophisticated editing and automated mixing. Several plug-in (software) effects are included, and you can purchase and install other plug-ins. Recording software can be updated at little cost. As for drawbacks, computers can crash and can be difficult to set up and optimize for audio work.

MIDI sequencing lets you record musical parts by entering notes slowly or one at a time if necessary, then play the performance at a regular tempo. You can edit notes to correct mistakes and change the instrument sounds after recording the performance. However, you are limited to the sounds of samples and sound modules unless you use MIDI/digital-audio software, which lets you add miked instruments to the mix.

The enclosed CD contains samples of several types of recordings. *Play CD tracks 1 and 2.* Track 2 demonstrates:

- Live stereo recording—orchestra
- Live stereo recording—rock group
- Live mix recording—jazz group
- Multitrack recording—pop group
- MIDI sequencing—synthesizer funk

Editing/Mastering

No matter which recording method you use, eventually you'll mix all your songs to a 2-track recorder. Then you may want to edit those recorded mixes. Editing is the process where you remove noises and count-offs between songs, put the songs in the desired order, and put a few seconds of silence between songs. This is done with a computer and editing software (a DAW, Figure 2.7).

The last step is to copy the edited program to CD-R or DVD-R. There's your final product, ready to duplicate.

Quality Levels of Recording Formats

Here is a list of various recording formats from lowest sound quality to highest:

1. Cassette (becoming obsolete): Has some hiss, distortion, and wow and flutter that can make wobbly pitch.
2. Analog tape recorded at 7–1/2 ips or with narrow width tape (1/4 inch, 1/2-inch): This medium is becoming obsolete. Analog tape recordings can sound very good but have some tape hiss.
3. MiniDisc: This format uses data compression that degrades the sound quality slightly. Details are in Chapter 9.
4. Digital recording at 16 bits, 44.1 kHz: This is CD quality. Media that offer 16-bit recording are DAT, CD, MDM, hard disk, and Flash memory.
5. Digital recording at 24 bits, 44.1 kHz: Media that offer 24-bit recording are DVD, some MDM models, hard disk, and Flash memory.
6. Digital recording at 24 bits, 96 kHz, or 192 kHz: Media that offer this format are DVD, hard disk, and Flash memory. Another option is high-speed analog tape recording, perhaps with noise reduction. Another choice is Super Audio CD (SACD). It is considered by many to be state-of-the-art, as is 24-bit/192 kHz recording.
7. MIDI: If the sequencer recording is played by the same sound modules or samplers that were used during recording, it sounds exactly like the original performance. MIDI sequencer recordings reach the public on CD, DVD, or SACD media.

No matter what type of recording you do, each stage contributes to the sound quality of the finished recording. A bad-sounding master CD can be caused by any weak link: low-quality microphones, bad mic placement, improperly set mixer controls, and so on. A great-sounding recording results when you get every stage right. This book will help you reach that goal.

3

SOUND, SIGNALS, AND STUDIO ACOUSTICS

When you make a recording, you deal with at least two kinds of invisible energy: sound waves and electrical signals. For example, a microphone converts sound into a signal. A signal is a varying voltage that carries information. In our case, it's musical information.

This chapter covers some characteristics of sound and audio signals. These facts will help you work with room acoustics, and will help you know what's going on inside your mixer as you adjust the controls. With this knowledge you can make better recordings.

Sound Wave Creation

To produce sound, most musical instruments vibrate against air molecules, which pick up the vibration and pass it along as sound waves. When these vibrations strike your ears, you hear sound.

To illustrate how sound waves are created, imagine a vibrating speaker cone in a guitar amp. When the cone moves out, it pushes the adjacent air molecules closer together. This forms a compression. When the cone moves in, it pulls the molecules farther apart, forming a rarefaction. As shown in Figure 3.1, the compressions have a higher

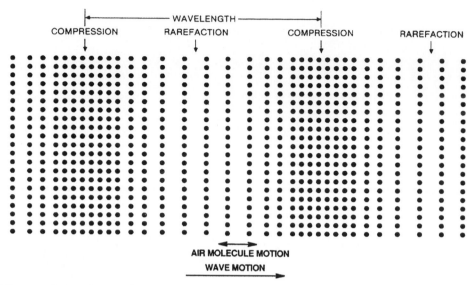

Figure 3.1 A sound wave.

pressure than normal atmospheric pressure; the rarefactions have a lower pressure than normal.

These disturbances pass from one molecule to the next in a spring-like motion—each molecule vibrates back and forth to pass the wave along. The sound waves travel outward from the sound source at 1130 feet per second (344 meters per second), which is the speed of sound in air at room temperature.

At some receiving point, such as an ear or a microphone, the air pressure varies up and down as the disturbances pass by. Figure 3.2 is a graph showing how sound pressure varies with time, like a wave. The high point of the graph is called a peak; the low point is called a trough. The horizontal center line of the graph is normal atmospheric pressure.

Sound waves tend to spread out as they travel away from the source. The compressions and rarefactions move out as expanding spheres. As the spherical waves expand, the sound pressure spreads over a larger area, so the pressure becomes weaker with distance from the source. This means that the farther you are from a sound source, the quieter the sound. Specifically, the sound pressure halves (drops 6 decibels; dB) each time the distance from the source doubles. This phenomenon is called the inverse square law.

Figure 3.2 Caption: Sound pressure vs. time of one cycle of a sound wave.

Figure 3.3 Three cycles of a wave.

Characteristics of Sound Waves

Figure 3.3 shows three waves in succession. One complete vibration from normal to high to low pressure and back to the starting point is called one cycle. The time it takes to complete one cycle—from the peak of one wave to the next—is called the period of the wave. One cycle is one period long.

Amplitude

The height of the wave is its amplitude. Loud sounds have high amplitudes (big pressure changes); quiet sounds have low amplitudes (small pressure changes). *Play track 3 on the enclosed CD to hear an example.*

Frequency

The sound source (in this case, the guitar-amp loudspeaker) vibrates back and forth many times a second. The number of cycles completed in one second is called frequency. The faster the speaker vibrates, the higher the frequency of the sound. Frequency is measured in hertz (Hz), which stands for cycles per second. One thousand hertz is called one kilohertz, abbreviated kHz.

The higher the frequency, the higher the perceived pitch of the sound. Low-frequency tones have a low pitch (like low E on a bass, which is 41 Hz). High-frequency tones have a high pitch (like four octaves above middle C, or 4186 Hz). *Track 4 on the enclosed CD illustrates this.* Doubling the frequency raises the pitch one octave.

Children can hear frequencies from 20 Hz to 20 kHz, and most adults with good hearing can hear up to 15 kHz or higher. Each musical instrument produces a range of frequencies, say, 41 Hz to 9 kHz for a string bass, or 196 Hz to 15 kHz for a violin.

Wavelength

When a sound wave travels through the air, the physical distance from one peak (compression) to the next is called a wavelength (refer to Figure 3.1). Low-pitched sounds have long wavelengths (several feet); high-pitched sounds have short wavelengths (a few inches or less). Wavelength is the speed of sound divided by frequency. So the wavelength of a 1000-Hz wave is 1.13 feet (0.344 m); 100 Hz is 11.3 feet (3.44 m), and 10 kHz is 1.35 inches (3.45 cm).

Phase and Phase Shift

The phase of any point on the wave is its degree of progression in the cycle—the beginning, the peak, the trough, or anywhere between. Phase is measured in degrees, with 360 degrees as one complete cycle. The beginning of a wave is 0 degrees; the peak is 90 degrees (one-quarter cycle), and the end is 360 degrees. Figure 3.4 shows the phases of various points on the wave.

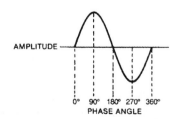

Figure 3.4 The phases of various points on a wave.

If there are two identical waves traveling together, but one is delayed with respect to the other, there is a phase shift between the two waves. The more delay, the more phase shift. Phase shift is measured in degrees. Figure 3.5 shows two waves separated by 90 degrees (one-quarter of a cycle) of phase shift. The dashed wave lags the solid wave by 90 degrees.

If you combine two identical sound waves, such as a sound and its reflection off a wall, the peaks of the two waves add together at certain points in the room. This doubles the sound pressure or amplitude, creating areas of louder sound at certain frequencies.

Phase Interference

When there is a 180-degree phase shift between two identical waves, the peak of one wave coincides with the trough of another (Figure 3.6). If these two waves are combined, they cancel out. This phenomenon is called phase cancellation or interference.

Suppose you have a signal with a wide range of frequencies, such as the singing voice. If you delay this signal and combine it with the original undelayed signal, some frequencies will be 180 degrees out of phase and will cancel. This makes a hollow, filtered tone quality.

Here's an example of how this can happen. Suppose you're recording a singer/guitarist with one mic near the singer and another mic near the guitar. Both mics pick up the singer. The singer's mic is close to the mouth, and you hear it with no delay in the signal. The guitar mic is farther from the mouth, so its voice signal is delayed. When you mix the two mics, you often hear a colored tone quality caused by phase cancellations between the two mics.

Suppose you're recording a stage play with a mic on a short stand on the floor. The mic picks up the direct sound from the actors, but it also

Figure 3.5 Two waves that are 90 degrees out of phase.

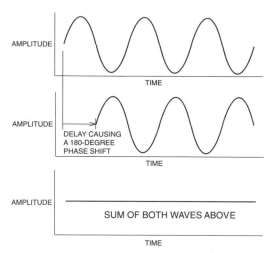

Figure 3.6 Phase interference: Adding two waves that are out of phase cancels the sound at that frequency.

picks up delayed reflections off the floor. Direct and delayed sounds combine at the mic, causing phase cancellations. You hear it as a hollow, filtered sound that changes when the actor walks while talking.

Harmonics

The type of wave shown in Figure 3.2 is called a sine wave. It is a pure tone of a single frequency, like a signal from a tone generator. In contrast, most musical tones have a complex waveform, which has more than one frequency component. All sounds are combinations of sine waves of different frequencies and amplitudes. Figure 3.7 shows sine waves of three frequencies combined to form a complex wave.

The lowest frequency in a complex wave is called the fundamental frequency. It determines the pitch of the sound. Higher frequencies in the complex wave are called overtones or upper partials. If the overtones are multiples of the fundamental frequency, they are called harmonics. For example, if the fundamental frequency is 200 Hz, the second harmonic is 400 Hz, and the third harmonic is 600 Hz.

The harmonics and their amplitudes help determine the tone quality or timbre of a sound, and help to identify the sound as a trumpet, piano, organ, voice, etc. *Play CD track 5.*

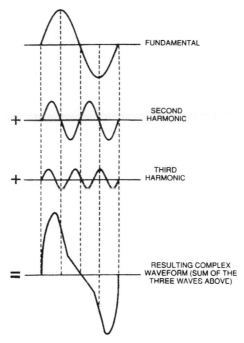

Figure 3.7 Adding fundamental and harmonic waveforms to form a complex waveform.

Noise (such as tape hiss) contains a wide band of frequencies and has an irregular, nonrepeating waveform.

Envelope

Another characteristic that identifies a sound is its envelope. When a note sounds, it doesn't last forever—it rises in volume, lasts a short time, then falls back to silence. This rise and fall in volume of one note is called the note's envelope. The envelope connects the peaks of successive waves that make up a note. Each musical instrument has a different envelope.

Most envelopes have four sections: attack, decay, sustain, and release (see Figure 3.8). During the attack, a note rises from silence to its maximum volume. Then it decays from maximum to some midrange level. This middle level is the sustain portion. During release, the note falls from its sustain level back to silence.

Percussive sounds, such as drum hits, are so short that they have only a rapid attack and decay. Other sounds, such as organ or violin

25

Figure 3.8 The four sections of the envelope of a note.

notes, last longer. They have slower attacks and longer sustains. Guitar plucks and cymbal crashes have quick attacks and slow releases. They hit hard then fade out slowly. *Play CD track 6.*

You can shorten a guitar string's decay or ringing by damping the string with the side of your hand. You can press a blanket against a kick drum head to damp the decay and get a tighter sound.

Behavior of Sound in Rooms

Because most music is recorded in rooms, you need to understand how room surfaces affect sound.

Echoes

Musical instruments make sound waves that travel outward in all directions. Some of the sound travels directly to your ears (or to a microphone) and is called direct sound. The rest strikes the walls, ceiling, floor, and furnishings of the recording room. At those surfaces, some of the sound is absorbed, some is transmitted through the surface, and the rest is reflected back into the room.

Because sound waves take time to travel (moving at about 1 foot per millisecond), the reflected sound reaches you after the direct sound. The reflection repeats the original sound after a short delay. If the sound is delayed about 50 msec or more, we call it an echo (Figure 3.9). In some concert halls we hear single echoes; in small rooms we often hear a short, rapid succession of echoes called flutter echoes. You can detect them by clapping your hand next to a wall. Flutter echoes happen when sound bounces back and forth between two parallel walls.

26

(A) Echo formation.

(B) Intensity vs. time of direct sound and its echoes.

Figure 3.9 Echoes. (A) Echo formation. (B) Intensity vs. time of direct sound and its echoes.

Reverberation

Sound reflects many times from all the surfaces in the room. These reflections sustain the sound of each note the musician plays. This persistence of sound in a room after the original sound has stopped is called reverberation (reverb). For example, reverberation is the sound you hear just after you shout in an empty gymnasium. The sound of your shout stays in the room and gradually dies away (decays). *Play CD track 10 to hear echoes and reverberation.*

Reverb is hundreds of echoes that gradually get quieter. The echoes follow each other so rapidly that they merge into a single continuous sound. Eventually, the room surfaces completely absorb the echoes. The timing of the echoes is random, and the echoes increase in number as they decay. Figure 3.10 shows how reverberation develops in a recording room.

Reverberation is a continuous fade-out of sound (HELLO-O-O-o-o), while an echo is a discrete repetition of a sound (HELLO hello hello hello).

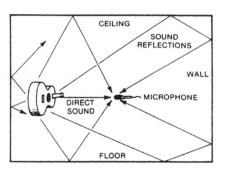

(A) Multiple sound reflections create reverberation.

(B) Intensity vs. time of direct sound, early reflections, and reverberation.

Figure 3.10 Reverberation. (A) Multiple sound reflections create reverberation. (B) Intensity vs. time of direct sound, early reflections, and reverberation.

Reverberation time (RT60) is the time it takes for reverb to decay 60 dB. Too long a reverb time makes a recording sound distant, muddy, and washed-out. That's why pop-music recordings are usually made in a fairly "dead" or nonreverberant studio, which has an RT60 of about 0.4 second or less. In contrast, classical music is recorded in "live," reverberant concert halls (RT60 about 1 to 3 seconds) because we want to hear reverb with classical music—it's part of the sound.

Reverberation comes to you from every direction because it is a pattern of many sound reflections off the walls, ceiling, and floor. Because we can tell where sounds come from, we can distinguish between the direct sound of an instrument coming to us from a single location and the reverberation coming to us from everywhere else. So we can ignore the reverb and concentrate on the sound source. In fact, we normally are not aware of reverberation.

But suppose you put a mic next to your ears, record an instrument in that room, and play back the recording. You'll hear a lot more reverb than what you heard live. What's going on? The reverb you recorded is not all around you. Instead, it's all up front between the speakers. So it's more audible; you can't discriminate against the reverb spatially. To reduce the amount of reverb in your recordings, you need to place mics close to instruments, and maybe add some sound-absorbing materials to the room.

Diffusion

When sound waves strike and reflect off a very bumpy or convex surface, they spread out or diffuse. This diffusion can be used to weaken sound reflections. Sound waves also spread out when they travel through a small opening.

Leakage

Sound from an instrument travels to the nearest mic, and also "leaks" into the mics intended for other instruments (Figure 3.11). This overlap of an instrument's sound into the mic of another instrument is called leakage (or bleed or spill).

It's very important to minimize leakage—to ensure that each mic picks up only its intended instrument. Suppose that you're miking a drum set and an acoustic piano. As the musicians play, you monitor what the mics pick up. When you turn up just the drum mics, the drums sound close-up or "tight." But when you also turn up the piano mic, the drum sound becomes distant or muddy. That's because the piano mic picks up drum leakage at a distance. *Play CD track 11 to hear an example of leakage.*

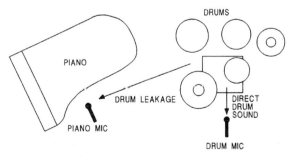

Figure 3.11 Example of leakage. The piano mic picks up leakage from the drums, changing the close drum sound to distant.

How to Tame Echoes, Reverb, and Leakage

Echoes, reverb, and leakage can make your recordings sound mushy and distant. There are two ways to prevent these problems: with recording techniques and with acoustic treatment.

Controlling Room Problems with Recording Techniques

Sometimes you can make clean recordings in an ordinary room—such as a club, living room, or basement—if you follow these suggestions:

- Mike close. Place each mic 1 to 6 inches from each instrument or voice. Then the mics will hear more of the instruments and less of the room reflections. You might want to use mini mics, which attach directly to instruments.

- Use directional mics—cardioid, supercardioid, or hypercardioid—which reject room acoustics.

- Record bass guitar and synth directly with a guitar cord or a direct box. Because you omit the microphone, you pick up no room acoustics. To get a good sound when recording electric guitar direct, record off the effects boxes or use a guitar-amp simulator.

- Overdub instruments one at a time rather than recording them all at once. You'll pick up a much cleaner sound. However, this loses the emotional interaction that occurs when all the musicians play together. You might record all the loud instruments at once: drums, bass, and electric guitar. Then overdub the quiet instruments: acoustic guitars, sax, piano, vocals.

- Record in a large room. This lets you spread the musicians farther apart, and weakens the sound reflections from the walls into the mics.

Controlling Room Problems with Acoustic Treatments

When should you apply acoustic treatment to a room or build a studio?

- You clap your hands next to a wall and you hear flutter echoes (a fluttering sound). These are caused by sounds bouncing back and forth between hard parallel walls.

- Your studio is a very live environment, such as a garage or concrete-block basement, so you hear too much room reverberation.
- Your studio is very small.
- You hear outside noises in your recordings.
- Bass-guitar amps and monitor speakers sound boomy.
- You want the freedom to mike several feet away without picking up noise or excess room reverb.
- You hear a lot of leakage in the mic signals.

If these conditions apply, check out the following suggestions on upgrading the acoustics of your studio.

Reverb and echoes are caused by sound reflections off room surfaces. So any surface that is highly sound-absorbent helps to reduce those problems.

To absorb high frequencies, use porous materials such as convoluted (bumpy) foam mattresses. They can be highly flammable, so cover them with flame-retardant treatment (such as at www.rosebrand.com). Nail or glue them to the walls, or mount them on frames. Thick foam works better than thin. Four-inch foam on the wall absorbs frequencies from about 200 to 800 Hz and up, depending on the angle at which sound strikes the foam. Leave some space between the foam panels. This helps to diffuse or spread out the sound in the room (Figure 3.12). Don't overdo the foam padding. A stuffed, dead room is uncomfortable to play in. Keep some reflections because they add "air" and liveliness to the sound.

Other high-frequency absorbers are sleeping bags, moving blankets, carpeting, curtains, and fiberglass insulation covered with muslin or burlap. If possible, space these materials several inches from the wall. The spacing helps them absorb mid-bass frequencies. A wide-range absorber is 4-inch pressed fiberglass board (Owens-Corning Type 703, $3 lb/ft^3$) covered in muslin or burlap.

Start with just a little absorption behind or above the musician you're recording. Add more absorbers, a few at a time, until your recordings sound as dead as you wish.

To absorb low frequencies, you can make bass traps. Here are four types:

1. Resonant tube trap: Take a 35- to 55-gallon rubber trashcan, stuff it with fiberglass insulation (wear a dust mask and gloves), and cover the open end with muslin or canvas. This tube trap absorbs frequencies near 1130/2H, where H is the height of the trashcan in feet.

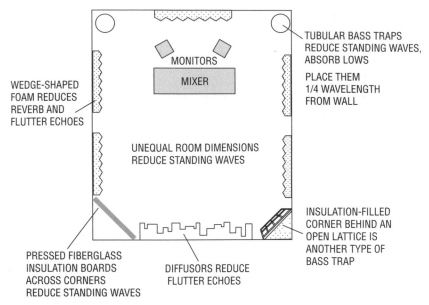

Figure 3.12 Acoustic treatments.

For example, a 3-foot-high trashcan absorbs 188 Hz. Placement is not critical.

2. Frictional tube trap: Make a 2-foot-diameter canvas bag, 8 feet tall, and fill it with fiberglass insulation. Hang one bag a few feet out from each room corner (Figure 3.12). The distance should be 1130/4F, where F is the frequency in hertz that you want to absorb. For example, to absorb 80 Hz, hang the bag 3.5 feet out from a room corner. (Thanks to David Moulton for those ideas.)

3. Lattice: Build a 3-foot-wide flat lattice frame of wood slats, the same height as the ceiling (Figure 3.12). Cover the frame with muslin or burlap. Put the frame diagonally across a room corner and fill the corner with R-30 insulation. Leave the foil on, foil side out toward the room. Put one assembly in each corner. (Thanks to Chips Davis for this idea.)

4. Insulation panel: Get 8 pieces of 2-foot × 4-foot × 4-inch rigid fiberglass insulation, type 705, from an insulation supplier. Cover each piece with muslin or burlap to contain the fibers. Place a piece across each room corner with the 2-foot edge touching the floor. Stack two

panels to make them 8 feet high. (Thanks to Ethan Winer for this idea.)

There are other ways to absorb bass. Wood paneling works well. It also helps to open closet doors and place couches and books a few inches from the walls. In a basement studio, nail acoustic tile to the ceiling joists with fiberglass insulation in the air space between tiles and ceiling.

You may not need any bass traps if you don't put any bass into the room. For example, don't turn up the bass-guitar amp—just record the bass direct and have the musicians wear headphones to hear the bass.

Controlling Standing Waves

Let's look at another acoustic problem: standing waves. If you play an amplified bass guitar through a speaker in a room, and do a bass run up the scale, you may hear some notes that boom out in the room. The room is resonating at those frequencies. These resonance frequencies, which are strongest below 300 Hz, occur in patterns called "standing waves." They can give a tubby or boomy coloration to musical instruments and monitor speakers.

Room resonances are worst in a cubical room. They are less of a problem if the room's length, width, and height are not multiples of each other. Table 3.1 shows several room dimensions in feet that reduce boomy-sounding standing waves.

For example, if the room width is 9.1 feet, and the ceiling is 8 feet high, the length should be 11.1 feet for best reduction of boominess.

Try to record in a large room because the room resonance frequencies are likely to be below the musical range. Use bass traps to absorb room resonances. Contrary to popular opinion, nonparallel walls don't prevent standing waves.

Making a Quieter Studio

The following tips will keep noises out of your recordings:

- Consider having the studio in a basement, because the earth blocks noises from the outside. The furnace or air-conditioning might be a problem, however.
- Turn off appliances and telephones while recording.
- Pause for ambulances and airplanes to pass.

Table 3.1 Room dimensions in feet to reduce standing waves

Height	Width	Length
8	9.1	11.1
8	9.4	11.8
8	10.1	11.3
8	10.2	12.3
8	11.6	16.8
8	11.8	13.6
8	12.8	18.6
8	13.0	21.0
10	11.4	13.9
10	11.7	14.7
10	12.6	14.1
10	12.8	15.4
10	14.5	21.0
10	14.7	17.0
10	16.0	23.3
10	16.2	26.2

- Close windows. Consider covering them with thick plywood.
- Close doors and seal with towels.
- Remove small objects that can rattle or buzz.
- Weather-strip doors all around, including underneath. (Leave the doors open for ventilation when not recording.)
- Replace hollow doors with solid doors.
- Block openings in the room with thick plywood and caulking.
- Put several layers of plywood and carpet on the floor above the studio, and put insulation in the air space between the studio ceiling and the floor above.
- Place microphones close to instruments and use directional microphones. This won't reduce noise in the studio, but it will reduce noise picked up by the microphones.

When building a new studio, you might want to make the walls of plastered concrete block because massive walls reduce sound transmission, or make the walls of gypsum board and staggered studs. Nail

Figure 3.13 Staggered-stud construction to reduce noise transmission.

gypsum board to 2 × 4 staggered studs on 2 × 6 footers as shown in Figure 3.13. Staggering the studs prevents sound transmission through the studs. Fill the air space between walls with insulation.

The ideal home-recording room for pop music is a large, well-sealed room with optimum dimensions. This room is in a quiet neighborhood. It should have some soft surfaces (carpet, acoustic-tile ceiling, drapes, couches), and some hard vibrating surfaces (wood paneling or gypsum board walls on studs).

Your home studio may not need acoustic treatment. Do some trial recordings to find out. But if your room could stand some improvement, the suggestions above should point you in the right direction.

For better results and a more professional appearance, consider buying some acoustic treatments from the following companies. Their Web sites are tubetrap.com, realtraps.com, acousticalsolutions.com, primacoustic.com, auralex.com, acousticsfirst.com, wallmate.net, illbruck-sonex.com, rpginc.com, and fstechnologies.com. Flame-retardant treatment for blankets and curtains is at www.rosebrand.com.

Signal Characteristics of Audio Devices

When a microphone converts sound to electricity, this electricity is called the signal. It has the same frequency and the same amplitude changes as the incoming sound wave.

When this signal passes through an audio device, the device may alter the signal. It might change the level of some frequencies or add unwanted sounds that are not in the original signal. Let's look at some of these effects.

Frequency Response

Suppose you have an audio device—a mic, mixer, effects unit, recorder, or speaker. You send a musical signal through the device. Usually the music contains some high and some low frequencies.

The device might respond differently to different frequencies. It might amplify the low notes and weaken the high notes. You can graph how the device responds to different frequencies by plotting its output level versus frequency. This graph is called a frequency response (Figure 3.14). The level in the graph is measured in dB, while frequency is measured in Hz. Generally, 1 dB is the smallest change in loudness that we can hear.

Suppose the level is the same at all frequencies. Then the graph forms a horizontal straight line and is called a "flat frequency response" (Figure 3.15). All the frequencies are reproduced at an equal level. In other words, the device passes all the frequencies without changing their relative levels. You get out the same amount of bass and treble that went in. A flat response does not affect the tonal balance of the incoming sound.

Many audio devices do not have a flat response across the audio band from 20 to 20,000 Hz. They have a limited range of frequencies that can be reproduced at an equal level (within a tolerance, such as ± 3 dB). In Figure 3.14, the frequency response shown by the solid line is 50 to 12,000 Hz ± 3 dB. That means the audio device passes all frequencies from 50 to 12,000 Hz at a nearly equal level—within 3 dB. It reproduces low sounds and high sounds equally well. The response is down 3 dB at 50 and 12,000 Hz, and is up 3 dB at 5000 Hz.

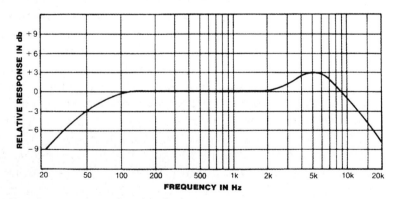

Figure 3.14 An example of a frequency response.

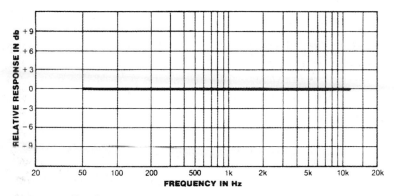

Figure 3.15 A flat frequency response.

Usually, the more extended or "wide" the frequency range is, the more natural and real the recording sounds. A wide, flat response gives accurate reproduction. A frequency response of 200 to 8000 Hz (± 3 dB) is narrow (poor fidelity); 80 to 12,000 Hz is wider (better fidelity), and 20 to 20,000 Hz is widest (best fidelity). *Play CD track 7.*

Also, the flatter the frequency response, the greater the fidelity or accuracy. A response deviation of ± 3 dB is good, ± 2 dB is better, and ± 1 dB is excellent. There are exceptions to this statement, which we'll look at in Chapter 10 in the section on equilizers.

When you turn a bass or treble knob on your guitar amp, mixer EQ, or stereo, you're changing the frequency response. If you turn up the bass, the low frequencies rise in level. If you turn up the treble, the high frequencies are emphasized. The ear interprets these effects as changes in tone quality—warmer, brighter, thinner, duller, and so on.

Figure 3.14 shows a non-flat frequency response. Toward the right side of this line, the response at high frequencies "rolls off" or declines. This shows that the upper harmonics are weak. The result is a dull sound. Toward the left side, the response at low frequencies rolls off. This means the fundamentals are weakened and the result is a thin sound.

The frequency response of an audio device might be made non-flat on purpose. For example, you might cut low frequencies with an equalizer to reduce breath pops from a microphone. Also, a microphone may sound best with a non-flat response, such as boosted high frequencies that add presence and sizzle.

Noise

Noise is another characteristic of audio signals. Every audio component produces a little noise—a rushing sound like wind in trees. Noise in a recording is undesirable unless it's part of the music.

You can make noise less audible by keeping the signal level in a device relatively high. If the level is low, you have to turn up the listening volume in order to hear the signal well. Turning up the volume of the signal also turns up the volume of the noise, so you hear noise along with the signal. But if the signal level is high, you don't have to turn up the listening level as much. Then the noise remains in the background.

Distortion

If you turn up the signal level too high, the signal distorts and you hear a gritty, grainy sound or clicks. This type of distortion is sometimes called "clipping," because the peaks of the signal are clipped off so they are flattened. To hear distortion, simply record a signal at a very high recording level (with the meters going well into the red area) and play it back. *Play CD track 9.* Digital recorders also produce quantization distortion at very low signal levels.

Optimum Signal Level

You want the signal level high enough to cover up the noise, but low enough to avoid distortion. Every audio component works best at a certain optimum signal level, and this is usually indicated by a 0 on a meter or lights that show the signal level.

Figure 3.16 shows the range of signal levels in an audio device. At the bottom is the noise floor of the device—the level of noise it produces with no signal. At the top is the distortion level—the point at which the signal distorts and sounds grungy. In between is a range in which the signal sounds clean. The idea is to maintain the signal around the 0 point on the average. With digital recorders, however, "0" is the maximum undistorted level.

Signal-to-Noise Ratio

The level difference in decibels between the signal level and the noise floor is called the "signal-to-noise ratio" or S/N (see Figure 3.16). The

Figure 3.16 The range of signal levels in an audio device.

higher the S/N, the cleaner the sound. An S/N of 60 dB is fair, 70 dB is good, and 80 dB or greater is excellent. *Play CD track 8.*

To illustrate S/N, imagine a person yelling a message over the sound of a train. The message yelled is the signal; the noise is the train. The louder the message, or the quieter the train, the greater the S/N. And the greater the S/N, the clearer the message.

Headroom

The level difference in decibels between the normal signal level and the distortion level is called "headroom" (see Figure 3.16). The greater the headroom, the greater the signal level the device can pass without running into distortion. If an audio device has a lot of headroom, it can pass high-level peaks without clipping them.

You want to set your mixer controls so that the signal has some headroom, is well above the noise floor, and is below distortion. Here's how.

During recording:

1. Set master faders and group faders to 0 (the shaded portion of fader travel).
2. Set one musician's mixer fader to 0.
3. Have the musician play the loudest part of the song.
4. Adjust the input trim to set the recording level to peak at about −3 dBFS maximum.
5. Repeat for the other faders and instruments.

During mixdown:

1. Set master faders and group faders to 0 (about three-quarters up).
2. Set up a mix with the channel faders.
3. Keeping the master and group faders near 0, adjust the channel faders' levels so that the mixer's stereo meters peak around 0. (The more faders in use, the lower the fader levels should be.)

At these settings, the signal levels in the mixer should be just about right. The signal should have no audible noise or distortion, and the mixer should have enough headroom so that loud peaks won't distort.

4

EQUIPPING YOUR STUDIO

You want to set up a recording system that's affordable, easy to use, and sonically excellent. With today's wide array of user-friendly sound tools, you can do just that. This chapter is a guide to equipment for a recording studio: what it does, what it costs, and how to set it up.

In this chapter we'll examine:

- Equipment and costs for budget studios
- Equipment for higher-end studios
- Details of each piece of equipment
- Cables and connectors
- Simple acoustic treatments
- Preventing hum and radio frequency interference

What is the bare-bones equipment you need to crank out a decent demo CD? How much does it cost? Thanks to the new breed of affordable equipment, you can put together a complete home recording studio for under $1200. That includes powered speakers, mics, recording software, and a sound card.

First you should know the general process for recording popular music:

1. Plug a mic or an instrument into a mic preamp. Connect the preamp into a track of a multitrack recorder. The preamp might be built into a mixer.

2. Record the instrument or vocal on one track.

3. Repeat this process with another instrument on another track, and so on—up to several tracks. Or record a group of instruments at once on multiple tracks.

4. Mix all these tracks with a mixer, combining them to 2 tracks of stereo music.

5. Record the stereo mix on a 2-track recorder. Repeat these steps for several songs.

6. Edit the mixes to put a few seconds of silence between the songs, and to put the songs in the desired order.

7. Copy the edited mixes onto a CD-R.

This process can be done in several ways; there are many equipment options. For example, you could record with a mini studio, MiniDisc recorder, hard-disk recorder, computer recording software, or a MIDI sequencer. I'll briefly explain what this equipment does.

Low-Cost Recording Equipment

A low-cost system includes microphones, powered monitor speakers, and a recording device.

Microphone

This device converts the sound of your voice (or any instrument) into an electrical signal that can be recorded. Microphone sound quality varies widely, so be sure to get one or more good mics costing at least $100 each. Condenser mics require phantom power, which is provided either by a phantom power supply or by XLR mic inputs in some mixers. Some condenser mics work on batteries. You'll also need a mic cable and at least one mic stand and boom costing about $35. If you want to record classical music in stereo, you'll need either a stereo mic, or a matched pair of high-quality condenser or ribbon mics of the same model number, plus a stereo mic-stand adapter.

Monitor System

Another important part of your studio is the monitor system—a pair of quality headphones or loudspeakers. You can use powered speakers, or use nonpowered speakers with a separate power amplifier. An essential tool, the monitor system tells you what you're doing to the recorded sound. The sound you hear over the monitors is approximately what the final listener will hear. Very good headphones are available for $100 and up, and good speakers cost about $400 a pair and up.

Recording Device

Four types of low-cost recording devices to choose from are a 2-track recorder, mini studio, 8-track recorder-mixer, and a computer with an audio interface and recording software. We'll look at each one.

2-Track Recorder

This device is suitable for on-location recording of classical music: an orchestra, symphonic band, string quartet, pipe organ, or soloist. Sometimes it can be used to record folk music or jazz. Some types of 2-track recorder are a hard-drive recorder, MiniDisc recorder, CD recorder, memory recorder, DAT, or laptop computer with recording software. Cost is $400 and up.

Mini Studio (4-Track Recorder/Mixer)

Also called a portable studio or pocket studio, this unit combines a 4-track recorder with a mixer—all in one portable chassis (Figure 4.1). It records in MP3 format to a memory card. A mini studio lets you record several instruments and vocals, then mix them to stereo or send them to your computer via USB. (A USB is a type of cable and ports for high-speed data transfer.) The sound quality is good enough to make demos, or to use as a musical scratchpad for ideas, but it is not quite good enough to release commercial albums. Costing about $300 and up, a mini studio can be a good choice for beginning recordists. Some manufacturers are Tascam, Boss, Digitech, Fostex, Korg, and Zoom (Samson).

Mini Studio features:

- Records on a memory card such as Compact Flash or SmartMedia.
- Sends the mix to your computer via a USB port for editing or CD burning.

Figure 4.1 Mini studio.

- Its internal MIDI sound module (synthesizer) plays back MIDI sequences. Includes MIDI files or rhythm patterns to jam with.
- Built-in effects.
- Built-in mic (in some models).
- Autopunch in/out.
- Battery or AC-adapter powered.
- Virtual tracks let you record multiple takes of a single performance, then select your favorite during mixdown.
- Guitar-amp modeling simulates various guitar amps; mic modeling simulates mic models.
- 2 mic inputs; records up to 2 tracks at a time. Plays back 4 tracks at once.

Digital Multitracker (8-Track Recorder-Mixer)

A step up from a mini studio, the digital multitracker combines a digital 8-track recorder with a mixer in a single chassis (Figure 4.2). It's convenient and portable. Plus, it offers CD sound quality and more tracks than the 4-track mini studios. The 8-track recording medium is a hard drive, Zip drive, MiniDisc, or Flash memory card. Other names for this device are stand-alone Digital Audio Workstation (DAW), portable digital studio, or recorder-mixer. The price is $400 to $1200.

Some features to look for:

Figure 4.2 Digital multitracker.

- Type of analog inputs: balanced or unbalanced. XLR and ¼-inch TRS (Tip-Ring-Sleeve) are balanced. RCA or ¼-inch TS (Tip-Sleeve) are unbalanced. Balanced inputs allow longer cable runs without picking up hum.
- Number of mic inputs: 2 to 8.
- Number of mixer channels: 8 and up.
- Number of simultaneous recording tracks: Two may be enough if you're recording only a track or two at a time, but a band may need to record all 8 tracks at once.
- Number of virtual tracks (recordings of separate takes of the same instrument): 8 and up.
- Automation (the mixer stores and resets your mixes).
- Phantom power for condenser mics.
- Number of effects processors (1 to 3).
- Midi Time Code (MTC) and Midi Machine Control (MMC), tempo map, and tap tempo.
- Backlit LCD display (bigger is better) with a waveform display.
- Built-in CD burner.
- A/D/A conversion: 16-bit is CD quality; 20- or 24-bit is better.
- Data compression (no data compression is preferred for higher sound quality).

Computer DAW

Another low-cost recording setup has three parts: a personal computer, an audio interface, and a digital-audio recording software (Figure 4.3).

HARD DRIVE

MONITOR

PCI SLOT, USB PORT, OR FIREWIRE PORT

COMPUTER

KEYBOARD

MOUSE

ALESIS FIREPORT

I/O INTERFACE (BREAKOUT BOX)

SOUND CARD

CONTROLLER WITH I/O

AUDIO INTERFACES (CHOOSE ONE)

Figure 4.3 Computer with a choice of audio interface and recording software.

MIDI/Digital-Audio software is the same, but also records MIDI sequences (explained later). The audio interface (such as a sound card) converts audio from your mic preamp or mixer into a signal that is recorded on the computer's hard disk as a magnetic pattern. Eight tracks or more can be recorded. The interface costs as little as $80 for a good sound card; pro-quality sound cards cost $200 or more. Low-cost recording software runs about $30 to $200. It's possible to make commercial-quality recordings with a computer recording system. Details are in Chapter 13, Computer Recording.

You mix the tracks with a mouse by adjusting the "virtual" controls that appear on your computer screen. Then you record the mix on your computer's hard drive.

Using a mouse can be fatiguing and can lead to repetitive stress syndrome. As an alternative to the mouse, you might buy a control surface ($400 and up). It looks like a mixer with faders, but it adjusts the virtual controls you see on the computer screen. That way you can use your computer for recording, and control the software with knobs and faders instead of a mouse.

After all your songs are mixed to two stereo tracks, you can use the software to edit the recording: remove noises and count-offs between songs, put the songs in order with a few seconds of silence between them, and match their perceived loudness. Then you use a CD-R burner to copy the edited program to a CD-R. There's your final product—ready to duplicate.

A computer studio costs about the same as a mini studio and is more powerful. It's a bargain. But because software requires computer skills, it's a little harder to learn and use than a hardware multitracker. Software recordings are at least CD quality—better than the MP3 recordings you get with a mini studio.

A computer studio can record MIDI (Musical Instrument Digital Interface) tracks as well as audio tracks. Using a piano-style keyboard or drum-machine pads, you play synthesized musical instruments—bass, drums, piano, etc. (Figure 4.4). Part of the recording software, called a MIDI sequencer, records the keystrokes that you play on the piano-style keyboard. When you play back the sequencer recording, it plays synthesized instruments from a sound card, sound module, synthesizer, or software synth.

As an alternative, use a keyboard workstation. This is a keyboard synth with a built-in sequencer and effects. The workstation lets you

Figure 4.4 MIDI sequencer recording with a computer.

create the entire musical backup on a stereo pair of channels, all in the keyboard. Then you record the audio from the keyboard into a computer. If you want to add a vocal, use MIDI/digital-audio software. Mix the tracks and burn a CD-R.

We've looked at several types of bare-bones recording setups. All can help you create quality demos. You can go much higher in price to get more features and better sound. For example, if you want to record an entire band playing all at once, with each instrument having its own mic, you'll need a system with more microphones, more tracks, and more headphones.

As we've seen, putting together a home studio or project studio doesn't have to cost much. As technology develops, better equipment is available at lower prices. That dream of owning your own studio is within reach.

Higher-Cost Recording Equipment

So far we've talked about devices that record 2 to 8 tracks. The next step up in price, quality, and flexibility are these:

- A hard-disk multitrack recorder, a separate mixer, and signal processors (Figure 4.5)
- A 16-, 24-, or 32-track recorder-mixer
- A high-end computer recording system

This equipment is good enough to record albums for commercial release. Let's look at the mixer first.

Figure 4.5 A studio using a separate multitrack recorder and mixer.

Mixer

A mixer (Figure 4.6) is an electronic device with an elaborate control panel. The mixer is connected to your multitrack recorder. You plug mics and electric instruments into the mixer, which amplifies their signals. While recording, you use the mixer to send those signals to the desired recorder tracks and to set recording levels. During mixdown, the mixer combines (mixes) the tracks to stereo. It also lets you adjust the sound quality of each track. The price is about $500 and up. A large, complex mixer is called a mixing console or board. Mixing consoles are explained in more detail in Chapter 11.

Hard-Disk Recorder (HD Recorder)

This device records up to 24 tracks on a built-in hard drive (Figure 4.7). Multiple recorders can be linked to get more tracks. Some examples are the Alesis ADAT HD24XR, Tascam MX-2424, Otari DR-100, iZ Technology RADAR, Fostex D-2424LV, and Mackie HDR 24/96.

METERS

OUTPUT
AND
MONITOR
SECTIONS

MASTER
FADERS

INPUT MODULES

Figure 4.6 A mixer.

49

Figure 4.7 A multitrack hard-disk recorder.

Figure 4.8 An effects unit.

Processors

Signal processors (Figure 4.8) are electronic devices (or software programs) that add special sonic effects such as reverberation, echo, chorus, and flanging. A compressor is a processor that acts like an automatic volume control. It turns down the vocal (or an instrument) when it is too loud—making it much easier to listen to.

HD Recorder-Mixer with 16 to 32 Tracks

As we said before, a multitrack HD recorder can be combined with a mixer in a single chassis, forming an HD recorder-mixer (Figure 4.9). It's also called a stand-alone DAW, portable digital studio, or digital multitracker. Effects and a CD burner are built in. Easy to use and connect, the recorder-mixer is a good tool for recording bands in the studio and in concert. Features to look for were given earlier in the above section Digital Multitracker (8-Track Recorder-Mixer). Some manufacturers are TASCAM, Akai, Korg, Fostex, Roland, Boss, and Yamaha.

High-End Recording Software and Hardware

The top DAW systems include elaborate recording software, and sometimes control surfaces and computer cards with DSP (Digital Signal Processing).

Figure 4.9 HD recorder-mixer.

Equipment Details

Now let's look more closely at each component of the recording studio.

Recorder

Here are some features found in all types of multitrack recorder.

Overdubbing

The musician plays along while listening to tracks already recorded and records a new part on an unused track. For example, suppose you've already recorded bass and drums and you want to add a guitar track. The performer listens to a headphone mix of the bass track, drum track, and his or her guitar signal. The musician plays the guitar while the bass and drum tracks play, and you record the guitar on an unused track.

Recording Two or More Tracks at Once

Recording several tracks simultaneously is useful for recording a live performance or an entire band at once. When you record in a studio, you often can record one or two tracks at a time. But when you record a live performance, you need to record all the tracks at once. Not all recorder-mixers let you do this.

Punch-In/Out

Use the punch-in and punch-out functions to fix mistakes. As the recorded track is playing, punch into record mode just before the mistake, play a new correct part that is recorded, and punch out of record mode

51

when you're finished. Most recorders accept a footswitch so you can punch-in with your foot while playing your instrument.

Bouncing Tracks

When you bounce (or ping-pong) tracks, you mix two or three tracks together and record the result on an unused track. Then you can erase the original tracks, freeing them for recording more instruments. This way you can record up to nine tracks with a 4-track machine. All recorder-mixers permit bouncing.

Pitch Control

Pitch control lets you adjust the recording speed up or down so you can match the pitch of recorded tracks to the pitch of new instruments to be recorded.

Locate

With this feature, certain locations in the recording can be stored in memory, such as the beginning of a verse or chorus. When you press "locate," the recorder goes to that location. You could use this feature to record repeatedly between two preset points, such as the beginning and end of a punch-in. A similar feature is return-to-zero, also called locate to zero, where the recorder goes to the beginning of the song. This feature makes it easy to practice a mix repeatedly.

Mixer

Described below are some of the features in mixers.

Mic Inputs

How many mics and mic inputs do you need? It depends on the instruments you want to record. If you want to mic a drum set, you might need 8 mics and 8 mic inputs, mixing those to 1 or 2 tracks. On the other hand, if you use MIDI instruments, you might need only one good mic for vocals and acoustic instruments. You can use one mic on several different instruments and vocals if you overdub them one at a time.

Mic input connectors are XLR or a 1/4-inch phone jack. An XLR connector looks like three small holes arranged in a triangle. If your mixer has such inputs, you can run long mic cables without picking up hum. (Hum is an unwanted low-pitched tone at 50 or 60 Hz caused by power wiring.) XLRs are found only in high-end units. Lower-cost recorder-mixers use 1/4-inch phone jacks (receptacles) for mic inputs, which are adequate for small studios. But to use them, you may need some female XLR-to-1/4-inch impedance-matching adapters. These are available at Radio Shack.

Insert Jacks

Insert jack connectors let you plug in a compressor in line with an input signal to reduce the dynamic range of that signal. A compressor is most often used on lead vocals.

Equalization

Equalization (EQ) means tone control. The simplest units have no EQ; you're stuck with the sound you get from your microphones. Most inexpensive units include a bass and treble control, one set per input. Fancier recorder-mixers have sweepable or semi-parametric EQ, which lets you continuously vary the frequency you want to adjust. This type of EQ offers the most control over the tone quality of each instrument you're recording.

Effects

A recording without effects sounds rather dead and plain. Effects such as reverberation, echo, and chorus can add sonic excitement to a recording. They are produced by devices called signal processors (see Figure 4.8) or by plug-ins, which are software effects used in a computer recording program.

The most essential effect is reverberation, a slow decay of sound such as you hear just after you shout in an empty gymnasium (HELLO-O-O-o-o-o . . .). Reverberation adds a sense of space; it can put your music in a concert hall, a small club, or a cathedral. This effect is usually produced by a digital reverb unit, available for $200 and up, or as a software plug-in.

Another popular effect is echo, a repetition of a sound (HELLO hello hello). It's made by a delay unit or delay plug-in, which also provides other effects such as chorus, doubling, and flanging.

A multieffects processor combines several effects in a single box. These effects can be heard one at a time or several at once. You can customize the sounds by pushing buttons to change the presets. (See Chapter 10 for more information on effects.)

Although effects are built into most recording software and recorder-mixers, an analog mixer needs to be used with external effects units. On the mixer is a set of connectors (labeled send and return) for hooking up an external effects unit, such as a reverb or delay device. A unit with one effects send lets you add one type of effect; a unit with two effects sends lets you add two different effects to create more sonic interest.

Microphones

Another essential item for the studio is a microphone. Good mics are needed for quality sound—and you get what you pay for. If you experiment with various types of microphones, you find big differences in fidelity. One or more microphones costing at least $100 each are recommended. It's false economy to use a cheap mic. You can't skimp here and expect to get quality sound. Any distortion or weird tone quality in the microphone may be difficult or impossible to remove later.

You may be able to borrow some good microphones, or use the ones you normally use for PA. Your ears should tell you if the fidelity is adequate for your purpose. Some people are happy to get any sound recorded; others settle for nothing less than professional sound quality.

Probably the most useful mic types for home recording are the cardioid condenser mic and cardioid dynamic mic. The cardioid pickup pattern helps reject room acoustics for a tighter sound. The condenser type is commonly used on cymbals, acoustic instruments, and studio vocals; dynamics are used typically on drums and guitar amps. (For more information on microphones, see Chapter 6.)

If you plan to record solo instruments or musical ensembles in stereo with two mics at a distance, you need two condenser mics of the same model number, or a stereo microphone. See Chapter 18 for details.

Phantom-Power Supply

A phantom-power supply powers the circuits in condenser mics. It uses the same cable that carries the mic's audio signal. You can omit the supply if your condenser mic has a battery, or if your mixer supplies phantom power.

Mic Preamp

This device amplifies a mic signal up to a higher voltage, called line level, which is needed by mixers and recorders. A stand-alone mic preamp provides a little cleaner sound than a mic preamp built into a mixer, but costs much more. Studios on a budget can do without it.

Direct Box

A direct box is a useful accessory for recorder-mixers with balanced XLR-type mic inputs. A direct box is a small device that connects an electric instrument (guitar, bass, synth) to a mixer's XLR-type mic input. It lets you record electric instruments directly into your mixer without a microphone. You can buy a direct box for as little as $50.

A direct box picks up a very clean sound, which may be undesirable for electric guitar. If you want to pick up the distortion of the guitar amp, use a microphone instead. Or use a guitar-amp modeling device or modeling plug-in.

Some recorder-mixers have "high-impedance" inputs meant for electric guitars. In this case, simply use a short guitar cord between your instrument and the mixer high-impedance input.

Monitor System

The monitor system lets you hear what you're recording and mixing. You can use a pair of high-quality headphones and a pair of loudspeakers and a power amplifier. The power amplifier strengthens the mixer's signal so it can drive loudspeakers. An alternative is a pair of powered monitors with built-in amplifiers.

The speakers should be accurate, high-fidelity types costing at least $200 each. Your home stereo might be good enough to serve, but skimping on a monitor system is not a smart move.

Nearfield studio monitor speakers (described in Chapter 5) are small, bookshelf-type speakers that are placed about 3 feet apart and 3 feet from you as you sit at your mixer.

If your monitor speakers are in the same room as your microphones, the mics pick up the sound of the speakers. This causes feedback or a muddy sound. In this case, it's better to monitor with headphones while recording.

If you're recording only yourself, one set of headphones is enough. But if you're recording another musician, you both need headphones. Many recorder-mixers have two headphone jacks for this purpose.

If you want to record or overdub several people at once, you need headphones for all of them. For example, if you're overdubbing three harmony vocalists, each one needs headphones to hear previously recorded tracks to sing with. To connect all these headphones, you could build a headphone junction box—an aluminum or plastic box that contains several headphone jacks. These are wired to a cable coming from your mixer's headphone jack. Or you could use a splitter cable, which makes two jacks out of one.

Rack and Patch Bay

A rack is an enclosure in which signal processors and other equipment are mounted. A patch bay or patch panel in a rack is a group of connectors that are wired to equipment inputs and outputs. Neither one is essential, but they are convenient. If you have a computer studio in which all the processing is done by software plug-ins, you may not need a patch bay.

Miscellaneous Equipment

Other equipment for your home studio includes mic cables, audio cables, USB or FireWire cables, power outlet strips, lighting, tables or studio furniture, mic pop filters, masking tape and a pen to label inputs and cables, contact cleaning fluid, DAT and MDM cleaning tapes, MIDI equipment stands, music stands, session forms, connector adapters, pen and paper, a flashlight, and so on.

Blank Recording Media

For your recorder you need some blank media to record on. Use the brand suggested by the recorder manufacturer. Listed below are the media used by various recorders:

Computer: Hard drive, CD-R, or CD-RW

HD recorder: A hard drive, which is built-in or removable

Digital multitracker (8-track recorder-mixer): A hard drive, Mini-Disc data disc, or a Flash memory card

Mini studio (4-track recorder-mixer): Flash memory card

Modular digital multitrack: S-VHS cassette (Alesis) or Hi-8 video cassette (Tascam).

MIDI Studio Equipment

MIDI studio equipment is covered in detail in Chapter 16. Here are some components in a typical MIDI studio that uses a piano-style synthesizer and a drum machine to make sounds (Figure 4.10).

MIDI Controller

A MIDI controller is a musical-performance device, such as a piano-style keyboard or drum pads, that puts out a MIDI signal when you play it. A MIDI signal is a string of numbers that tells which notes you played, when you played them, and so on.

Synthesizer

A synthesizer (synth) is a device or software that simulates the sound of real musical instruments or generates original sounds. There are several types. A stand-alone or hardware synthesizer has a piano-style keyboard and sound generators. When you play this instrument, it produces both a MIDI and an audio signal. Other types of synthesizers are a synth chip on a sound card, a stand-alone sound module, or a software synth.

Figure 4.10 One type of MIDI studio.

Sequencer

A sequencer is a device, or a computer program, that records the MIDI signal of your performance into computer memory for later editing and playback. It also records notes that you enter on a musical scale. The notes can be edited. A sequence is a computer .mid file of the notes you played, their timing, their volume, etc.

Sampler

A sampler is a device or a software plug-in that records and plays single notes, or short musical phrases, of real musical instruments. The sampler stores the samples in computer memory or on CD-Rs and floppy disks. To play the recorded samples, you trigger them with a MIDI controller or sequencer.

Sound Module

A sound module is a device that plays prerecorded samples or synthe-sized sounds when triggered by a MIDI controller or sequencer. It does not record samples.

Drum Machine

A drum machine is a device that simulates a drummer. It's a sequencer that records a drum performance done on its built-in pads. Each pad plays a different drum sound. When you play back your recorded per-formance, the samples play. This simulates a drum set and percussion. Many recorder-mixers have a drum machine built in.

Line Mixer

A line mixer is a small mixer that combines the signals of synths, sound modules, samplers, and drum machines. A line mixer has no mic preamps. If you want to use a mic with your MIDI studio, you also need a mic preamp or a mixer with mic preamps. If all your synthesizers are inside your computer, you don't need a line mixer.

MIDI Interface

A MIDI interface is a circuit card or device that connects to a computer, and has a MIDI IN connector and MIDI OUT connector. It converts a MIDI signal into computer data, and vice versa, so that you can record, edit, and play back a performance done on a MIDI controller. There are four types of MIDI interface: a MIDI card, a sound card with MIDI connectors, an I/O interface, and a control surface.

MIDI Software

Listed below are some types of MIDI software.

- Sequencer software ($30 to $1500): Records a performance done on a MIDI controller, or records notes that you enter on a musical scale. The notes can be edited.
- MIDI/digital-audio software ($30 to $1500): Also called digital-audio sequencer software. This software lets you record MIDI sequences (from a MIDI controller) and digital audio tracks (from mics) on a hard disk. The sequences and digital audio play at the same time in sync. Both can be edited.
- Notation software ($80 to $600): Displays your performance as a score on screen. You can edit the notes, type in lyrics and chords, and print out the score. Some recording software has notation included.
- Editor/librarian software ($200): Edits synth patches and stores them on disk.

Setting Up Your Studio

Once you have your equipment, you need to connect it together with cables, and possibly install equipment racks and acoustic treatment. Let's consider each step.

Cables

Cables carry electric signals from one audio component to another. They are usually made of one or two insulated conductors (wires) surrounded by a fine-wire mesh shield that reduces hum. Outside the shield is a plastic or rubber insulating jacket. On both ends of each cable is a connector (which comes in various types).

Cables are either balanced or unbalanced. A balanced line is a cable that uses two wires (conductors) to carry the signal surrounded by a shield (see Figure 4.11). Each wire has equal impedance to ground. An unbalanced line has a single conductor surrounded by a shield (see Figure 4.12). The balanced line rejects hum better than an unbalanced line,

Figure 4.11   A 2-conductor shielded, balanced line.

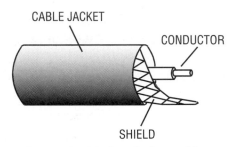

Figure 4.12   A 1-conductor shielded, unbalanced line.

but an unbalanced line less than 10 feet long usually provides adequate hum rejection and costs less.

A cable carries one of these four signal levels or voltages:

- Mic level (about 2 millivolts, or 0.002 volt)
- Electric guitar or keyboard level (about 0.1 volt)
- Line level (0.316 volt for unbalanced equipment, 1.23 volts for balanced equipment)
- Speaker level (about 20 volts)

The term "0.316 volt" also is known as "−10 dBV"; the term "1.23 volts" also is known as +4 dBu (see Appendix A).

Equipment Connectors

Recording equipment also has balanced or unbalanced connectors built into the chassis. Be sure your cable connectors match your equipment connectors.

Balanced equipment connectors include:

- 3-pin (XLR-type) connector (Figure 4.13)
- 1/4-inch TRS phone jack (Figure 4.14)

Unbalanced equipment connectors:

- 1/4-inch TS (Tip-Sleeve) phone jack (Figure 4.14)
- Phono jack (RCA connector) (Figure 4.15)

A jack is a receptacle; a plug inserts into a jack.

Cable Connectors

There are several types of cable connectors used in audio. Figure 4.16 shows a 1/4-inch mono phone plug (or TS phone plug) used with cables for unbalanced microphones, synthesizers, and electric instruments. The

Figure 4.13 A 3-pin XLR-type connector used in balanced equipment.

Figure 4.14 A 1/4-inch phone jack used in balanced and unbalanced equipment.

Figure 4.15 A phono (RCA) jack used in unbalanced equipment.

61

Figure 4.16 A mono (TS) 1/4-inch phone plug.

Figure 4.17 An RCA (phono) plug.

tip terminal is soldered to the cable's center conductor; the sleeve terminal is soldered to the cable shield.

Figure 4.17 shows an RCA or phono plug, used to connect unbalanced line-level signals. The center pin is soldered to the cable's center conductor; the cup terminal is soldered to the cable shield.

Figure 4.18 shows a 3-pin pro audio connector (XLR type). It is used with cables for balanced mics and balanced recording equipment. The female connector (with holes; Figure 4.18A) plugs into equipment outputs. The male connector (with pins; Figure 4.18B) plugs into equipment inputs. Pin 1 is soldered to the cable shield, pin 2 is soldered to the "hot" red or white lead, and pin 3 is soldered to the remaining lead. This wiring applies to both female and male connectors.

Figure 4.19 shows a stereo (TRS) phone plug used with stereo headphones and with some balanced line-level cables. For headphones, the tip terminal is soldered to the left-channel lead; the ring terminal is soldered to the right-channel lead; and the sleeve terminal is soldered to the common lead. For balanced line-level cables, the sleeve terminal is soldered to the shield; the tip terminal is soldered to the hot red or white lead; and the ring terminal is soldered to the remaining lead.

Some mixers have insert jacks that are stereo phone jacks; each jack accepts a stereo phone plug. Tip is the send signal to an audio device input, ring is the return signal from the device output, and sleeve is ground.

Figure 4.18 A 3-pin pro audio connector (XLR-type). (A) female. (B) male.

Figure 4.19 A stereo (TRS) phone plug.

If you have unbalanced microphone inputs (1/4-inch diameter holes) on your recorder or mixer, use a balanced cable (with XLRs on both ends) from mic to input. This reduces hum. At the mixer input, plug an impedance-matching adapter into the mic cable and into the mixer input. Sold at Radio Shack, the adapter has a female XLR connector on one end and a 1/4-inch phone plug on the other (see Figure 4.20).

Cable Types

Cables are also classified according to their function. In a studio, you'll use several types of cables: power, mic, MIDI, speaker, USB, FireWire, S/PDIF, Tascam TDIF, and Alesis Lightpipe cables. You'll also use guitar cords and patch cords, which are also cables.

A power cable (an AC extension cord or a power cord on a device) is made of three heavy-gauge wires surrounded by an insulating jacket. The wires are thick to handle lots of current.

A mic cable is usually 2-conductor shielded. It has two wires to carry the signal, surrounded by a fine-wire cylinder or shield that reduces hum

Figure 4.20 An impedance-matching adapter.

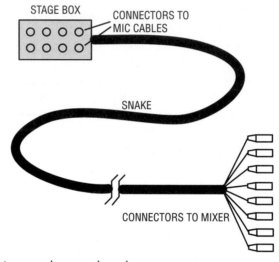

Figure 4.21 A stage box and snake.

pickup. On one end of the cable is a connector that plugs into the microphone, usually a female XLR-type. On the other end is either a 1/4-inch phone plug or a male XLR-type connector that plugs into your mixer.

Rather than running several mic cables to your recorder-mixer, you might consider using a snake—a box with multiple mic connectors—all wired to a thick multiconductor cable (Figure 4.21). A snake is especially convenient if you're running long cables to recording equipment. It's essential for most on-location recording.

Professional balanced equipment is interconnected with mic cable: 2-conductor shielded cable having a female XLR on one end and a male XLR on the other. Professional patch bays use balanced cables with TRS phone plugs.

A MIDI cable uses a 5-pin DIN connector on each end of a 2-conductor shielded cable. The cable connects MIDI OUT to MIDI IN or MIDI THRU to MIDI IN.

A speaker cable connects the power amp to each loudspeaker. Speaker cables are normally made of lamp cord (zip cord). To avoid wasting power, speaker cables should be as short as possible and should be heavy gauge (between 12- and 16-gauge). Number 12 gauge is thicker than 14; 14 is thicker than 16.

A USB cable or a FireWire cable connects a peripheral device (like an audio interface) to a computer. USB and FireWire are covered in detail in Chapter 13.

An S/PDIF cable transfers a digital signal from one device's S/PDIF output to another device's S/PDIF input. It uses a shielded unbalanced cable (ideally a 75-ohm RG59 video cable) with an RCA plug on each end.

A Tascam TDIF cable is a multiconductor cable with a 25-pin D-sub connector on both ends. It's used to connect multiple digital-audio signals from Tascam multitrack recorders to digital mixers or computer TDIF interfaces.

An Alesis Lightpipe cable is an optical cable with a Toslink connector on both ends. This cable is used to connect 8 channels of digital-audio signals from an Alesis multitrack recorder to a digital mixer or computer Lightpipe interface.

A guitar cord is made of 1-conductor shielded cable with a 1/4-inch phone plug on each end. It connects between a direct box and an electric musical instrument: guitar, bass, synthesizer, or drum machine.

Patch cords connect your recorder-mixer to external devices: an effects unit, a 2-track recorder, and a power amplifier. They also connect an analog mixer to the analog inputs and outputs of a multitrack recorder, usually as a snake that combines several cables. An unbalanced patch cord is made of 1-conductor shielded cable with either a 1/4-inch phone plug or a phono (RCA) connector on each end. A stereo patch cord is two patch cords joined side by side.

Rack/Patch Bay

You might want to mount your signal processors in a rack—a wooden or metal enclosure with mounting holes for equipment (Figure 4.22). You also might want to install a patch panel or patch bay: a group of connectors that are wired to equipment inputs and outputs. Using a patch bay and patch cords, you can change equipment connections easily. You

Figure 4.22 A rack and patch panel.

also can bypass or patch around defective equipment. Note that patch bays increase the chance of hum pickup slightly because of the additional cables and connectors.

Figure 4.23 shows some typical patch-panel assignments.

Equipment Connections

The instruction manuals of your equipment tell how to connect each component to the others. In general, use cables that are as short as possible to reduce hum, but that are long enough to let you make changes.

Be sure to label all your cables on both ends according to what they plug into; for example, TRACK 6 OUT or REVERB IN. If you label your cables, when you change connections temporarily, or the cable becomes unplugged, you'll know where to plug it back in. A piece of masking tape folded over the end of the cable makes a stay-put label.

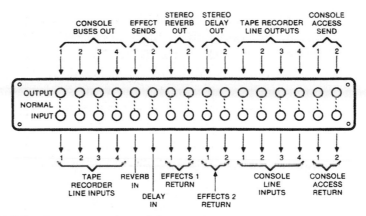

Figure 4.23 Some typical patch-bay assignments.

Figure 4.24 Typical connections for recording-studio equipment.

Typically, you follow this procedure to connect equipment (see Figure 4.24):

1. Plug the AC power cords of audio equipment and electric musical instruments into AC outlet strips fed from the same circuit breaker. Plug the power amplifier into its own outlet on the same breaker so that it receives plenty of current.

2. Connect mics and direct boxes to mic cables.

3. Connect mic cables either to the snake junction box or directly into mixer mic inputs. Connect the snake connectors into mixer mic

inputs. If your mixer has phone-jack mic inputs, you may need to use an impedance-matching adapter (female XLR to phone) between the mic cable and the mic input jack (Figure 4.20).

4. Set the output volume of synthesizers and drum machines about three-quarters up. Connect the synthesizer's and drum machine's audio outputs to mixer line inputs. If this causes hum, use a direct box.

5. If the mixer is a stand-alone unit (not part of a recorder-mixer), connect the mixer 2-track or tape outputs to the inputs of a 2-track recorder (DAT, computer audio interface, or MiniDisc recorder).

6. Connect the 2-track recorder outputs to the mixer's 2-track or tape inputs.

7. Connect the mixer's monitor outputs to the power-amp inputs. Connect the power-amp outputs to loudspeakers. Or if you are using powered (active) monitors, connect the mixer monitor outputs to the monitor-speaker inputs.

8. If the mixer does not have internal effects, connect the mixer aux-send connectors to effects inputs (not shown). Connect the effects outputs to the mixer aux-return or bus-in connectors.

9. If you're using a separate mixer and multitrack recorder, connect mixer bus 1 to recorder track 1 IN; connect bus 2 to track 2, and so on. Also connect the recorder's track 1 OUT to the mixer's line input 1; connect the track 2 OUT to the mixer's line input 2, and so on. As an alternative, connect insert jacks to multitrack inputs and outputs. At each insert plug, connect the tip (send) terminal to a track input and connect the ring (return) terminal to the same track's output.

10. If you have several headphones for musicians, connect the cue output to a small amplifier to drive their headphones. Or if the mixer's headphone signal is powerful enough, connect it to a box with several headphone jacks wired in parallel.

Semi-pro studio equipment with unbalanced connectors (usually phone or RCA) operates at a level called −10 or −10 dBV. Pro studio equipment with balanced connectors (XLR or TRS) works at +4 or +4 dBu. Check your equipment manuals to determine their input and output levels. When you connect devices that run at different levels, set the +4/−10 switch on each unit to match the levels. If there is no such switch on either device, connect between them a +4/−10 converter box such as the Ebtech Line Level Shifter (www.ebtechaudio.com). Or try the cables shown in Appendix A, Figures A.3 and A.4.

Figure 4.25 shows a typical layout for a DAW recording studio that uses the same equipment just described.

A recorder-mixer studio can be quite simple. It omits the external multitrack recorder, computer, computer monitor and keyboard, audio interface, and outboard effects.

Acoustic Treatment

Along with the recording equipment, you need a quiet room to record in—the larger the better. A basement, living room, or bedroom can work. Many garages have been turned into studios.

If the room has a lot of hard surfaces, you might need to place some material in the studio to absorb sound reflections. This treatment reduces the reverberation in the room and gives a clearer recorded sound. For budget or improvised studios, the acoustic treatment has to be limited. Try surrounding the instrument and its mic with thick blankets or sleeping bags hung a few feet away. Maybe hang them on ropes tied between room fixtures. Carpet the floor and nail some convoluted (bumpy) mattress foam to wood-paneled walls (make sure it is fire-proofed). Put muslin-covered 2-ft × 4-ft pieces of 705 fiberglass insulation across each

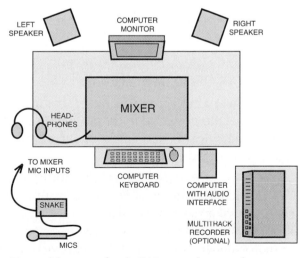

Figure 4.25 Typical layout of a DAW recording studio.

corner to absorb bass. You might prefer to use some commercially made sound absorbers.

Add absorbers spaced evenly around the walls until your recordings sound reasonably dry (free of audible room reverberation). There's more on acoustic treatments in Chapter 3.

Note: By using close miking, direct boxes, and overdubs, you might be able to make good recordings in a room without any acoustic treatment.

Hum Prevention

You patch in a piece of audio equipment, and there it is—HUM! It's a low-pitched tone or buzz. This annoying sound is a tone at 60 Hz (50 Hz in Europe) and multiples of that frequency.

Hum is caused mainly by

- Cables picking up magnetic and electrostatic hum fields radiated by power wiring—especially if the cable shield connection is broken.
- Ground loops. A ground loop is a circuit made of ground wires. It can occur when two pieces of equipment are connected to ground through a power cord, and also are connected to each other through a cable shield. The ground voltage may be slightly different at each piece of equipment, so a 50- or 60-Hz hum signal flows between the components along the cable shield.

These are the most important points to remember about hum prevention:

- To prevent ground loops, plug all equipment into outlet strips powered by the same breaker.
- Some power amps create hum if they don't get enough AC current. So connect the power amp (or powered speakers) AC plug to its own wall outlet socket—the same outlet that feeds the outlet strips for the recording equipment.
- If possible, use balanced cables going into balanced equipment. Balanced cables have XLR or TRS connectors and two conductors surrounded by a shield. Ideally, the shield should be connected to the chassis ground (not the signal ground) at both ends of the cable.
- Transformer-isolate unbalanced connections. If that is not an option, use the cable assemblies shown in Figures 3 and 4 in Appendix A.

- Don't use dimmers to change the studio lighting levels. Use multi-way incandescent bulbs instead.

Even if your system is wired properly, a hum or buzz may appear when you make a connection. Follow these tips to stop the hum:

- If the hum is coming from a direct box, flip its ground-lift switch.
- Check cables and connectors for broken leads and shields.
- Unplug all equipment from each other. Start by listening just to the powered monitor speakers. Connect a component to the system one at a time, and see when the hum starts.
- Remove audio cables from your devices and monitor each device by itself. It may be defective.
- Turn down the volume on your power amp (or powered speakers), and feed them a higher-level signal.
- Use a direct box instead of a guitar cord between instrument and mixer.
- To stop a ground loop when connecting two devices, connect between them a 1:1 isolation transformer or hum eliminator (such as Jensen or Ebtech). See Figures 3 and 4 in Appendix A.
- Add ground-lift adapters to line-level balanced cables at the male XLR end. **Caution**: a lifted (disconnected or "telescoping") shield can act as an RF antenna. To prevent RFI pickup, solder a 0.001-microfarad (μF) capacitor between the lifted shield and XLR pin 1.
- Make sure that the snake box is not touching metal.
- Do not connect XLR pin 1 to the connector shell. Tighten the mic-connector screws.
- Try another mic.
- If you hear a hum or buzz from an electric guitar, have the player move to a different location or aim in a different direction. You might also attach a wire between the player's body and the guitar strings near the tailpiece to ground the player's body.
- Turn down the high-frequency EQ on a buzzing bass guitar track.
- To reduce buzzing between notes on an electric-guitar track, apply a noise gate.
- Route mic cables and patch cords away from power cords; separate them vertically where they cross. Also keep recording equipment

71

and cables away from computer monitors, power amplifiers, and power transformers.

- See Rane's excellent article on sound system interconnections at www.rane.com.

By following all these tips, you should be able to connect audio equipment without introducing any hum. Good luck!

Reducing Radio Frequency Interference

Radio frequency interference (RFI) is heard as buzzing, clicks, radio programs, or "hash" in the audio signal. It's caused by CB transmitters, computers, lightning, radar, radio and TV transmitters, industrial machines, auto ignitions, stage lighting, and other sources. Many of the following techniques are the same used to reduce hum from other sources. To reduce RFI:

- If you think that a speaker cable, mic cable, or patch cord is picking up RFI, wrap the cable several times around an RFI choke (available at Radio Shack). Put the choke near the device that is receiving audio.
- Install high-quality RFI filters in the AC power outlets. The cheap types available from local electronics shops are generally ineffective.
- If a cable shield is floating (disconnected) at one end, solder a 0.001-μF capacitor between XLR pin 1 and the shield.
- If a MIC is picking up RFI, solder a 0.047 μF capacitor between pin 1 and 2, and between pin 1 and 3, in the female XLR connector of the MIC cable.
- Periodically clean connector contacts with Caig Labs De-Oxit, or at least unplug and plug them in several times.

This chapter briefly covered the equipment and connectors for a recording studio. The rest of this book explains each piece of equipment in detail and tells how to use it for best results.

5

MONITORING

One of the most exciting moments in recording comes when the finished mix is played over the studio monitor speakers. The sound is so clear you can hear every detail, and so powerful you can feel the deep bass throbbing in your chest.

You use the monitor system to listen to the output signals of the console or the recorders. It consists of the console monitor mixer, the power amplifiers, loudspeakers, and the listening room. The power amplifier boosts the electrical power of the console signal to a sufficient level to drive a loudspeaker. The speaker converts the electrical signal into sound, and the listening-room acoustics affect the sound from the speaker.

A quality monitor system is a must if you want your mixes to sound good. The power amp and speakers tell you what you're doing to the recorded sound. According to what you hear, you adjust the mix and judge your mic techniques. Clearly, the monitor system affects the settings of many controls on your mixer, as well as your mic selection and placement. And all those settings affect the sound you're recording. So, using inadequate monitors can result in a poor-sounding product coming out of your studio.

It's important to use accurate speakers that have a flat frequency response. If your monitors are weak in the bass, you will tend to boost the bass in the mix until it sounds right over those monitors. But when that mix is played over speakers with a flatter response, it will sound too

73

bassy because you boosted the bass on your mixer. So, using monitors with weak bass results in bassy recordings; using monitors with exaggerated treble results in dull recordings, and so on. In general, colorations in the monitors will be inverted in your mixdown recording.

That's why it's so important to use an accurate monitor system—one with a wide, smooth frequency response. Such a system lets you hear exactly what you recorded.

Speaker Requirements

The requirements for an accurate studio monitor are these:

- **Wide, smooth frequency response**. To ensure accurate tonal reproduction, the on-axis response of the direct sound should be ±4 dB or less from 40 Hz to 15 kHz. The low-frequency response of a small monitor speaker should extend to at least 70 Hz.

- **Uniform off-axis response**. The high-frequency output of a speaker tends to diminish off-axis. Ideally the response at 30 degrees off-axis should be only a few decibels down from the response on-axis. That way, a producer and engineer sitting side-by-side will hear the same tonal balance. Also, the tonal balance will not change as the engineer moves around at the console.

- **Good transient response**. This is the ability of the speaker to accurately follow the attack and decay of musical sounds. If a speaker has good transient response, the bass guitar sounds tight, not boomy and drum hits have sharp impact. Some speakers are designed so that the woofer and tweeter signals are aligned in time. This aids transient response.

- **Clarity and detail**. You should be able to hear small differences in the sonic character of instruments, and to sort them out in a complex musical passage.

- **Low distortion**. Low distortion is necessary because it lets you listen to the speaker for a long time without your ears hurting. A good spec might be: Total harmonic distortion under 3% from 40 Hz to 20 kHz at 90 dBSPL (sound pressure level).

- **Sensitivity**. Sensitivity is the sound pressure level a speaker produces at 1 meter (m) when driven with 1 watt (W) of pink noise. Pink noise is random noise with equal energy per octave. This noise is either band-limited to the range of the speaker or is a one-third-octave band centered at 1 kHz. Sensitivity is measured in dB/W/m (dB sound pressure level per 1 W at 1 m). A spec of 93 dB/W/m is considered high; 85 dB/W/m is low. The higher the sensitivity, the less amplifier power you need to get adequate loudness.

- **High output capability**. This is the ability of a speaker to play loudly without burning out. You often need to monitor at high levels to hear quiet details in the music. Plus, when you record musicians who play loudly in the studio, it can be a letdown for them to hear a quiet playback. So you may need a maximum output of 110 dBSPL.

This formula calculates the maximum output of a speaker (how loud it can play):

$$dBSPL = 10 \log(P) + S$$

where dBSPL is the sound pressure level at 1 m, P is the continuous power rating of the speaker in watts, and S is the sensitivity rating in dB/W/m.

For example, if a speaker is rated at 100 W maximum continuous power, and its sensitivity is 94 dBSPL/W/m, its maximum output SPL is $10 \log(100) + 94 = 114$ dBSPL (at 1 m from the speaker). The level at 2 m will be about 4 to 6 dB less.

Nearfield™ Monitors

Many professional recording studios use large monitor speakers that have deep bass. However, they are expensive, heavy, and difficult to install, and they are affected by the acoustics of the control room. If you want to avoid this hassle and expense, consider using a pair of Nearfield monitor speakers (Figure 5.1). A Nearfield monitor is a small, wide-range speaker typically using a cone woofer and dome-shaped tweeter. You place a pair of them about 3 or 4 feet apart, on stands just behind the console, about 3 or 4 feet from you. Nearfields are far more popular than large wall-mounted speakers.

This technique, developed by audio consultant Ed Long, is called Nearfield monitoring. Because the speakers are close to your ears, you hear mainly the direct sound of the speakers and tend to ignore the room acoustics. Plus, Nearfield monitors sound very clear and provide sharp stereo imaging. Some units have bass or treble tone controls built in to compensate for the effects of speaker placement and room surfaces.

Figure 5.1 A Nearfield monitor speaker.

Nearfield monitors have enough bass to sound full when placed far from walls. Although most Nearfields lack deep bass, they can be supplemented with a subwoofer to reproduce the complete audio spectrum. Or you can check the mix occasionally with headphones that have deep bass.

Some Nearfields are in a satellite-subwoofer format. The two satellite speakers are small units, typically including a 4-inch woofer and 3/4-inch dome tweeter. The satellites are too small to produce deep bass, but that is handled by the subwoofer—a single cabinet with one or two large woofer cones. Typically, the subwoofer (sub) produces frequencies from 100 Hz down to 40 Hz or below. Because we do not localize sounds below about 100 Hz, all the sound seems to come from the satellite speakers. The sub-satellite system is more complicated to set up than two larger speakers, but offers deeper bass.

Powered (Active) Monitors

Some monitors have a power amplifier built in. You feed them a line-level signal (labeled MONITOR OUT) from your mixing console. Most powered monitors are bi-amplified: they have one amplifier for the woofer and another for the tweeter. The advantages of bi-amplification include:

- Distortion frequencies caused by clipping the woofer power amplifier will not reach the tweeter, so there is less likelihood of tweeter burnout if the amplifier clips. In addition, clipping distortion in the woofer amplifier is made less audible.
- Intermodulation distortion is reduced.
- Peak power output is greater than that of a single amplifier of equivalent power.
- Direct coupling of amplifiers to speakers improves transient response—especially at low frequencies.
- Bi-amping reduces the inductive and capacitive loading of the power amplifier.
- The full power of the tweeter amp is available regardless of the power required by the woofer amp.

The Power Amplifier

If your monitor speakers are not powered, you need a power amplifier (Figure 5.2). It boosts your mixer's line-level signal to a higher power in order to drive the speakers.

How many watts of power do you need? The monitor speaker's data sheet gives this information. Look for the specification called "Recommended amplifier power." A power amp of 50 W per channel continuous is about the minimum for Nearfield monitors; 150 W is better. Too much power is better than too little, because an underpowered system is likely to clip or distort. This creates high frequencies that can damage tweeters.

A good monitor power amp has distortion under 0.05% at full power. It should have a high damping factor—at least 100—to keep the bass tight. The amp should be reliable. Look for separate level controls for left and right channels. The amplifier should have a clip or peak light that flashes when the amp is distorting.

Figure 5.2 A power amplifier.

Speaker Cables and Polarity

When you connect the power amp to the speakers, use good wiring practice. Long or thin cables waste amplifier power by heating. So put the power amp(s) close to the speakers and use short cables with thick conductors—at least 16 gauge. The low resistance of these cables helps the power amplifier to damp the speaker motion and tighten the bass.

If you wire the two speakers in opposite polarity, one speaker's cone moves out while the other speaker's cone moves in. This causes vague stereo imaging, weak bass, and a strange sense of pressure on your ears. Be sure to wire the speakers in the same polarity as follows: In both channels, connect the amplifier positive (+ or red) terminal to the speaker positive (+ or red) terminal. Setting the correct polarity is also called "speaker phasing."

Control-Room Acoustics

The acoustics of the control room affect the sound of the speakers. Sound waves leaving each speaker strike the room surfaces. At those surfaces, some frequencies are absorbed, while other frequencies are reflected. At your ears, the sound waves reflected from the room surfaces combine with the direct sound from the speakers. Reflections that arrive within 20 to 65 milliseconds (msec) after the direct sound blend with the direct sound and affect the tonal balance you hear.

Suppose the walls are covered with carpet or blankets so that they absorb only the high frequencies. Then the walls will reflect mainly the low frequencies. When you listen to a speaker playing in such a room, you hear the direct sound from the speaker plus the bassy wall reflections. The combined sound will be bass-heavy. Now suppose the walls are made of wood paneling mounted on studs. Such a vibrating surface absorbs lows and reflects highs. The sound you hear probably will be thin and overly bright.

Clearly, the room surfaces should reflect (or absorb) all frequencies about equally to avoid coloring the sound of the speakers. Equal absorption (±25%) from about 250 to 4000 Hz is usually adequate. As described in Chapter 3, you can use flexible panels or bass traps to absorb lows, in combination with fibrous materials or foam to absorb highs, or use thick fibrous material spaced from the wall and ceiling.

Room resonances or standing waves can cause some bass notes to blare out and cause other notes to disappear. Be sure to control these resonances as suggested in Chapter 3.

Room acoustics also affect the decay-in-time of the sound coming from the speakers. Whenever the speakers play a note that ends suddenly, the sound of that note continues to bounce around the room. This causes echoes and reverberation that prolong the sound. This long decay of sound is not part of the recording. So the control room should be relatively dead; that is, it should have a short reverberation time. A typical living room has a reverb time of about 0.4 seconds; the control room should, too, so the engineer will hear about the same amount of room reverb that a home listener will hear. A totally dead room is uncomfortable to listen in.

Using Nearfield monitors makes the room acoustics less important, but it still helps to treat the room acoustics. To prevent sound reflections from the wall behind the speakers, apply muslin-covered fiberglass insulation or acoustic foam. This treatment improves the monitors' sound. Stereo imaging and depth are greatly improved, the sound is clearer, and the frequency response is flatter. The treatment will reduce boominess and ringing, and make transients sharper. Also, your recordings will translate better to other speakers.

If your control room is separate from the studio, the control room should be built to keep out sound from the studio. You want to hear only the sound from the monitors, not the live sound from the musicians. In a home studio, you can achieve isolation simply by putting the control-room equipment in a room far removed from the studio, with the doors closed.

A control room built next to the studio needs good isolation. Use double-wall construction with staggered studs (see Figure 5.3). Put

Figure 5.3 Staggered-stud construction to reduce noise transmission.

fiberglass insulation between the two walls. The door between the two rooms should be solid wood and should be weatherstripped all around—including underneath. Use a double-pane window (mounted in rubber) between the control room and studio.

In some home or project studios, the control room is the same room as the studio. Because no isolation is used, the cost of building the studio is much less. The engineer records while listening with headphones, and does the critical monitoring during playback and mixdown.

Speaker Placement

Once you have acquired the speakers and worked on the room acoustics, you can install the speakers. Mount them at ear height so the mixer doesn't block their sound. To prevent sound reflections off the mixing console, place the speakers on stands behind the console's meter bridge, rather than putting them on top. For best stereo imaging, align the speaker drivers vertically and mount the speakers symmetrically with respect to the side walls. Place the two speakers as far apart as you're sitting from them; aim them toward you, and sit exactly between them (Figure 5.4). To get the smoothest low-frequency response, put the speakers near the shorter wall, and sit forward of the halfway point in the room.

Try to position the monitors several feet from the nearest wall. Wall reflections can degrade the frequency response and stereo imaging. The closer to the wall the monitors are, the more bass you hear. In small rooms you might have to place the monitors against the wall, which will exaggerate the bass. But some monitors have a low-frequency attenuation switch to compensate.

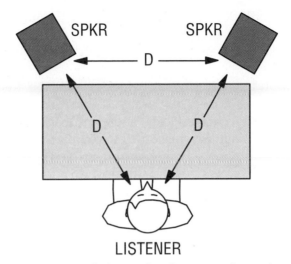

Figure 5.4 The recommended speaker/listener relationship for best stereo imaging.

Using the Monitors

You've treated the room acoustics, and you've connected and placed the speakers as described earlier. Now it's time to adjust the stereo balance.

1. Play a mono musical signal into an input channel on your mixer, and assign it to the stereo output channels 1 and 2.

2. Adjust the input channel pan pot so that the signal reads the same on the stereo output channel 1 and 2 meters.

3. Place the two speakers the same distance from you.

4. Sit at the mixer exactly midway between the speakers. If you sit off-center, you will hear the image shifted toward one side. Listen to the image of the sound between the speaker pair. You should localize it midway between the monitors; that is, straight ahead.

5. If necessary, center the image by adjusting the left or right volume control on your power amp.

When you do a mixdown, try to keep the listening level around 85 dBSPL—a fairly loud home listening level. As discovered by Fletcher and Munson, we hear less bass in a program that is played quietly than in the same program played loudly. If you mix a program while monitoring at, say, 100 dBSPL, the same program will sound weak in the bass when heard at a lower listening level—which is likely in the home. So, programs meant to be heard at 85 dBSPL should be mixed and monitored at that level.

Loud monitoring also exaggerates the frequencies around 4 kHz. A recording mixed loud may sound punchy, but the same recording heard at a low volume will sound dull and lifeless.

Here's another reason to avoid extreme monitor levels: **Loud sustained sound can damage your hearing or cause temporary hearing loss at certain frequencies.** If you must do a loud playback for the musicians (who are used to high SPLs in the studio), protect your ears by wearing earplugs or leaving the room.

You can get a low-cost sound level meter from Radio Shack. Play a musical program at 0 VU or 0 dB on the mixer meters and adjust the monitor level to obtain an average reading of 85 dBSPL on the sound level meter. Mark the monitor-level setting.

Before doing a mix, you may want to play some familiar commercial CDs over your monitors to remind yourself what a good tonal balance sounds like. Listen to the amount of bass, midrange, and treble, and try to match those in your mixes. But listen to several CDs, because they vary.

While mixing, monitor the program alternately in stereo and mono to make sure there are no out-of-phase signals that cancel certain frequencies in mono. Also beware of center-channel buildup: Instruments or vocals that are panned to center in the stereo mix sound 3 dB louder when monitored in mono than they do in stereo. That is, the balance changes in mono—the center instruments are a little too loud. To prevent this, don't pan tracks hard left and hard right. Bring in the side images a little so they will be louder in mono.

You'll mix the tracks to sound good on your accurate monitors. But also check the mix on small inexpensive speakers to see whether anything is missing or whether the mix changes drastically. Make sure that bass instruments are recorded with enough edge or harmonics to be audible on the smaller speakers. It's a good idea to make a cassette or CD copy of the mix for listening in a car, boom box, or compact stereo.

Headphones

Many home studios have the mixer and monitors in the same room as the musicians. In this case, you monitor with quality headphones while recording and overdubbing, then monitor with speakers when you mix. If you're monitoring as the musicians are playing, block out their sound by using closed-cup headphones or in-the-ear earphones. You may find that headphones provide adequate isolation if the music you're recording is quiet. Some popular headphones for studio monitoring are the Sony MDR-7506, AKG K240, and Sennheiser HD 280 Pro.

Compared to speakers, headphones have several advantages:

* They cost much less.
* There is no coloration from room acoustics.
* The tone quality is the same in different environments.
* They are convenient for on-location monitoring.
* It's easy to hear small changes in the mix.
* Transients are sharper due to the absence of room reflections.

Headphones have several disadvantages:

* They become uncomfortable after long listening sessions.
* Cheap headphones have inaccurate tone quality.
* Headphones don't project bass notes through your body.
* The bass response varies due to changing headphone pressure.
* The sound is in your head rather than out front.
* You hear no room reverberation, so you may add in too much or too little artificial reverb.
* It's difficult to judge the stereo spread. Over headphones, panned signals tend not to sound as far off center as the same signals heard over speakers. The same is true of stereo recordings made with a coincident pair of mics.

Because speakers sound different from headphones, it's best to do mixes over speakers.

The Cue System

The cue system is a monitor system for musicians to use as they're recording. It includes some of the aux knobs in your mixer, a small power amplifier, a headphone connector box, and headphones. Musicians often can't hear each other well in the studio, but they can listen over headphones to hear each other in a good balance. Also, they can listen to previously recorded tracks while overdubbing.

Headphones for a cue system should be durable and comfortable. They should be closed-cup to avoid leakage into microphones. This is an ideal situation; open-air phones may work well enough. Also, the cue "phones" should have a smooth response to reduce listening fatigue, and should play loud without burning out. Make sure they are all the same model so each musician hears the same thing. A built-in volume control is convenient.

A suggested cue system is shown in Figure 5.5. Connect a power amp to an aux-send or monitor output of your mixer. The amp drives several resistor-isolated headphones, which are in parallel.

If your mixer has a strong signal at its headphone jack, you can get by with four headphone jacks in a small metal box wired in parallel and connected to the mixer's headphone jack.

Although some consoles can provide several independent cue mixes, the ideal situation is to set up individual cue mixers near each musician. Then they can set their own cue mix and listening level. The inputs of these mixers are fed from the console output buses.

Suppose a vocalist sings into a microphone and hears that mic's signal over the cue headphones. If the singer's voice and the headphone's sound are opposite in polarity, the voice partially cancels or sounds funny in the headphones. Make sure that the voice and headphones are the same polarity.

Here's how. While talking into a mic and listening to it on headphones, reverse the ground and signal leads to the headphones connec-

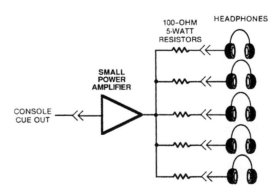

Figure 5.5 A cue system.

tor. The position that gives the fullest, most solid sound in the head-phones is correct.

All the headphones in your studio should be the same model, so that everyone will hear with correct polarity.

Conclusion

Ultimately, what you hear from the monitors influences your recording techniques and affects the quality of your recordings. So take the time to plan and adjust the control-room acoustics. Choose and place the speakers carefully. Monitor at proper levels and listen on several systems. You'll be rewarded with a monitor system you can trust.

MICROPHONES

What microphone is best for recording an orchestra? What's a good snare mic? Should the microphone be a condenser or dynamic, omni or cardioid?

You can answer these questions more easily once you know the types of microphones and understand their specs. First, it always pays to get a high-quality microphone—which costs at least $100. The mic is a source of your recorded signal. If that signal is noisy, distorted, or tonally colored, you'll be stuck with those flaws through the whole recording process. Better get it right up front.

Even if you have a MIDI studio and get all your sounds from samples or synthesizers, you still might need a good microphone for sampling, or to record vocals, sax, acoustic guitar, and so on.

A microphone is a transducer—a device that changes one form of energy into another. Specifically, a mic changes sound into an electrical signal. Your mixer amplifies and modifies this signal.

Transducer Types

Mics for recording can be grouped into three types depending on how they convert sound to electricity: dynamic, ribbon, or condenser.

A dynamic mic capsule, or transducer, is shown in Figure 6.1. A coil of wire attached to a diaphragm is suspended in a magnetic field. When

Figure 6.1 A dynamic transducer.

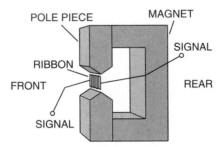

Figure 6.2 A ribbon transducer.

sound waves vibrate the diaphragm, the coil vibrates in the magnetic field and generates an electrical signal similar to the incoming sound wave. Another name for a dynamic mic is moving-coil mic, but this term is seldom used.

In a ribbon mic capsule, a thin metal foil or ribbon is suspended in a magnetic field (Figure 6.2). Sound waves vibrate the ribbon in the field and generate an electrical signal.

A condenser or capacitor mic capsule has a conductive diaphragm and a metal backplate placed very close together (Figure 6.3). They are charged with static electricity to form two plates of a capacitor. When sound waves strike the diaphragm, it vibrates. This varies the spacing

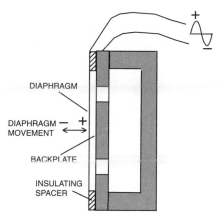

DIAPHRAGM

DIAPHRAGM —
MOVEMENT

BACKPLATE

INSULATING
SPACER

Figure 6.3 A condenser transducer.

between the plates. In turn, this varies the capacitance and generates a signal similar to the incoming sound wave. Because of its lower diaphragm mass and higher damping, a condenser mic responds faster than a dynamic mic to rapidly changing sound waves (transients).

Two types of condenser mic are true condenser and electret condenser. In a true condenser mic (externally biased mic), the diaphragm and backplate are charged with a voltage from a circuit built into the mic. In an electret condenser mic, the diaphragm and backplate are charged by an electret material, which is in the diaphragm or on the backplate. Electrets and true condensers can sound equally good, although some engineers prefer true condensers, which tend to cost more.

A condenser mic needs a power supply to operate, such as a battery or phantom power supply. Phantom power is 12 to 48 volts DC applied to pins 2 and 3 of the mic connector through two equal resistors. The microphone receives phantom power and sends audio signals on the same two conductors. Many mixing consoles supply phantom power at their mic input connectors. You simply plug the mic into the mixer to power it. Dynamics and ribbons need no power supply. You can plug these types of mics into a phantom supply without damage, unless either signal conductor is accidentally shorted to the mic housing. Figure 6.4 shows a cutaway view of a typical dynamic vocal mic and condenser instrument mic.

89

Figure 6.4 Inside a typical dynamic vocal mic and condenser instrument mic.

General Traits of Each Transducer Type

Condenser

- Wide, smooth frequency response
- Detailed sound, extended highs
- Omni type has excellent low-frequency response
- Transient attacks sound sharp and clear
- Preferred for acoustic instruments, cymbals, studio vocals
- Can be miniaturized

Dynamic

- Tends to have rougher response, but still quite usable
- Rugged and reliable
- Handles heat, cold, and high humidity
- Handles high volume without distortion
- Preferred for guitar amps and drums
- If flat response, can take the "edge" off woodwinds and brass

Ribbon

- Prized for its warm, smooth tone quality
- Delicate
- Complements digital recording

There are exceptions to the tendencies listed above. Some dynamics have a smooth, wide-range frequency response. Some condensers are rugged and handle high SPLs. It depends on the specs of the particular mic. *Track 13 on the enclosed CD demonstrates the sound of each transducer type.*

Polar Pattern

Microphones also differ in the way they respond to sounds coming from different directions. An omnidirectional microphone is equally sensitive to sounds arriving from all directions. A unidirectional mic is most sensitive to sound arriving from one direction—in front of the mic—but softens sounds entering the sides or rear of the mic. A bidirectional mic is most sensitive to sounds arriving from two directions—in front of and behind the mic—but rejects sounds entering the sides.

There are three types of unidirectional patterns: cardioid, supercardioid, and hypercardioid. A mic with a cardioid pattern is sensitive to sounds arriving from a broad angle in front of the mic. It is about 6 dB less sensitive at the sides, and about 15 to 25 dB less sensitive in the rear. The supercardioid pattern is 8.7 dB less sensitive at the sides and has two areas of least pickup at 125 degrees away from the front. The hypercardioid pattern is 12 dB less sensitive at the sides and has two areas of least pickup at 110 degrees away from the front.

To hear how a cardioid pickup pattern works, talk into a cardioid mic from all sides while listening to its output. Your reproduced voice is loudest when you talk into the front of the mic, and softest when you talk into the rear. *Play CD track 14.*

The super- and hypercardioid reject sound from the sides more than the cardioid. They are more directional, but they pick up more sound from the rear than the cardioid does.

A microphone's polar pattern is a graph of its sensitivity versus the angle at which sound comes into it. The polar pattern is plotted on polar graph paper. Sensitivity is plotted as distance from the origin. Figure 6.5 shows various polar patterns.

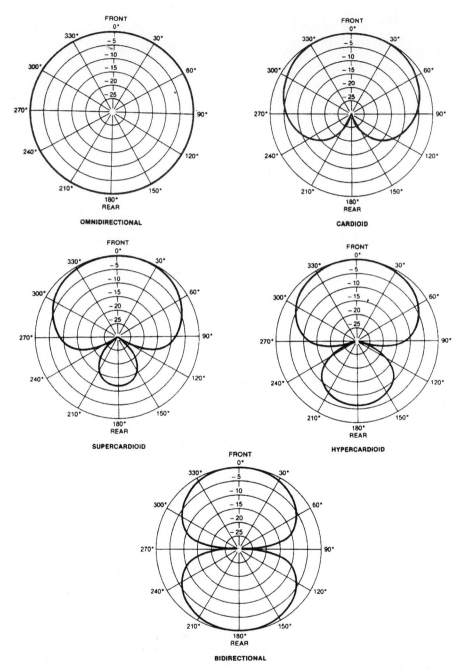

Figure 6.5 Various polar patterns. Sensitivity is plotted vs. angle of sound incidence.

Traits of Various Polar Patterns

Omnidirectional

- All-around pickup
- Most pickup of room reverberation (*play CD track 14*)
- Not much isolation unless you mike close
- Low sensitivity to pops (explosive breath sounds)
- Low handling noise
- No up-close bass boost (proximity effect)
- Extended low-frequency response in condenser mics—great for pipe organ or bass drum in an orchestra or symphonic band
- Lower cost in general

Unidirectional (cardioid, supercardioid, hypercardioid)

- Selective pickup
- Rejection of room acoustics, background noise, and leakage
- Good isolation—good separation between tracks
- Up-close bass boost (except in mics that have holes in the handle)
- Better gain-before-feedback in a sound-reinforcement system
- Coincident or near-coincident stereo miking (explained in Chapter 7)

Cardioid

- Broad-angle pickup of sources in front of the mic
- Maximum rejection of sound approaching the rear of the mic

Supercardioid

- Maximum difference between front hemisphere and rear hemisphere pickup (good for stage-floor miking)
- More isolation than a cardioid
- Less reverb pickup than a cardioid

Hypercardioid

- Maximum side rejection in a unidirectional mic
- Maximum isolation—maximum rejection of reverberation, leakage, feedback, and background noise

Bidirectional

- Front and rear pickup, with side sounds rejected (for across-table interviews or two-part vocal groups, for example)
- Maximum isolation of an orchestral section when miked overhead
- Blumlein stereo miking (two bidirectional mics crossed at 90 degrees)

In a good mic, the polar pattern should be about the same from 200 Hz to 10 kHz. If not, you'll hear off-axis coloration: the mic will have a different tone quality on and off axis. Small-diaphragm mics tend to have less off-axis coloration than large-diaphragm mics.

You can get either the condenser or dynamic type with any kind of polar pattern (except bidirectional dynamic). Ribbon mics are either bidirectional or hypercardioid. Some condenser mics come with switchable patterns. Note that the shape of a mic does not indicate its polar pattern.

If a mic is end-addressed, you aim the end of the mic at the sound source. If a mic is side-addressed, you aim the side of the mic at the sound source. Figure 6.6 shows a typical side-addressed condenser mic with switchable polar patterns.

Boundary mics that mount on a surface have a pattern that is half-omni (hemispherical), half-supercardioid, or half-cardioid (like an apple sliced in half through its stem). The boundary mounting makes the mic more directional so it picks up less room acoustics.

Frequency Response

As with other audio components, a microphone's frequency response is the range of frequencies that it will reproduce at an equal level (within a tolerance, such as ±3 dB).

The following is a list of sound sources and the microphone frequency response that is adequate to record the source with high fidelity. A wider range response works, too.

- Most instruments: 80 Hz to 15 kHz
- Bass instruments: 40 Hz to 9 kHz
- Brass and voice: 80 Hz to 12 kHz
- Piano: 40 Hz to 12 kHz
- Cymbals and some percussion: 300 Hz to 15 or 20 kHz
- Orchestra or symphonic band: 40 Hz to 15 kHz

DIAPHRAGM

GRILLE

ELECTRONICS

Figure 6.6 A typical multi-pattern mic that is side-addressed.

If possible, use a mic with a response that rolls off below the lowest fundamental frequency of the instrument you're recording. For example, the frequency of the low-E string on an acoustic guitar is about 82 Hz. A mic used on the acoustic guitar should roll off below that frequency to avoid picking up low-frequency noise such as rumble from trucks and air conditioning. Some mics have a built-in low-cut switch for this purpose. Or you can filter out the unneeded lows at your mixer.

A frequency-response curve is a graph of the mic's output level in dB at various frequencies. The output level at 1 kHz is placed at the 0 dB line on the graph, and the levels at other frequencies are so many decibels above or below that reference level.

The shape of the response curve suggests how the mic sounds at a certain distance from the sound source. (If the distance is not specified, it's probably 2 to 3 feet.) For example, a mic with a wide, flat response reproduces the fundamental frequencies and harmonics in the same proportion as the sound source. So a flat-response mic tends to provide accurate, natural reproduction at that distance.

A rising high end or a "presence peak" around 5 to 10 kHz sounds more crisp and articulate because it emphasizes the higher harmonics (Figure 6.7). *Play CD track 15.* Sometimes this type of response is called tailored or contoured. It's popular for guitar amps and drums because it adds punch and emphasizes attack. Some microphones have switches that alter the frequency response.

Most uni- and bidirectional mics boost the bass when used within a few inches of a sound source. You've heard how the sound gets bassy when a vocalist sings right into the mic. This low-frequency boost related to close mic placement is called the proximity effect, and it's often plotted on the frequency-response graph. Omni mics have no proximity effect; they sound tonally the same at any distance.

The warmth created by proximity effect adds a pleasing fullness to drums. In most recording situations, though, the proximity effect lends an unnatural boomy or bassy sound to the instrument or voice picked up by the mic. Some mics—multiple-D or variable-D types—are designed to reduce it. These types have holes or slots in the mic handle. Some mics have a bass-rolloff switch to compensate for the bass boost. Or you can roll off the excess bass with your mixer's equalizer until the sound is natural. By doing so, you also reduce low-frequency leakage picked up by the microphone.

Note that mic placement can greatly affect the recorded tone quality. A flat-response mic does not always guarantee a natural sound because mic placement has such a strong influence. Tonal effects of mic placement are covered in Chapter 7.

Figure 6.7 An example of the frequency response of a microphone with proximity effect and a presence peak around 5 kHz.

Impedance (Z)

This spec is the mic's effective output resistance at 1 kHz. A mic impedance between 150 and 600 ohms is low; 1000 to 4000 ohms is medium; and
above 25 kilohms is high.

Always use low-impedance mics. If you do, you can run long mic
cables without picking up hum or losing high frequencies. The input
impedance of a mixer mic input is about 1500 ohms. If it were the same
impedance as the mic, about 250 ohms, the mic would "load down" when
you plug it in. Loading down a mic makes it lose level, distort, or sound
thin. To prevent this, a mic input has an impedance much higher than
that of the microphone. But it's still called a low-Z input.

More information on impedance is in Appendix E.

Maximum SPL

To understand this spec, first we need to understand sound pressure level
(SPL). It is a measure of the intensity of a sound. The quietest sound we
can hear, the threshold of hearing, is 0 dBSPL. Normal conversation at 1
foot measures about 70 dBSPL; painfully loud sound is above 120 dBSPL.

If the maximum SPL spec is 125 dBSPL, the mic starts to distort
when the instrument being miked is putting out 125 dBSPL at the mic.
A maximum SPL spec of 120 dB is good, 135 dB is very good, and 150 dB
is excellent.

Dynamic mics tend not to distort, even with very loud sounds. Some
condensers are just as good. Some have a pad you can switch in to
prevent distortion in the mic circuitry. Because a mic pad reduces signal-
to-noise ratio (S/N), use it only if the mic distorts.

Sensitivity

This spec tells how much output voltage a mic produces when driven by
a certain SPL. A high-sensitivity mic puts out a stronger signal (higher
voltage) than a low-sensitivity mic when both are exposed to an equally
loud sound.

A low-sensitivity mic needs more mixer gain than a high-sensitivity
mic. More gain usually results in more noise. When you record quiet
music at a distance (classical guitar, string quartet), use a mic of high sensitivity to override mixer noise. When you record loud music or mike
close, sensitivity matters little because the mic signal level is well above

the mixer noise floor. That is, the S/N is high. Listed below are typical sensitivity specs for three transducer types:

- Condenser: 5.6 mV/Pa (high sensitivity)
- Dynamic: 1.8 mV/Pa (medium sensitivity)
- Ribbon or small dynamic: 1.1 mV/Pa (low sensitivity)

The louder the sound source, the higher the signal voltage the mic puts out. A very loud instrument, such as a kick drum or guitar amp, can cause a microphone to generate a signal strong enough to overload the mic preamp in your mixer. That's why most mixers have pads or input-gain controls—to prevent preamp overload from hot mic signals.

Self-Noise

Self-noise or equivalent noise level is the electrical noise or hiss a mic produces. It's the dBSPL of a sound source that would produce the same output voltage that the noise does.

Usually the self-noise spec is A-weighted. That means the noise was measured through a filter that makes the measurement correlate more closely with the annoyance value. The filter rolls off low and high frequencies to simulate the frequency response of the ear.

An A-weighted self-noise spec of 14 dBSPL or less is excellent (quiet); 21 dB is very good, 28 dB is good; and 35 dB is fair—not good enough for quality recording.

Because a dynamic mic has no active electronics to generate noise, it has very low self-noise (hiss) compared to a condenser mic. So most spec sheets for dynamic mics do not specify self-noise.

Signal-to-Noise Ratio

This is the difference in decibels between the mic's sensitivity and its self-noise. The higher the SPL of the sound source at the mic, the higher the S/N. Given an SPL of 94 dB, an S/N spec of 74 dB is excellent; 64 dB is good. The higher the S/N ratio, the cleaner (more noise-free) the signal, and the greater the "reach" of the microphone.

Reach is the clear pickup of quiet, distant sounds due to high S/N. Reach is not specified in data sheets because any mic can pick up a source at any distance if the source is loud enough. For example, even a cheap mic can reach several miles if the sound source is a thunderclap.

Polarity

The polarity spec relates the polarity of the electrical output signal to the acoustic input signal. The standard is "pin 2 hot." That is, the mic produces a positive voltage at pin 2 with respect to pin 3 when the sound pressure pushes the diaphragm in (positive pressure).

Be sure that your mic cables do not reverse polarity. On both ends of each cable, the wiring should be pin 1 shield, pin 2 red, pin 3 white or black. Or the wiring on both ends should be pin 1 shield, pin 2 white, pin 3 black.

If some mic cables are correct polarity and some are reversed, and you mix their mics to mono, the bass may cancel.

Microphone Types

The following sections describe several types of recording mics.

Large-Diaphragm Condenser Microphone

This is a condenser microphone, usually side-addressed, with a diaphragm 1 inch or larger in diameter (Figure 6.6). It generally has very good low-frequency response and low self-noise. Common uses are studio vocals and acoustic instruments. Examples: AKG C12 VR, C414, C2000B and C3000B; Audio-Technica AT2020/3035/4040, Audix SCX25, Blue Blueberry, M-Audio Luna, CAD Equitek Series and M177, DPA 4041, Lawson L47MP MKII and L251, Manley Gold Reference, Neumann U87, U47 and TLM 103; Soundelux Elux 251, Shure KSM Series, MXL V67G, V69, 900, 2001 and 2003; Rode NT1A, Studio Projects B and C Series, Samson CL7 and C01, Nady SCM 950 and 100, M-Audio Luna and Nova, and Behringer B1 and B2.

Small-Diaphragm Condenser Microphone

This is a stick-shaped or "pencil" cardioid condenser microphone, usually end-addressed, with a diaphragm under 1 inch in diameter (Figure 6.4). It generally has very good transient response and detail, making it a fine choice for close miking acoustic instruments—especially cymbals, acoustic guitar, and piano. Examples: AKG C 451 B; Audio-Technica AT 3031 and AT 4051a; Audix SCX1, ADX50, and ADX51; CAD Equitek e60; M-Audio Pulsar; Samson C02; Crown CM-700; DPA 4006; Neumann KM

184; Sennheiser e614 and MKH50; Shure KSM109/SL, KSM137/Sl and SM81; MXL 600 and 603S; Behringer B5 with cardioid capsule; and Studio Projects C4.

Dynamic Instrument Microphone

This is a stick-shaped dynamic microphone, end-addressed (Figure 6.4). Although it may have a flat response, it generally has a presence peak and some low-frequency rolloff to prevent boominess when used up close. It's often used on drums and guitar amps. Examples: Shure SM57, AKG D112 (kick drum), Audio-Technica AT AE2500 (kick), Electro-Voice N/D868 (kick), Audix D1 through D6 and I-5, and Sennheiser MD421, e604 and e602 (kick).

Live-Vocal Microphone

This unidirectional mic is shaped like an ice-cream cone because of its large grille used to reduce breath pops. It can be a condenser, dynamic, or ribbon type, and it usually has a presence peak and some low-frequency rolloff. Examples: Shure SM58 and Beta 58, Shure SM85 and SM87, AKG D3800, Audix OM5, Beyerdynamic M88 TG, Crown CM-200A, EV N-Dym Series, and Neumann KMS 105.

Ribbon Microphone

This mic can be side- or end-addressed. It generally is used wherever you want a warm, smooth tone quality (sometimes with reduced highs). Examples: models by Beyerdynamic, Coles, Royer, and AEA.

Boundary Microphone

Boundary mics are designed to be used on surfaces. Tape them to the underside of a piano lid, or tape them to the wall for pickup of room ambience. They can be used on hard baffles between instruments, or on panels to make the mics directional. A boundary mic uses a mini condenser mic capsule mounted very near a sound-reflecting plate or boundary (Figure 6.8). Because of this construction, the mic picks up direct sound and reflected sound at the same time, in-phase at all frequencies. So you get a smooth response free of phase cancellations. A conventional mic near a surface sounds colored; a boundary mic on a surface sounds

Figure 6.8 Typical PZM construction.

natural. Examples: AKG C 562 BL, Audio-Technica AT 841a, Beyerdynamic MPC 22, Crown PZM-30D and PZM-6D, and Shure Beta 91.

Other benefits are a wide, smooth frequency response free of phase cancellations, excellent clarity and reach, and the same tone quality anywhere around the mic. The polar pattern is half-omni or hemispherical. Some boundary mics have a half-cardioid or half-supercardioid polar pattern. They work great on a conference table, or near the front edge of a stage floor to pick up drama or musicals.

Miniature Microphone

Mini condenser mics can be attached to drum rims, flutes, horns, acoustic guitars, fiddles, and so on. Their tone quality is about as good as larger studio microphones and the price is relatively low. With these tiny units you can mike a band in concert without cluttering the stage with boom stands (Figure 6.9), or you can mike a whole drum set with two or three of these. Although you lose individual control of each drum in the mix, the cost is low and the sound is quite good with some bass and treble boost. Compared to large mics, mini mics tend to have more noise (hiss)

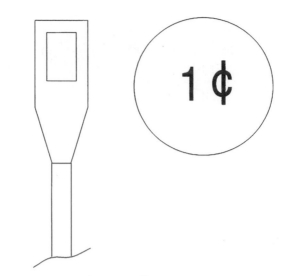

Figure 6.9 A mini mic is the size of a penny.

in distant-miking applications. A lavalier mic is a mini mic worn on the chest to pick up speech from a newscaster or a wandering lecturer. Examples: AKG Micro Mic Series, Shure Beta 98S, Audix M1245 and Micro-D, Countryman Isomax B6, Crown GLM-100, DPA 4060, and Sennheiser e608.

Stereo Microphone

A stereo microphone combines two directional mic capsules in a single housing for convenient stereo recording (Figure 6.10). Simply place the mic a suitable distance and height from the sound source, and you'll get a stereo recording with little fuss. Examples: AKG C426 B Comb, Audio-Technica AT825 and AT822, Crown SASS-P MKII, Neumann SM69, Shure VP88, Nady RSM-2, AEA R88, and Royer SF-12.

Because there is no spacing between the mic capsules, there also is no delay or phase shift between their signals. Coincident stereo microphones are mono-compatible—the frequency response is the same in mono and stereo—because there are no phase cancellations if the two channels are combined.

Figure 6.10 A stereo microphone.

Digital Microphone

This condenser microphone has a built-in analog-to-digital converter. It is usually side-addressed, has a large diaphragm, has a flat response, and very low self-noise. Its output is a digital signal, which is immune to picking up hum. Examples: Beyerdynamic MCD 100, and Neumann Solution-D.

Headworn Microphone

This microphone is used for a live performance which might be recorded. It is a small condenser mic worn on the head, either omni- or unidirectional. The headworn mic allows the performer freedom of movement on stage. Some models provide excellent gain before feedback and isolation. Examples: AKG C 420, Audio-Technica ATM73, Countryman Isomax E6, and Crown CM-311A.

Microphone Selection

Table 6.1 is a guide to choosing a mic based on your requirements.

Suppose you want to record a grand piano playing with several other instruments. You need the microphone to reduce leakage. Table 6.1 recommends a unidirectional mic or an omni mic up close. For this particular piano, you also want a natural sound, for which the table suggests

Table 6.1 Mic Application Guide

Requirement	Characteristic
Natural, smooth tone quality	Flat frequency response
Bright, present tone quality	Bright, present tone quality
Extended lows	Omni condenser or dynamic with good low-frequency response
Extended highs (detailed sound)	Condenser
Reduced "edge" or detail	Dynamic
Boosted bass up close	Directional mic
Flat bass response up close	Omni mic, or directional mic with sound holes in the handle
Reduced pickup of leakage, feedback, and room acoustics	Directional mics
Enhanced pickup of room acoustics	Omni mics
Miking close to a surface, even coverage of moving sources or large sources, inconspicuous mic	Boundary mic
Coincident or near-coincident stereo (see Chapter 18)	Stereo mic
Extra ruggedness	Dynamic mic
Reduced handling noise	Omni mic, or unidirectional with shock mount
Reduced breath popping	Omni mic, or unidirectional with pop filter
Distortion-free pickup of very loud sounds	Condenser with high maximum SPL spec, or dynamic
Low self-noise, high sensitivity, noise-free pickup of quiet sounds	Large-diaphragm condenser mic

a mic with a flat response. You want a detailed sound, so a condenser mic is the choice. A microphone with all these characteristics is a flat-response, unidirectional condenser mic. If you're miking close to a surface (the piano lid), a boundary mic is recommended.

Now suppose you're recording an acoustic guitar on stage, and the guitarist roams around. This is a moving sound source, for which the table recommends a mini mic attached to the guitar. Feedback and leakage are not a problem because you're miking close, so you can use an omni mic. Thus, an omni condenser mic is a good choice for this application.

For a home studio, a suggested first choice is a cardioid condenser mic with a flat frequency response. This type of mic is especially good for studio vocals, cymbals, percussion, and acoustic instruments. Remember that the mic needs a power supply to operate, such as a battery or phantom power supply.

Your second choice of microphone for a home studio is a cardioid dynamic microphone with a presence peak in the frequency response. This type is good for drums and guitar amps. I recommend cardioid over omni for a home studio. The cardioid pattern rejects the leakage, background noise, and room reverb often found in home studios. An omni mic, however, can do that, too, if you mike close enough. Also, omni mics tend to provide a more natural sound at lower cost, and they have no proximity effect.

Mic Accessories

There are many devices used with microphones to route their signals or to make them more useful. These include pop filters, stands and booms, shock mounts, cables and connectors, stage boxes and snakes, and splitters.

Pop Filter

A much needed accessory for a vocalist's microphone is a pop filter or windscreen. It usually is a foam "sock" that you put over the mic. Some microphones have pop filters or ball-shaped grilles built in.

Why is it needed? When a vocalist sings a word starting with "p," "b," or "t" sounds, a turbulent puff of air is forced from the mouth. A microphone placed close to the mouth is hit by this air puff, resulting in a thump or little explosion called a pop. The windscreen reduces this problem.

The best type of pop filter is a nylon screen in a hoop, or a perforated-metal disk, placed a few inches from the mic.

You can also reduce pop by placing the mic above or to the side of the mouth, or by using an omni mic. *CD track 16 demonstrates how a pop filter or mic placement can prevent breath pops.*

Stands and Booms

Stands and booms hold the microphones and let you position them as desired. A mic stand has a heavy metal base that supports a vertical pipe. At the top of the pipe is a rotating clutch that lets you adjust the height of a smaller telescoping pipe inside the large one. The top of the small pipe has a standard 5/8-inch 27 thread, which screws into a mic stand adapter.

A boom is a long horizontal pipe that attaches to the vertical pipe. The angle and length of the boom are adjustable. The end of the boom is threaded to accept a mic stand adapter, and the opposite end is weighted to balance the weight of the microphone.

Shock Mount

A shock mount holds a mic in a resilient suspension to isolate the mic from mechanical vibrations, such as floor thumps and mic-stand bumps.

Many mics have an internal shock mount which isolates the mic capsule from its housing; this reduces handling noise as well as stand thumps.

Cables and Connectors

Mic cables carry the electrical signal from the mic to the mixing console, mic preamp, or recorder. With low-impedance mics, you can use hundreds of feet of cable without hum pickup or high-frequency loss. Some mics have a permanently attached cable for convenience and low cost; others have a connector in the handle to accept a separate mic cable. The second method is preferred for serious recording because if the cable breaks, you have to repair or replace only the cable, not the whole microphone.

Mic cables are made of one or two insulated conductors surrounded by a fine-wire mesh shield to keep out electrostatic hum. If you hear a loud buzz when you plug in a microphone, check that the shield is securely soldered in place.

After acquiring a microphone, you may need to wire its 2-conductor shielded cable to a 3-pin XLR audio connector. Here are the solder connections:

1. Pin 1: Shield
2. Pin 2: "Hot" or "in-polarity" lead (usually red or white)
3. Pin 3: "Cold" or "out-of-polarity" lead (usually black)

If the mic output is a 3-pin XLR, but your recorder or mixer mic input is an unbalanced phone jack, a different wiring is needed:

1. Phone-plug tip (the short terminal): Hot lead
2. Phone-plug sleeve (the long terminal): Shield and cold lead

Wind your mic cables onto a large spool, which can be found in the electrical section of hardware stores. Plug the cables together as you wind them.

Snake

It is messy and time-consuming to run mic cables from several mics all the way to a mixer. Instead, you can plug all your mics into a stage box with several connectors (Figure 6.11). The snake—a thick multi-conductor cable—carries the signals to the mixer. At the mixer end, the cable divides into several mic connectors that plug into the mixer.

Splitter

When you record a band in concert, you might want to feed each mic's signal to your recording mixer and to the band's PA and monitor mixers. A mic splitter does the job. For each microphone channel, it has one XLR input for a microphone, a "direct" XLR output wired to the input, and

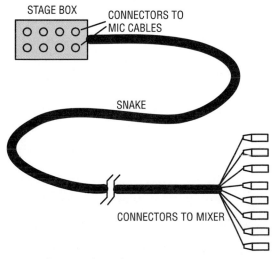

Figure 6.11 A stage box and snake.

one or more transformer-isolated XLR outputs with a ground-lift switch. The mixer that provides phantom power must be connected to the direct XLR output.

Summary

We talked about some mic types, specs, and accessories. You should have a better idea about what kind of microphone to choose for your own applications.

Mic manufacturers are happy to send you free catalogs and application notes, which are also available on mic company Web sites. Mic dealers also may have this literature.

Remember, you can use any microphone on any instrument if it sounds good to you. Just try it and see if you like it. To make high-quality recordings, though, you need good mics with a smooth, wide-range frequency response, low noise, and low distortion.

MICROPHONE TECHNIQUE BASICS

Suppose you're going to mike a singer, a sax, or a guitar. Which mic should you choose? Where should you place it?

Your mic technique has a powerful effect on the sound of your recordings. In this chapter we'll look at some general principles of miking that apply to all situations. Chapter 8 covers common mic techniques for specific instruments.

Which Mic Should I Use?

Is there a "right" mic to use on a piano, a kick drum, or a guitar amp? No. Every microphone sounds different, and you choose the one that gives you the sound you want. Still, it helps to know about two main characteristics of mics that affect the sound: frequency response and polar pattern.

Most condenser mics have an extended high-frequency response— they reproduce sounds up to 15 or 20 kHz. This makes them great for cymbals or other instruments that need a detailed sound, such as acoustic guitar, strings, piano, and voice. Dynamic moving-coil microphones have a response good enough for drums, guitar amps, horns, and woodwinds. Loud drums and guitar amps sound dull if recorded with a flat-response mic; a mic with a presence peak (a boost around 5 kHz) gives more edge or punch.

Suppose you are choosing a microphone for a particular instrument. In general, the frequency response of the mic should cover at least the frequencies produced by that instrument. For example, an acoustic guitar produces fundamental frequencies from 82 Hz to about 1 kHz, and produces harmonics from about 1 to 15 kHz. So a mic used on an acoustic guitar should have a frequency response of at least 82 Hz to 15 kHz if you want to record the guitar accurately. Table 7.1 shows the frequency ranges of various instruments.

The polar pattern of a mic affects how much leakage and ambience it picks up. Leakage is unwanted sound from instruments other than the one at which the mic is aimed. Ambience is the acoustics of the record-

Table 7.1 Frequency Ranges of Various Musical Instruments

Instrument	Fundamentals (Hz)	Harmonics (kHz)
Flute	261–2349	3–8
Oboe	261–1568	2–12
Clarinet	165–1568	2–10
Bassoon	62–587	1–7
Trumpet	165–988	1–7.5
French horn	87–880	1–6
Trombone	73–587	1–7.5
Tuba	49–587	1–4
Snare drum	100–200	1–20
Kick drum	30–147	1–6
Cymbals	300–587	1–15
Violin	196–3136	4–15
Viola	131–1175	2–8.5
Cello	65–698	1–6.5
Acoustic bass	41–294	1–5
Electric bass	41–300	1–7
Acoustic guitar	82–988	1–15
Electric guitar	82–1319	1–3.5 (through amp)
Electric guitar	82–1319	1–15 (direct)
Piano	28–4196	5–8
Bass (voice)	87–392	1–12
Tenor (voice)	131–494	1–12
Alto (voice)	175–698	2–12
Soprano (voice)	247–1175	2–12

ing room—its early reflections and reverb. The more leakage and ambience you pick up, the more distant the instrument sounds.

An omni mic picks up more ambience and leakage than a directional mic when both are the same distance from an instrument. So an omni tends to sound more distant. To compensate, you have to mike closer with an omni.

How Many Mics?

The number of mics you need varies with what you're recording. If you want to record an overall acoustic blend of the instruments and room ambience, use just two microphones or a stereo mic (Figure 7.1). This method works great on an orchestra, symphonic band, choir, string quartet, pipe organ, small folk group, or a piano/voice recital. Stereo miking is covered in detail later in this chapter.

To record a pop-music group, you mike each instrument or instrumental section. Then you adjust the mixer volume control for each mic to control the balance between instruments (Figure 7.2).

To get the clearest sound, don't use two mics when one will do the job. Sometimes you can pick up two or more sound sources with one mic (Figure 7.3). You could mike a brass section of four players with one mic on four players, or with two mics on every two players. Or mike a choir

Figure 7.1   Overall miking of a musical ensemble with two distant microphones.

Figure 7.2 Individual miking with multiple close mics and a mixer.

Figure 7.3 Multiple miking with several sound sources on each microphone.

in a studio in four groups: put one mic on the basses, one on the sopranos, and so on.

Picking up more than one instrument with one mic has a problem: during mixdown, you can't adjust the balance among instruments recorded on the same track. You have to balance the instruments before recording them. Monitor the mic, and listen to see if any instrument is too quiet. If so, move it closer to the mic.

How Close Should I Place the Mic?

Once you've chosen a mic for an instrument, how close should the mic be? Mike a few inches away to get a tight, present sound; mike farther away for a distant, spacious sound. (Try it to hear the effect.) *Play CD track 17.* The farther a mic is from the instrument, the more ambience, leakage, and background noise it picks up. So mike close to reject these unwanted sounds. Mike farther away to add a live, loose, airy feel to overdubs of drums, lead-guitar solos, horns, etc.

Close miking sounds close; distant miking sounds distant. Here's why. If you put a mic close to an instrument, the sound at the mic is loud. So you need to turn up the mic gain on your mixer only a little to get a full recording level. And because the gain is low, you pick up very little reverb, leakage, and background noise (Figure 7.4A).

If you put a mic far from an instrument, the sound at the mic is quiet. You'll need to turn up the mic gain a lot to get a full recording level. And because the gain is high, you pick up a lot of reverb, leakage, and background noise (Figure 7.4B).

If the mic is very far away—maybe 10 feet—it's called an ambience mic or room mic. It picks up mostly room reverb. A popular mic for ambience is a boundary microphone taped to the wall. You mix it with the

(A) A close microphone picks up mainly direct sound, resulting in a close sound quality.

(B) A distant microphone picks up mainly reflected sound, resulting in a distant sound quality.

Figure 7.4 (A) A close microphone picks up mainly direct sound, which results in a close sound quality. (B) A distant microphone picks up mainly reflected sound, which results in a distant sound quality.

usual close mics to add a sense of space. Use two for stereo. When you record a live concert, you might want to place ambience mics over the audience, aiming at them from the front of the hall, to pick up the crowd reaction and the hall acoustics.

Classical music is always recorded at a distance (about 4 to 20 feet away) so that the mics will pick up reverb from the concert hall. It's a desirable part of the sound.

Leakage

Suppose you're close-miking a drum set and a piano at the same time (Figure 7.5). When you listen to the drum mics alone, you hear a close, clear sound. But when you mix in the piano mic, that nice, tight drum sound degrades into a distant, muddy sound. That's because the drum sound leaked into the piano mic, which picked up a distant drum sound from across the room. *CD track 11 is an example of leakage.*

There are many ways to reduce leakage:

- Mike each instrument closely. That way the sound level at each mic is high. Then you can turn down the mixer gain of each mic, which reduces leakage at the same time.
- Overdub each instrument one at a time.
- Record direct.
- Filter out frequencies above and below the range of each instrument.
- Use directional mics (cardioid, etc.) instead of omni mics.

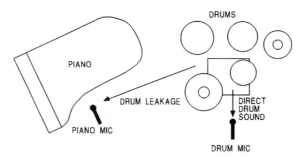

Figure 7.5 Example of leakage. The piano mic picks up leakage from the drums, which changes the close drum sound to distant.

- Record in a large, fairly dead studio. In such a room, leakage reflected from the walls is weak.
- Put portable walls (goboes) between instruments.

Don't Mike Too Close

Miking too close can color the recorded tone quality of an instrument. If you mike very close, you might hear a bassy or honky tone instead of a natural sound.

Why? Most musical instruments are designed to sound best at a distance, at least 1-1/2 feet away. The sound of an instrument needs some space to develop. A mic placed a foot or two away tends to pick up a well-balanced, natural sound. That is, it picks up a blend of all the parts of the instrument that contribute to its character or timbre.

Think of a musical instrument as a loudspeaker with a woofer, midrange, and tweeter. If you place a mic a few feet away, it will pick up the sound of the loudspeaker accurately. But if you place the mic close to the woofer, the sound will be bassy. Similarly, if you mike close to an instrument, you emphasize the part of the instrument that the microphone is near. The tone quality picked up very close may not reflect the tone quality of the entire instrument.

Suppose you place a mic next to the sound hole of an acoustic guitar, which resonates around 80 to 100 Hz. A microphone placed there hears this bassy resonance, giving a boomy recorded timbre that does not exist at a greater miking distance. To make the guitar sound more natural when miked close to the sound hole, you need to roll off the excess bass on your mixer, or use a mic with a bass rolloff in its frequency response.

The sax projects highs from the bell, but projects mids and lows from the tone holes. So if you mike close to the bell, you miss the warmth and body from the tone holes. All that's left at the bell is a harsh tone quality. You might like that sound, but if not, move the mic out and up to pick up the entire instrument. If leakage forces you to mike close, change the mic or use equalization (EQ).

Usually, you get a natural sound if you put the mic as far from the source as the source is big. That way, the mic picks up all the sound-radiating parts of the instrument about equally. For example, if the body of an acoustic guitar is 18 inches long, place the mic 18 inches away to get a natural tonal balance. If this sounds too distant or hollow, move in a little closer.

Where Should I Place the Mic?

Suppose you have a mic placed a certain distance from an instrument. If you move the mic left, right, up, or down, you change the recorded tone quality. In one spot, the instrument might sound bassy; in another spot, it might sound natural, and so on. So, to find a good mic position, simply place the mic in different locations—and monitor the results—until you find one that sounds good to you.

Here's another way to do the same thing. Close one ear with your finger, listen to the instrument with the other ear, and move around until you find a spot that sounds good. Put the mic there. Then make a recording and see if it sounds the same as what you heard live. Don't try this with kick drums or screaming guitar amps!

Why does moving the mic change the tone quality? A musical instrument radiates a different tone quality in each direction. Also, each part of the instrument produces a different tone quality. For example, Figure 7.6 shows the tonal balances picked up at various spots near a guitar. *CD track 18 illustrates the effect of mic placement on guitar tonal balance. CD track 19 demonstrates close and distant stereo miking of the acoustic guitar.*

Figure 7.6 Microphone placement affects the recorded tonal balance.

Other instruments work the same way. A trumpet radiates strong highs directly out of the bell, but does not project them to the sides. So a trumpet sounds bright when miked on-axis to the bell and sounds more natural or mellow when miked off to one side. A grand piano miked one foot over the middle strings sounds fairly natural, under the soundboard sounds bassy and dull, and in a sound hole it sounds mid-rangey.

It pays to experiment with all sorts of mic positions until you find a sound you like. There is no one right way to place the mics because you place them to get the tonal balance you want.

On-Surface Techniques

Sometimes you're forced to place a mic near a hard reflecting surface. Examples:

- Recording drama or opera with the mics near the stage floor.
- Recording an instrument that has hard surfaces around it.
- Recording a piano with the mic close to the lid.

In these cases, you'll often pick up an unnatural, filtered tone quality. Here's why. Sound travels to the microphone via two paths: directly from the sound source, and reflected off the nearby surface (Figure 7.7).

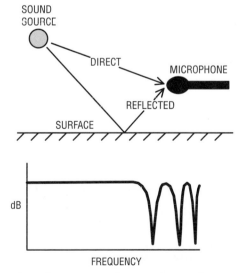

Figure 7.7 A mic placed near a surface picks up direct sound and a delayed reflection, which gives a comb-filter frequency response.

117

Because of its longer travel path, the reflected sound is delayed compared to the direct sound. The direct and delayed sound waves combine at the mic, which causes phase cancellations of various frequencies. The series of peaks and dips in the response is called a comb-filter effect, and it sounds like mild flanging.

Boundary mics solve the problem. In a boundary mic, the diaphragm is very close to the reflecting surface so that there is no delay in the reflected sound. Direct and reflected sounds add in-phase over the audible range of frequencies, resulting in a flat response (Figure 7.8). *Play CD track 20.*

You might tape an omni boundary mic to the underside of a piano lid, to a hard-surfaced panel, or to a wall for ambience pickup. A uni-directional boundary mic works great on a stage floor to pick up drama. A group of these mics will clearly pick up people at a conference table.

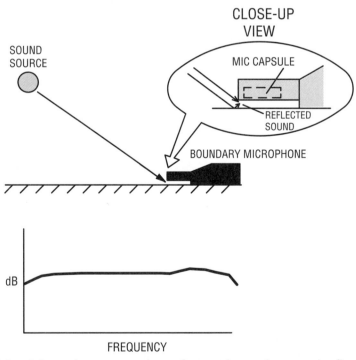

Figure 7.8 A boundary mic on the surface picks up direct and reflected sounds in phase.

The Three-to-One Rule

Let's say you're miking several instruments, each with its own mic. If you place the musicians too close together, the sound will be blurred. But if you spread out the musicians and mike them close, the sound will be clearer.

Specifically, try to space the mics at least three times the mic-to-source distance (as in Figure 7.9). This is called the 3:1 rule. For example, if two mics are each placed 1 foot from their sound sources, the mics should be at least 3 feet apart. This will prevent the blurred, colored sound caused by phase cancellations between mics. *Play CD track 21.*

The mics can be closer together than 3:1 if you use two cardioid mics aiming in opposite directions.

Suppose you're recording a singer/guitarist. There's a mic on the singer and a mic on the acoustic guitar. The vocal picked up by the guitar mic is delayed because the vocal sound travels a longer path to that mic. The two vocal signals in the mix—direct and delayed—interfere with each other and make a hollow sound.

Try these solutions:

- Mike the voice and guitar very close. Roll off the excess bass with your mixer's EQ. Maybe use a pickup on the guitar instead of a mic.
- Place two bidirectional mics so the tops of their grilles touch. This gets rid of any delay between their signals. Aim the "dead" side of

Figure 7.9 The 3:1 rule of microphone placement avoids phase interference between microphone signals.

the vocal mic at the guitar; aim the dead side of the guitar mic at the mouth.

- Use just one mic midway between the mouth and guitar. Adjust the balance by changing the mic's height.
- Delay the vocal mic signal by about 1 msec. Then the signals of the two mics will be more in-phase, preventing phase cancellations when they are mixed to the same channel.

Off-Axis Coloration

Some mics have off-axis coloration—a dull or colored effect on sound sources that are not directly in front of the mic. Try to aim the mic at sound sources that put out high frequencies, such as cymbals. When you pick up a large source such as an orchestra, use a mic that has the same response over a wide angle. Such a mic has similar polar patterns at middle and high frequencies. Most large-diaphragm mics have more off-axis coloration than smaller mics (under 1 inch).

Stereo Mic Techniques

Stereo mic techniques capture the sound of a musical group as a whole, using only two or three microphones. When you play back a stereo recording, you hear phantom images of the instruments in various spots between the speakers. These image locations—left to right, front to back—correspond to the instrument locations during the recording session.

Stereo miking is the preferred way to record classical-music ensembles and soloists. In the studio, you can stereo-mike a piano, drum set cymbals, vibraphone, harmony singers, or other large sound sources.

Goals of Stereo Miking

One goal is accurate localization. That is, instruments in the center of the group are reproduced midway between the two speakers. Instruments at the sides of the group are heard from the left or right speaker. Instruments halfway to one side are heard halfway to one side, and so on.

Figure 7.10 shows three stereo localization effects. Figure 7.10A shows some instrument positions in an orchestra: left, left-center, center, right-center, right. In Figure 7.10B, the reproduced images of these instru-

(A) Orchestra instrument locations (top view).

(B) Images localized accurately between speakers (the listener's perception).

(C) Narrow stage effect.

(D) Exaggerated separation effect.

Figure 7.10 Stereo localization effects. (A) Orchestra instrument locations (top view). (B) Images localized accurately between speakers (the listener's perception). (C) Narrow stage effect. (D) Exaggerated separation effect.

ments are accurately localized between the speakers. The stereo spread, or stage width, extends from speaker to speaker. (You might want to record a string quartet with a narrower spread.)

If you space or angle the mics too close together, you get a narrow stage effect (Figure 7.10C). If you space or angle the mics too far apart, you hear exaggerated separation (Figure 7.10D). That is, instruments halfway to one side are heard near the left or right speaker.

To judge stereo effects, you have to sit exactly between your monitor speakers (the same distance from each). Sit as far from the speakers as the spacing between them. Then the speakers appear to be 60 degrees apart. This is about the same angle an orchestra fills when viewed from a typical ideal seat in the audience (say, tenth row center). If you sit off-center, the images shift toward the side on which you're sitting and are less sharp.

Types of Stereo Mic Techniques

To make a stereo recording, you use one of these basic techniques:

- Coincident pair (XY or MS)
- Spaced pair (AB)
- Near-coincident pair (ORTF, etc.)
- Baffled pair (sphere, OSS, SASS, PZM wedge, etc.)

Let's look at each technique.

Coincident Pair

With this method, you mount two directional mics with grilles touching, diaphragms one above the other, and angled apart (Figure 7.11). For example, mount two cardioid mics with one grille above the other, and angle them 120 degrees apart. You can use other patterns too: supercardioid, hypercardioid, or bidirectional. The wider the angle between mics, the wider the stereo spread.

How does this technique make images you can localize? A directional mic is most sensitive to sounds in front of the mic (on-axis) and progressively less sensitive to sounds arriving off-axis. That is, a directional mic puts out a high-level signal from the sound source it's aimed at, and produces lower-level signals from other sound sources.

The coincident pair uses two directional mics that are angled symmetrically from the center line (Figure 7.11). Instruments in the center of the group make the same signal from each mic. During playback, you hear a phantom image of the center instruments midway between your

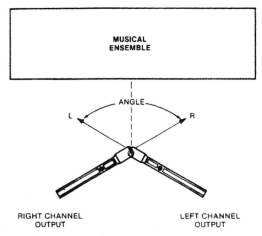

Figure 7.11 Coincident-pair technique.

speakers. That's because identical signals in each channel produce an image in the center.

If an instrument is off-center to the right, it is more on-axis to the right-aiming mic than to the left-aiming mic. So the right mic will produce a higher level signal than the left mic. During playback of this recording, the right speaker will play at a higher level than the left speaker. This reproduces the image off-center to the right—where the instrument was during recording.

The coincident pair codes instrument positions into level differences between channels. During playback, the brain decodes these level differences back into corresponding image locations. A pan pot in a mixing console works on the same principle. If one channel is 15 to 20 dB louder than the other, the image shifts all the way to the louder speaker.

Suppose we want the right side of the orchestra to be reproduced at the right speaker. That means the far-right musicians must produce a signal level 20 dB higher from the right mic than from the left mic. This happens when the mics are angled far enough apart. The correct angle depends on the polar pattern.

Instruments partway off center produce interchannel level differences less than 20 dB, so you hear them partway off center.

Listening tests have shown that coincident cardioid mics tend to reproduce the musical group with a narrow stereo spread. That is, the group does not spread all the way between speakers.

A coincident-pair method with excellent localization is the Blumlein array. It uses two bidirectional mics angled 90 degrees apart and facing the left and right sides of the group.

A special form of the coincident-pair technique is Mid-Side or MS (Figure 7.12). In this method, a cardioid or omni mic faces the middle of the orchestra. A matrix circuit sums and differences the cardioid mic with a bidirectional mic aiming to the sides. This produces left- and right-channel signals. You can remotely control the stereo spread by changing the ratio of the mid signal to the side signal. This remote control is useful at live concerts, where you can't physically adjust the mics during the concert. MS localization can be accurate.

To make coincident recordings sound more spacious, boost the bass 4 dB (+2 dB at 600 Hz) in the L–R or side signal.

A recording made with coincident mics is mono-compatible. That is, the frequency response is the same in mono or stereo. Because the mics occupy almost the same point in space, there is no time or phase difference between their signals. And when you combine them to mono, there

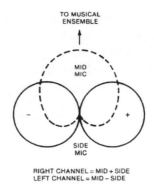

Figure 7.12 Mid-side (MS) technique.

Figure 7.13 Spaced-pair technique.

are no phase cancellations to degrade the frequency response. If you expect that your recordings will be heard in mono (say, on TV), then you'll probably want to use coincident methods.

Spaced Pair

Here, you mount two identical mics several feet apart and aim them straight ahead (Figure 7.13). The mics can have any polar pattern, but omni is most popular for this method. The greater the spacing between mics, the greater the stereo spread.

How does this method work? Instruments in the center of the group make the same signal from each mic. When you play back this recording, you hear a phantom image of the center instruments midway between your speakers.

124

If an instrument is off-center, it is closer to one mic than the other, so its sound reaches the closer microphone before it reaches the other one. Both mics make about the same signal, except that one mic signal is delayed compared with the other.

If you send a signal to two speakers with one channel delayed, the sound image shifts off center. With a spaced-pair recording, off-center instruments produce a delay in one mic channel, so they are reproduced off center.

The spaced pair codes instrument positions into time differences between channels. During playback, the brain decodes these time differences back into corresponding image locations.

A delay of 1.2 msec is enough to shift an image all the way to one speaker. You can use this fact when you set up the mics. Suppose you want to hear the right side of the orchestra from the right speaker. The sound from the right-side musicians must reach the right mic about 1.2 msec before it reaches the left mic. To make this happen, space the mics about 2 to 3 feet apart. This spacing makes the correct delay to place right-side instruments at the right speaker. Instruments partway off center make interchannel delays less than 1.2 msec, so they are reproduced partway off center.

If the spacing between mics is, say, 12 feet, then instruments that are slightly off center produce delays between channels that are greater than 1.2 msec. This places their images at the left or right speaker. I call this "exaggerated separation" or a "ping-pong" effect (Figure 7.10D).

On the other hand, if the mics are too close together, the delays produced will be too small to provide much stereo spread. Also, the mics will tend to emphasize instruments in the center because the mics are closest to them.

To record a good musical balance of an orchestra, you need to space the mics about 10 or 12 feet apart. But then you get too much separation. You could place a third mic midway between the outer pair and mix its output to both channels. That way, you pick up a good balance, and you hear an accurate stereo spread.

The spaced-pair method tends to make off-center images unfocused or hard to localize. Why? Spaced-pair recordings have time differences between channels. Stereo images produced solely by time differences are unfocused. You still hear the center instruments clearly in the center, but off-center instruments are hard to pinpoint. Spaced-pair miking is a good choice if you want the sonic images to be diffuse or blended, instead of sharply focused.

Another flaw of spaced mics: If you mix both mics to mono, you may get phase cancellations of various frequencies. This may or may not be audible.

Spaced mics, however, give a "warm" sense of ambience, in which the concert-hall reverb seems to surround the instruments and, sometimes, the listener. Here's why: The two channels of recorded reverb are incoherent; that is, they have random phase relationships. Incoherent signals from stereo speakers sound diffuse and spacious. Because spaced mics pick up reverb incoherently, it sounds diffuse and spacious. The simulated spaciousness caused by this phasiness is not necessarily realistic, but it is pleasant to many listeners.

Another advantage of the spaced pair is that you can use omni mics. An omni condenser mic has deeper bass than a uni condenser mic.

Near-Coincident Pair

In this method, you angle apart two directional mics, and space their grilles a few inches apart horizontally (Figure 7.14). Even a few inches of spacing increases the stereo spread and adds a sense of ambient warmth or air to the recording. The greater the angle or spacing between mics, the greater the stereo spread.

How does this method work? Angling directional mics produces level differences between channels. Spacing mics produces time differences. The level differences and time differences combine to create the stereo effect.

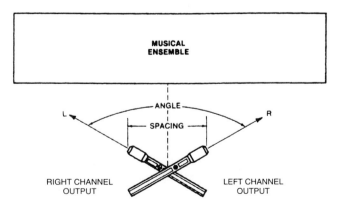

Figure 7.14 Near-coincident pair technique.

If the angling or spacing is too great, you get exaggerated separation. If the angling or spacing is too small, you'll hear a narrow stereo spread.

A common near-coincident method is the ORTF system, which uses two cardioids angled 110 degrees apart and spaced 7 inches (17 cm) horizontally. Usually this method gives accurate localization. That is, instruments at the sides of the orchestra are reproduced at or very near the speakers, and instruments halfway to one side are reproduced about halfway to one side.

Baffled Omni Pair

This method uses two omni mics, usually ear-spaced, and separated by either a hard or soft baffle (Figure 7.15). To create stereo, it uses time differences at low frequencies and level differences at high frequencies. The spacing between mics creates time differences. The baffle creates a sound shadow (reduced high frequencies) at the mic farthest from the source. Between the two channels, there are spectral differences—differences in frequency response.

Some examples of baffled-omni pairs are the Schoeps or Neumann sphere microphones, the Jecklin Disk, and the Crown SASS-P MKII stereo microphone.

Figure 7.15 Baffled-omni technique.

Comparing the Four Techniques

1. Coincident pair:
   - Uses two directional mics angled apart with grilles touching.
   - Level differences between channels produce the stereo effect.
   - Images are sharp.
   - Stereo spread ranges from narrow to accurate.
   - Signals are mono compatible.

2. Spaced pair:
   - Uses two mics spaced several feet apart, aiming straight ahead.
   - Time differences between channels produce the stereo effect.
   - Off-center images are diffuse.
   - Stereo spread tends to be exaggerated unless a third center mic is used, or unless spacing is under 2 to 3 feet.
   - Provides a warm sense of ambience.
   - Tends not to be mono compatible, but there are exceptions.
   - Good low-frequency response if you use omni condensers.

3. Near-coincident pair:
   - Uses two directional mics angled apart and spaced a few inches apart horizontally.
   - Level and time differences between channels produce the stereo effect.
   - Images are sharp.
   - Stereo spread tends to be accurate. Provides a greater sense of air than coincident methods.
   - Tends not to be mono compatible.

Play CD track 22 to hear a comparison of the coincident, near-coincident and spaced-pair techniques.

4. Baffled omni pair:
   - Uses two omni mics, usually ear-spaced, with a baffle between them.
   - Level, time, and spectral differences produce the stereo effect.

- Images are sharp.
- Stereo spread tends to be accurate.
- Good low-frequency response.
- Good imaging with headphones.
- Provides more air than coincident methods.
- Tends not to be mono compatible, but there are exceptions.

Hardware

A handy device is a stereo mic adapter or stereo bar. It mounts two mics on a single stand, and lets you adjust the angle and spacing. You might prefer to use a stereo mic instead of two mics. It has two mic capsules in a single housing for convenience.

How to Test Imaging

Here's a way to check the stereo imaging of a mic technique.

1. Set up the stereo mic array in front of a stage.
2. Record yourself speaking from various locations on stage where the instruments will be—center, half-right, far right, half-left, far left. Announce your position.
3. Play back the recording over speakers.

You'll hear how accurately the technique translated your positions, and you'll hear how sharp the images are.

We looked at several mic arrays to record in stereo. Each has its pros and cons. Which method you choose depends on the sonic compromises you're willing to make.

8

MICROPHONE TECHNIQUES

This chapter describes some ways to select and place mics for musical instruments and vocals. These techniques are popular, but they're just suggested starting points. Feel free to experiment.

Before you mike an instrument, listen to it live in the studio, so you know what sound you're starting with. You might want to duplicate that sound through your monitor speakers.

Electric Guitar

Let's start by looking at the chain of guitar, effects, amplifier, and speaker. At each point in the chain where you record, you'll get a different sound (Figure 8.1).

1. The electric guitar puts out an electrical signal that sounds clean and clear.

2. This signal might go through some effects boxes, such as distortion, wah wah, compression, chorus, or stereo effects.

3. Then the signal goes through a guitar amp, which boosts the signal and adds distortion. At the amplifier output (preamp out or external speaker jack), the sound is very bright and edgy.

4. The distorted amp signal is played by the speaker in the amp. Because the speaker rolls off above 4 kHz, it takes the edge off the distortion and makes it more pleasant.

Figure 8.1 Three places to record the electric guitar.

You can record the electric guitar in many ways (Figure 8.1):

• With a mic in front of the guitar amp
• With a direct box
• Both miked and direct
• Through a signal processor or stomp box

The song you're recording will tell you what method it wants. Just mike the amp when you want a rough, raw sound with tube distortion and speaker coloration. Rock 'n' roll or heavy metal usually sounds best with a miked amp. If you record through a direct box, the sound is clean and clear, with crisp highs and deep lows. That might work for quiet jazz or R&B. Use whatever sounds right for the particular song you're recording.

First, try to kill any hum you hear from the guitar amp. Turn up the guitar's volume and treble controls so that the guitar signal overrides hum and noise picked up by the guitar cable. Ask the guitarist to move around, or rotate, to find a spot in the room where hum disappears. Flip the polarity switch on the amp to the lowest-hum position. To remove buzzes between guitar notes, try a noise gate, or ask the player to keep his or her hands on the strings.

Miking the Amp

Small practice amplifiers tend to be better for recording than large, noisy stage amps. If you use a small one, place it on a chair to avoid picking up sound reflections from the floor (unless you like that effect).

A common mic for the guitar amp is a cardioid dynamic type with a "presence peak" in its frequency response (a boost around 5 kHz). The cardioid pattern reduces leakage (off-mic sounds from other instruments). The dynamic type handles loud sounds without distorting, and the presence peak adds "bite." Of course, you can use any mic that sounds good to you.

As a starting point, try miking the amp about an inch from the grille near a speaker cone, slightly off-center—where the cone meets the dome.

The closer you mike the amp, the bassier the tone. The farther off-center the mic is, the duller the tone. Often, distant miking sounds great when you overdub a lead guitar solo played through a stack of speakers in a live room. Try a boundary mic on the floor or on the wall, several feet away.

Recording Direct

Now let's look at recording direct (also known as direct injection or DI). The electric guitar produces an electrical signal that you can plug into your mixer. You bypass the mic and guitar amp, so the sound is clean and clear. Just remember that amp distortion is desirable in some songs.

Mixer mic inputs tend to have an impedance (Z) around 1500 ohms. But a guitar pickup is several thousand ohms. So if you plug a high-Z electric guitar directly into a mic input, the input will load down the pickup and give a thin or dull sound.

To get around this loading problem, use a direct box between the guitar and your mixer (Figure 8.2). The DI box has a high-Z input and low-Z output, thanks to a built-in transformer or circuit. Some mixers

Figure 8.2 Typical direct box.

have a high-Z input jack built in, so you can plug the electric guitar or bass directly into this jack.

The direct box should have a ground-lift switch to prevent ground loops and hum. Set it to the position where you monitor the least hum. You might try a mix of direct sound and miked sound. *Play track 23 on the enclosed CD to hear demonstrations of electric-guitar recording methods.*

Electric Guitar Effects

If you want to record the guitarist's effects, connect the output of the effects boxes into the direct-box input. Many players have a rack of signal processors that creates their unique sound, and they just give you their direct feed. Be open to their suggestions, and be diplomatic about changing the sound. If they are studio players, they often have a better handle on effects than you might as the engineer.

You might want a "fat" or spacious lead-guitar sound. Here are some ways to get it:

- Send the guitar signal through a digital delay set to 20 to 30 msec. Pan guitar left, delay right. Adjust levels for nearly equal loudness from each speaker. (Watch out for phase cancellations in mono.)
- Send the guitar signal through a pitch-shifter, set for about 10 cents of pitch bending. Pan guitar left, pitch-shifted guitar right. (A cent is 1/100 of an equal-temperament semitone. There are 100 cents in a half-tone or semitone interval of pitch).
- Record two guitarists playing identical parts, and pan them left and right. This works great for rhythm-guitar parts in heavy metal.
- Double the guitar. Have the player re-record the same part on an unused track while listening to the original part. Pan the original part left and pan the new part right.
- Add stereo reverb or stereo chorus.

Some guitar processors add many effects to an electric guitar, such as distortion, EQ, chorus, and compression. An example is the Line 6 Pod. You simply plug the electric guitar into the processor, adjust it for the desired sound, and record the signal direct. You wind up with a fully produced sound with a minimum of effort.

Re-amping is a technique that lets you work on the amp's sound during mixdown rather than during recording. Record the guitar direct, then feed that track's signal into a guitar processor or miked guitar amp during mixdown. Use a low- to high-Z transformer between the track output and the processor or amp input. Record the processor or amp on an open track. In a digital audio workstation (DAW), you can start with a track of a direct-recorded guitar, then insert a guitar-amp modeling plug-in.

Electric Bass

BWAM, dik diddy bum. Do your bass tracks sound that clear? Or are they more muddy, like, "Bwuh, dip dubba duh"? Here's how to record the electric bass so it's clean and easy to hear in a mix.

As always, first you work on the sound of the instrument itself. Put on new strings if the old ones sound dull. Adjust the pickup screws (if any) for equal output from each string. Also adjust the intonation and tuning.

Usually, you record the electric bass direct for the cleanest possible sound. A direct pickup gives deeper lows than a miked amp, but the amp gives more midrange punch. You might want to mix the direct and miked sound. Use a condenser or dynamic mic with a good low-frequency response, placed 1 to 6 inches from the speaker.

When mixing a direct signal and a mic signal, make sure they are in-phase with each other. To do this, set them to equal levels and reverse the polarity of the direct signal or the mic signal. The polarity that gives the most bass is correct.

Have the musician play some scales to see if any notes are louder than the rest. You might set a parametric equalizer to soften these notes, or use a compressor.

The bass guitar should be fairly constant in level (a dynamic range of about 6 dB) to be audible throughout the song, and to avoid clipping the recording on loud peaks. To do this, run the bass guitar through a compressor. Set the compression ratio to about 4:1; set the attack time fairly slow (8 to 20 msec) to preserve the attack transient; and set the release time fairly fast (1/4 to 1/2 second). If the release time is too fast, you get harmonic distortion.

EQ can make the bass guitar clearer. Try cutting around 60 to 80 Hz, or at 400 Hz. A boost at 2 to 2.5 kHz adds edge or slap, and a boost at 700 to 900 Hz adds "growl" and harmonic clarity. If you boost the lows around 100 Hz, try boosting at a lower frequency in the kick drum's EQ to keep their sounds distinct. A fretless bass will probably need different EQ or less EQ than a fretted bass.

Here are some ways to make the bass sound clean and well defined:

- Record the bass direct.
- Use no reverb or echo on the bass.
- Have the bass player turn down the bass amp in the studio, just loud enough to play adequately. This reduces muddy-sounding bass leakage into other mics.
- Better yet, don't use the amp. Instead, have the musicians monitor the bass (and each other) with headphones.
- Have the bass player try new strings or a different guitar. Some guitars are better for recording than others. Use roundwound strings for a bright tone or flatwounds for a rounder tone.
- Ask the bass player to use the treble pickup near the bridge.
- Be sure to record the bass with enough edge or harmonics so the bass will be audible on small, cheap speakers.
- Try a bass-guitar signal processor such as the Bass Rockman (discontinued), Zoom, or DigiTech.

If the bass part is full and sustained, it's probably best to go for a mellow sound without much pluck. Let the kick drum define the rhythmic pattern. But if both the bass and kick are rhythmic and work independently, then you should hear the plucks. Listen to the song first, then get a bass sound appropriate for the music. A sharp, twangy timbre is not always right for a ballad; a full, round tone will get lost in a fusion piece.

Often, a musician plays bass lines on a synth or sound module. The module is triggered from a keyboard, a sequencer, or a bass guitar plugged into a pitch-to-MIDI converter. Connect the module output to your mixer line in.

Two effects boxes for the electric bass are the octave box and the bass chorus. The octave box takes the bass signal and drops it an octave in pitch. That is, it divides the bass signal's fundamental frequency in two. You put 82 Hz in; you get 41 Hz out. This gives an extra deep, growly sound. So does a 5-string bass.

A bass chorus gives a wavy, shimmering effect. Like a conventional chorus box, it detunes the signal and combines the detuned signal with the direct signal. Also, it removes the lowest frequencies from the detuned signal, so that the chorus effect doesn't thin out the sound.

Synthesizer, Drum Machine, and Electric Piano

For the most clarity, you usually DI a synth, MIDI sound module, drum machine, or electric piano. Set the volume on the instrument about three-quarters up to get a strong signal. Try to get the sound you want from patch settings rather than EQ.

Plug the instrument into a phone jack input on your mixer, or use a direct box. If you connect to a phone jack and hear hum, you probably have a ground loop. Here are some fixes:

- Power your mixer and the instrument from the same outlet strip. If necessary, use a thick extension cord between the outlet strip and the instrument.
- Use a direct box instead of a guitar cord, and set the ground-lift switch to the position where you monitor the least hum.
- To reduce hum from a low-cost synth, use battery power instead of an AC adapter.

A synth can sound dry and sterile. To get a livelier, funkier sound, you might run the synth signal into a power amp and speakers, and mike the speakers a few feet away.

If the keyboard player has several keyboards plugged into a keyboard mixer, you may want to record a premixed signal from that mixer's output. Record both outputs of stereo keyboards.

Leslie Organ Speaker

This glorious device has a rotating dual-horn on top for highs and a woofer on the bottom for lows. Only one horn of the two makes sound; the other is for weight balance. The swirling, grungy sound comes from the phasiness and Doppler effect of the rotating horn, and from the distorted tube electronics that drive the speaker. Here are a few ways to record it (Figure 8.3):

- In mono: Mike the top and bottom separately, 3 inches to 1 foot away. Aim the mics into the louvers. In the top mic's signal, roll off the lows below 150 Hz.
- In stereo: Record the rotating horn in stereo with a mic on either side. Problem: The horn will sound like it's rotating twice as fast, because the mics will pick up the horn twice per rotation.

137

Figure 8.3 Miking a Leslie organ speaker.

- In stereo: Record the top horn with a stereo mic or a pair of mics out front. Put a mic with a good low end on the bottom speaker, and pan it to center.

When you record the Leslie, watch out for wind noise from the rotating horn and buzz from the motor. Mike farther away if you monitor these noises.

Rather than recording an actual Leslie speaker and Hammond B3 organ, you might prefer to use a software emulation of those instruments. CDs and soft synths are available that simulate great sound with samples, which can be triggered by MIDI sequencers or MIDI controllers (covered in Chapter 16 on MIDI).

Drum Set

The first step is to make the drums sound good live in the studio. If the set sounds poor, you'll have a hard time making it sound great in the control room! You might put the drum set on a riser 1–1/2 feet high to reduce bass leakage and to provide better eye contact between the drummer and the rest of the band. To reduce drum leakage into other mics, you could surround the set with goboes—padded thick-wood panels about 4 feet tall. For more isolation, place the set in a drum booth, a small padded room with windows. It's also common to overdub the set in a live room.

Tuning

One secret to creating a good drum sound lies in careful tuning. It's easier to record a killer sound if you tune the set to sound right in the studio before miking it.

First let's consider drum heads. Plain heads have the most ring or sustain, while heads with sound dots or hydraulic heads dampen the ring. Thin heads are best for recording because they have crisp attack and long sustain. Old heads become dull, so use new heads.

When you tune the toms, first take off the heads and remove the damping mechanism, which can rattle. Put just the top head on and hand-tighten the lugs. Then, using a drum key, tighten opposite pairs of lugs one at a time, one full turn. After you tighten all the lugs, repeat the process, tightening one-half turn. Then press on the head to stretch it. Continue tightening a half-turn at a time until you reach the pitch you want. You'll get the most pleasing tone when the heads are tuned within the range of the shell resonance.

To reduce ugly overtones, try to keep the tension the same around the head. While touching the center of the head, tap with a drumstick on the head near each lug. Adjust tension for equal pitch around the drum. If you want a downward pitch bend after the head is struck, loosen one lug.

Keep the bottom head off the drum for the most projection and the broadest range of tuning. In this case, pack the bottom lugs with felt to prevent rattles. But you may want to add the bottom head for extra control of the sound. Projection is best if the bottom head is tighter than the top head—say, tuned a fourth above the top head. There will be a muted attack, an "open" tone, and some note bending. If you tune the bottom head looser than the top, the tone will be more "closed," with good attack.

With the kick drum (bass drum), a loose head gives lots of slap and attack, and almost no tone. The opposite is true for a tight head. Tune the head to complement the style of music. For more attack or click, use a hard beater.

Tune the snare drum with the snares off. A loose batter head or top head gives a deep, fat sound. A tight batter head sounds bright and crisp. With the snare head or bottom head loose, the tone is deep with little snare buzz, while a tight snare head yields a crisp snare response. Set the snare tension just to the point where the snare wires begin to "choke" the sound, then back off a little.

Damping and Noise Control

Usually the heads should ring without any damping. But if the toms or snare drum rings too much, put some plastic damping rings on them. Or tape some gauze pads, tissues, or folded handkerchiefs to the edge of the heads. Put masking tape on three sides of the pad so that the untaped edge is free to vibrate and dampen the head motion. Don't overdo the damping, or the drum set will sound like cardboard boxes.

Oil the kick drum pedal to prevent squeaks. Tape rattling hardware in place.

Sometimes a snare drum buzzes in sympathetic vibration with a bass-guitar passage or a tom-tom fill. Try to control the buzz by wedging a thick cotton wad between the snares and the drum stand. Or tune the snare to a different pitch than the toms.

Drum Miking

Now you're ready to mike the set. For a tight sound, place a mic near each drum head. For a more open, airy sound, use fewer mics or mix in some room mics placed several feet away. Typical room mics are omni condensers or boundary mics. Figure 8.4 shows typical mic placements for a rock drum set. Let's look at each part of the kit.

Snare

The most popular type of mic for the snare is a cardioid dynamic with a presence peak. The cardioid pattern reduces leakage; its proximity effect boosts the bass for a fatter sound. The presence peak adds attack. You might prefer a cardioid condenser for its sharp transient response.

Bring the mic in from the front of the set on a boom. Place the mic even with the rim, 1 or 2 inches above the head (Figure 8.5). Angle the mic down to aim where the drummer hits, or attach a mini condenser mic to the side of the snare drum so it "looks at" the top head over the rim.

Some engineers mike both the top and bottom heads of the snare drum, with the microphones in opposite polarity. A mic under the snare drum gives a zippy sound; a mic over the snare drum gives a fuller sound. You might prefer to use just a top mic, and move it around until it picks up both the top head and snares. The sound is full with the mic

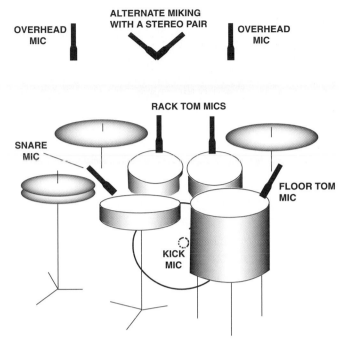

Figure 8.4 Typical mic placements for a rock drum set.

Figure 8.5 Snare-drum miking.

near the top head, and thins out and becomes brighter as you move the mic toward the rim and down the side of the drum.

Whenever the hi-hat closes, it makes a puff of air that can "pop" the snare-drum mic. Place the snare mic so the air puff doesn't hit it. To prevent hi-hat leakage into the snare mic:

141

- Mike the snare closely.
- Bring the snare boom in under the hi-hat, and aim the snare mic away from the hi-hat.
- Use a piece of foam or pillow to block sound from the hi-hat.
- Use a de-esser on the snare.
- Overdub the hi-hat.

Hi-Hat

Try a cardioid condenser mic about 6 inches over the cymbal edge that's farthest from the drummer (Figure 8.6). To avoid the air puff just mentioned, don't mike the hi-hat off its side; mike it from above aiming down. This also reduces snare leakage. You may not need a hi-hat microphone, especially if you use room mics. Usually the overhead mics pick up enough hi-hat.

Tom-Toms

You can mike the toms individually, or put a mic between each pair of toms. The first option sounds more bassy. Place a cardioid dynamic about

Figure 8.6 Hi-hat miking.

142

Figure 8.7 Tom-tom miking.

1 inch over the drumhead and 1 inch in from the rim, angled down about 45 degrees toward the head (Figure 8.7). Again, the cardioid's proximity effect gives a full sound. Another way is to clip mini condenser mics to the toms, peeking over the top rim of each drum.

If the tom mics pick up too much of the cymbals, aim the "dead" rear of the tom mics at the cymbals. If you use a supercardioid or hypercardioid mic, aim the null of best rejection at the cymbals.

Another way to reduce cymbal leakage is to remove the bottom heads from the toms and mike them inside a few inches from the head, off-center. This also keeps the mics out of the drummer's way. The sound picked up inside the tom-tom has less attack and more tone than the sound picked up outside.

Kick Drum

Place a blanket or folded towel inside the drum, pressing against the beater head to dampen the vibration and tighten the beat. The blanket shortens the decay portion of the kick-drum envelope. To emphasize the attack, use a wood or plastic beater—not felt—and tune the drum low.

143

A popular mic for kick drum is a large-diameter, cardioid dynamic type with an extended low-frequency response. Some mics are designed specifically for the kick drum, such as the AKG D112, Audio-Technica AT AE2500, Electro-Voice N/D868, and Shure Beta 52A.

For starters, place the kick mic inside on a boom, a few inches from where the beater hits (Figure 8.8). Mic placement close to the beater picks up a hard beater sound; off-center placement picks up more skin tone, and farther away picks up a boomier shell sound.

Other miking tips: Hang a mini omni condenser mic inside near the beater, or place an omni condenser a few inches from the beater. These mics respond to very deep frequencies and have sharp transient response, which helps the attack.

How should the recorded kick drum sound? Well, they don't call it kick drum for nothing. THUNK! You should hear a powerful low-end thump plus an attack transient.

Kick drum often needs a fair amount of EQ to sound good. Typically you cut several decibels around 400 Hz to remove the "papery" sound, boost 60 to 80 Hz if the kick sounds thin, and boost around 3 to 5 kHz to add click or snap. Don't overdo the high-frequency boost; usually you don't want too much "point" on the kick sound.

Cymbals

To capture all the crisp "ping" of the cymbals, a good mic choice is a cardioid condenser with an extended high-frequency response, flat or rising. Place the overhead mics about 2 to 3 feet above the cymbal edges; closer miking picks up a low-frequency ring. The cymbal edges radiate the most highs. Place the cymbal mics to pick up all the cymbals equally. If your recording will be heard in mono, or for sharper imaging, you might want to mount the mic grilles together and angle the mics apart (Figure 8.4).

Figure 8.8 Kick-drum miking.

144

Or use a stereo mic. Also try a near-coincident pair aimed at the high-hat and floor tom.

Recorded cymbals should sound crisp and smooth, not muffled or harsh.

Room Mics

Besides the close-up drum mics, you might want to use a distant pair of room mics when you record drum overdubs. Place the mics about 10 or 20 feet from the set to pick up room reverb. When mixed with the close-up mics, the room mics give an open, airy sound to the drums. Popular room mics are omni condensers or boundary mics taped to the control-room window. You might compress the room mics for special effect. If you don't have enough tracks for room microphones, try raising the overhead mics.

Boundary Mic Techniques

Boundary mics let you pick up the set in unusual ways. You can strap one on the drummer's chest to pick up the set as the drummer hears it. Tape them to hard-surfaced goboes surrounding the drummer. Put them on the floor under the toms and near the kick drum, or hang a pair over the cymbals. Try a supercardioid boundary mic in the kick drum.

Recording with Two to Four Mics

Sometimes you can mike the set simply. Place a stereo mic (or two mics) overhead and put another mic in the kick. If necessary, add a snare-drum mic (Figure 8.9). This method works well for acoustic jazz, and often for rock. If you want the toms to sound fuller, boost the lows in the overhead mics. Also try two mics about 18 inches apart angled down at the set from just over the drummer's head.

Another setup is shown in Figure 8.10. It uses only one mini omni condenser mic and one kick-drum mic. This method sometimes works well on small drum sets. Clip a mini omni condenser mic to the snare-drum rim about 4 inches above the rim, in the center of the set, aiming at the hi-hat. Also mike the kick drum.

The mini mic will pick up the snare, hi-hat, and toms all around it, and will pick up the cymbals from underneath. Move the mic closer or farther from the toms, and raise or lower the cymbals, until you hear a

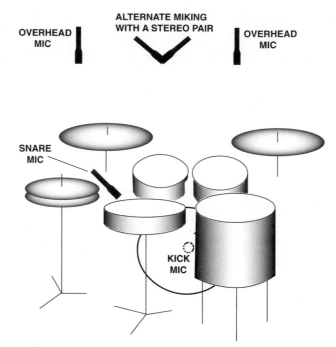

Figure 8.9 Miking a drum set with four mics.

Figure 8.10 Miking a drum set with a mini omni mic.

pleasing balance. Add a little bass and treble. You'll be surprised at the good sound and even coverage you can get with this simple setup.

Want a stereo effect? Mount one mic 4 inches above the snare drum rim between the hi hat, snare drum, and rack tom. Adjust position for best balance. Mount another mic four inches above the floor-tom rim, on the side farthest from the drummer (Figure 8.11). Pan the mini mics left and right.

Track 24 on the enclosed CD demonstrates several methods of miking a drum set.

Drum Recording Tips

After you set up all the mics, ask the drummer to play. Listen for rattles and leakage by soloing each microphone. Try not to spend much time getting a sound; otherwise you waste the other musicians' time and wear out the drummer.

To keep the drum sound tight, turn off mics not in use in a particular tune, use a noise gate on each drum mic, or overdub the drums.

One effect for the snare drum is gated reverb. It's a short splash of bright-sounding reverberation, which is rapidly cut off by a noise gate or expander. Many effects units have a gated-reverb program.

Another trick is recording "hot." Using an analog multitrack (or its plug-in equivalent), record the drums at a high level so they distort just a little. It's also common to compress the kick.

Figure 8.11 Miking a drum set with two mini omni mics.

A drummer might use drum pads, or drum triggers, fed into a sound module. Record directly off the module. You might want to mike the cymbals anyway for best sound.

If you're recording a drum machine and it sounds too mechanical, add some real drums. The machine can play a steady background while the drummer plays fills.

When miking drums on stage for PA, you don't need a forest of unsightly mic stands and booms. Instead, you can use short mic holders that clip onto drum rims and cymbal stands, or use mini condenser mics.

In a typical rock mix, the drums either are the loudest element, or are slightly quieter than the lead vocal. The kick drum is almost as loud as the snare. If you don't want a wimpy mix, keep those drums up front!

Try these EQ settings to enhance the recorded sound of the drums:

- Snare and rack toms: Fat at 200 Hz, crack at 5 kHz. Cut around 400 Hz on toms for clarity. If the sound is too tubby, cut around 200 Hz.

- Floor toms: Fullness at 80 to 100 Hz.

- Cymbals: Sizzle at 10 kHz or higher. Roll off the lows below 500 Hz to reduce low-frequency leakage.

- Kick drum: Boost at 3 to 5 kHz for click. Filter out highs above 9 kHz to reduce leakage from cymbals. To remove the "cardboard" sound, cut at 300 to 600 Hz.

A typical track assignment for drums might be

1. Kick
2. Snare
3 & 4. Toms in stereo
5 & 6. Overheads in stereo

Try these tricks to come up with unusual drum sounds:

- Record with a cheap dynamic or crystal mic, maybe in a can.

- Run the drums through extreme processing: compression, gating, distortion, pitch shifting, tremolo, and so on.

- Substitute other objects for drums, cymbals, drumsticks, and brushes.

- Move a mic around a cymbal or drumhead while recording it.

- Put the drums in a reverberant room or hallway.

- Try the preverb effect described in Chapter 10.

Instead of recording an acoustic drum set, you might use an electronic drum set or CDs of drum samples. Copy the samples into a sampler or sampling software, then trigger them with a MIDI sequencer or MIDI controller.

Percussion

Let's move on to percussion, such as the cowbell, triangle, tambourine, or bell tree. A good mic for metal percussion is a condenser type because it has sharp transient response. Mike at least 1 foot away so the mic doesn't distort.

You can pick up congas, bongos, and timbales with a single mic between the pair, a few inches over the top rim, aimed at the heads. Or put a mic on each drum. It often helps to mike these drums top and bottom, with the bottom mic in opposite polarity. A cardioid dynamic with a presence peak gives a full sound with a clear attack.

For xylophones and vibraphones, place two cardioid mics 1–1/2 feet above the instrument, aiming down. Cross the mics 135 degrees apart or place them about 2 feet apart. You'll get a balanced pickup of the whole instrument.

Acoustic Guitar

The acoustic guitar has a delicate timbre that you can capture through careful mic selection and placement. First prepare the acoustic guitar for recording. To reduce finger squeaks, try commercial string lubricant, a household cleaner/waxer, talcum powder on fingers, or smooth-wound strings. Ask the guitarist to play louder; this increases the "music-to-squeak" ratio!

Replace old strings with new ones a few days before the session. Experiment with different kinds of guitars, picks, and finger picking to get a sound that's right for the song.

For acoustic guitar, a popular mic is a condenser with a smooth, extended frequency response from 80 Hz up. This kind of mic has a clear, detailed sound. You can hear each string being plucked in a strummed chord. Usually the sound picked up is as crisp as the real thing.

Now let's look at some mic positions. To record a classical guitar solo in a recital hall, mike about 3 to 6 feet away to pick up room reverb. Try a stereo pair (Figure 8.12A), such as XY, ORTF, MS, or spaced pair

Figure 8.12 Some mic techniques for acoustic guitar.

(described in Chapter 7). If you record a classical guitar solo in a dead studio, mike about 1.5 to 2 feet away and add artificial reverb.

When you record pop, folk, or rock music, try a spot about 6 to 12 inches from where the fingerboard joins the guitar body—at about the 12th fret (Figure 8.12B). That's a good starting point for capturing the acoustic guitar accurately. Still, you need to experiment and use your ears. Close to the bridge, the sound is woody and mellow.

In general, close miking gives more isolation, but tends to sound harsh and aggressive. Distant miking lets the instrument "breathe"; you hear a gentler, more open sound.

Another spot to try: Tape a mini omni mic onto the body near the bottom of the sound hole, and roll off the excess bass. This spot gives good isolation (Figure 8.12C).

The guitar will sound more real if you record in stereo. Try one mic near the bridge, and another near the 12th fret (Figures 8.12D and E). Pan part-way left right. Another way to record stereo is with an XY pair of cardioid mics about 6 inches from the end of the fingerboard, mixed with a 3-foot-spaced pair of omni mics about 3 feet away.

Is feedback or leakage a problem? Mike close to the sound hole (Figure 8.12F). The tone there is very bassy, so turn down the low-frequency EQ on your mixer until the sound is natural. Also cut a few decibels around 3 kHz to reduce harshness.

You get the most isolation with a contact pickup. It attaches to the guitar, usually under the bridge. The sound of a pickup is something like an electric guitar. You can mix a mic with a pickup to add air and string noise to the sound of the pickup. That way, you get good isolation and good tone quality.

Singer/Guitarist

Normally you overdub the guitar and vocal separately. But if you have to record both at once, the vocal might sound filtered or hollow because of phase cancellations between the vocal mic and guitar mic. This can happen whenever two mics pick up the same source at approximately equal levels, at different distances, and both mixed to the same channel. Try one of these methods to solve the problem:

- Angle the vocal mic up and angle the guitar mic down to isolate the two sources. Follow the 3:1 rule described in Chapter 7.
- Use a pickup or mini mic on the guitar.
- Delay the vocal mic about 1 msec. This keeps the two mic signals in phase, preventing phase cancellations. Some multitrack recorders have a track-delay feature for this purpose.
- Use a coincident pair of figure-eight (bidirectional) mics crossed at 90 degrees. Aim the front of one mic at the voice; aim the front of the other mic at the guitar.
- Use a stereo mic or stereo pair about 1 foot out front; raise or lower the mics to adjust the voice/guitar balance.

Grand Piano

This magnificent instrument is a challenge to record well. First have the piano tuned, and oil the pedals to reduce squeaks. You can prevent thumps by stuffing some foam or cloth under the pedal mechanism.

For a classical-music solo, record in a reverberant room such as a recital hall or concert hall. Reverb is part of the sound. Set the piano lid on the long stick. Use condenser mics with a flat response. Place a stereo mic, or a stereo pair of cardioid mics, about 7 feet away and 7 feet high, up to 9 feet away and 9 feet high (Figure 8.13). Move the mics closer to reduce reverb, farther to increase it. When using a pair of omni mics, place them 1.3 to 2 feet apart, 3 to 6.5 feet from the piano, and 4 to 5 feet high (Figure 8.14). You might need to mix in a pair of hall mics: Try cardioids aiming away from the piano about 25 feet away.

Figure 8.13 Suggested grand-piano miking for classical music (using cardioid mics).

Figure 8.14 Suggested grand-piano miking for classical music (using omnidirectional mics).

When recording a piano concerto, give the piano a spot mic about 3 feet away. Put the mic in a shock mount.

Pop music demands close miking. Close mics pick up less room acoustics and leakage, and give a clear sound that cuts through the mix. Try not to mike the strings closer than 8 inches, or else you'll emphasize the strings closest to the mics. You want equal coverage of all the notes the pianist plays.

One popular method uses two spaced mics inside the piano. Use omni or cardioid condensers, ideally in shock mounts. Put the lid on the long stick. If you can, remove the lid to reduce boominess. Center one mic over the treble strings and one over the bass strings. Typically, both mics are 8 to 12 inches over the strings and 8 inches horizontally from the hammers (Figure 8.15, top, bass and treble mics). Aim the mics straight down or angle them to aim at the hammers. Pan the mics partly

Figure 8.15   Suggested grand-piano miking for popular music.

left and right for stereo. As an alternative, try two ear-spaced omni condensers or an ORTF pair about 12 to 18 inches above the strings. *Track 25 on the enclosed CD demonstrates some mic techniques for grand piano.*

The spaced mics might have phase cancellations when mixed to mono, so you might want to try coincident miking (Figure 8.15, top, stereo pair). Boom-mount a stereo mic, or an XY pair of cardioids crossed at 120 degrees. Miking close to the hammers sounds percussive; toward the tail has more tone.

For more clarity and attack, boost EQ around 10 kHz or use a mic with a rising high-frequency response.

Boundary mics work well, too. If you want to pick up the piano in mono, tape a boundary mic to the underside of the raised lid, in the center of the strings, near the hammers. Use two for stereo over the bass and treble strings. Put the bass mic near the tail of the piano to equalize the mic distances to the hammers (Figure 8.15, bottom). If leakage is a problem, close the lid and cut EQ around 250 Hz to reduce boominess.

If your studio lacks a piano, consider using a software emulation of a piano. Some programs provide high-quality samples of piano notes that can be played with a sequencer or a MIDI controller. Examples: Steinberg Grand VST 2.0 ($199 at www.steinberg.net) and Maxim Digital Audio Piano (freeware at www.mda-vst.com).

Upright Piano

Here are some ways to mike an upright piano (Figure 8.16):

(A) Remove the panel in front of the piano to expose the strings over the keyboard. Place one mic near the bass strings and one near the treble strings about 8 inches away. Record in stereo and pan the signals left and right for the desired piano width. If you can spare only one mic for the piano, just cover the treble strings.

(B) Remove the top lid and upper panel. Put a stereo pair of mics about 1 foot in front and 1 foot over the top. If the piano is against a wall, angle the piano about 17 degrees from the wall to reduce tubby resonances.

(C) Aim the soundboard into the room. Mike the bass and treble sides of the soundboard a few inches away. In this spot, the mics pick up less pedal thumps and other noises. Try cardioid dynamic mics with a presence peak.

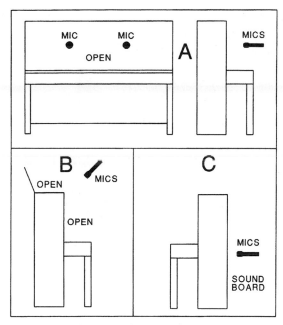

Figure 8.16 Some mic techniques for upright piano.

Acoustic Bass

The acoustic bass (string bass, double bass) puts out frequencies as low as 41 Hz, so use a mic with an extended low-frequency response. As always, closer miking improves isolation, while distant miking tends to sound more natural. Try these techniques (Figure 8.17):

- 4 to 18 inches in front of the bridge, on the side toward the G string (top string), a few inches above the bridge.
- For more fullness, move the mic toward the f-hole. Move the mic upward for more definition.
- 18 to 24 inches above the treble f-hole.
- Mix a pickup with a mic, or use a pickup alone and EQ it to sound good.

If you need more isolation, place a cardioid dynamic mic near the treble f-hole and roll off the excess bass on your mixer. Or try a cardioid dynamic mic 6 inches from the strings, 4 inches below the bridge, pointing at the base of the bridge (Figure 8.17). Mix with a pickup.

155

ACOUSTIC BASS
SIDE VIEW

18-24 IN. ABOVE
TREBLE F-HOLE

MIC OUT FRONT

4-18 IN.

MIC IN FOAM
BEHIND BRIDGE

CARDIOID DYNAMIC

Figure 8.17 Some mic techniques for the acoustic bass.

Here are some methods that isolate the bass and let the player move around. They work well for PA:

- Wrap a mini omni condenser mic in foam rubber (or in a foam windscreen) and mount it in the bridge aiming up (Figure 8.17).
- Tape a mini omni mic to the bridge.
- Wrap a regular cardioid mic in foam padding (except the front grille) and squeeze it behind the bridge (Figure 8.17) or tailpiece.
- For best isolation, try a direct feed from a pickup. This method adds clarity and deep bass, but probably will need some EQ. You might mix the pickup with a microphone.

Banjo

Try a flat-response mic about 1 foot away (Figure 8.18). If you need more isolation, mike closer and roll off some bass. The banjo sounds pleasantly mellow when miked toward the edge of the head, near the resonator

BANJO

Figure 8.18 Four methods for miking a banjo.

holes (if the banjo has them). Cloth stuffed inside will reduce feedback in PA situations.

For the most isolation, tape a mini omni condenser mic to the head about 1 inch in from the bottom edge, or on the tailpiece, or on the bridge. You can wedge a pickup between the strings below the bridge and the banjo head. Put the pickup flat against the head surface.

Mandolin, Dobro, Bouzouki, and Lap Dulcimer

Mike these about 8 to 12 inches away with a condenser mic. If you need more lows and more isolation, mike close to an f-hole. You can tape a mini omni condenser mic near an f-hole and tweak EQ for the best sound.

Hammered Dulcimer

Place a flat-response condenser mic about 2 feet over the center of the soundboard (Figure 8.19A). On stage, place a cardioid dynamic or condenser 6 to 12 inches over the middle of the top end (Figure 8.19B). For the best gain-before-feedback in a PA system, mix in a mini omni condenser mic (or a cardioid with bass rolloff) very near the sound hole (Figure 8.19C).

Fiddle (Violin)

Listen to the fiddle itself to make sure it sounds good. Correct any instrument problems before miking.

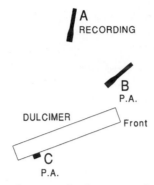

Figure 8.19 Some mic techniques for hammered dulcimer.

Figure 8.20 Three fiddle-miking methods.

First try a flat-response condenser mic (omni or cardioid) about 2 feet over the bridge. This distant miking gives an airy, silky sound. Close miking (about 1 foot, Figure 8.20) sounds more aggressive, which is desirable in old-time or bluegrass music. Aim the mic toward the f-holes for warmth or toward the fingerboard for clarity. Try miking the fiddle from the side if you want to reduce the midrange (around 1 to 2 kHz). If the ceiling is low, nail a square yard of acoustic foam up there to prevent reflections.

If you have to mike close—say, for a singing fiddler—aim the mic horizontally at the mouth about 6 inches away (Figure 8.20), or aim the mic at the player's chin from 1 foot above the fiddle. The mic will pick up both the singer and the fiddle.

If you need more isolation, try a mini omni mic. Wrap its cable in foam rubber (or a windscreen) 1–1/2 inches from the capsule. Wedge the foam under the tailpiece, and position the mic capsule halfway between the tailpiece and bridge, a half inch over the body (Figure 8.21). If necessary, cut a little at 3 kHz to reduce harshness and boost around 200 Hz for warmth. Another option is to clip the mic to the tailpiece and mount it over an f-hole.

A good spot for a pickup is on the left side of the top (player's view), on the player's side of the bridge.

To record a classical violin solo, try a stereo mic (or a stereo pair) 5 to 15 feet away in a reverberant room.

String Section

Place the strings in a large, live room and mike them at distance to pick up a natural acoustic sound. A common mic choice is a condenser with a flat response. First try a stereo mic or stereo pair of mics about 4 to 20 feet behind the conductor, raised about 15 feet.

MINI OMNI
MIC PLACEMENTS
ON THE FIDDLE

OVER
F-HOLE,
CLIPPED TO
TAILPIECE

BETWEEN
TAILPIECE
AND BRIDGE

Figure 8.21 Two ways to close-mike a fiddle for isolation.

If the room is noisy or too dead, or the balance is poor, you'll need to mike close and add digital reverb. Try one mic on every two to four violins, 6 feet off the floor, aiming down. Same for the violas. Mike the cello about 1 to 2 feet from the bridge, to the right side between the bridge and f-hole. When you mix the strings to stereo, pan them evenly between the monitor speakers. Spread them left, center, and right to make a "curtain of sound." If you can spare only one track for the strings, use a stereoizer effect during mixdown.

String Quartet

Record a quartet in stereo using a stereo mic or a pair of mics. Place them about 6 to 10 feet away to capture the room ambience. The monitored instruments should not spread all the way between speakers. If you want to narrow the stereo stage, angle or space the mics closer together.

Bluegrass Band and Old-Time String Band

Suppose you're recording a group that has a good acoustic balance. Try a stereo mic or stereo pair of mics about 3 feet away and 6 feet high (lower if the group is seated). Move the players toward or away from the mics to adjust their balance.

You'll have more control if you mike all the instruments up close and mix them. This also gives a more "commercial" sound. The production style aims for a natural timbre on all the instruments, either with no effects or with slight reverb.

Harp

Use a condenser mic with a flat response. If the harp is playing with an orchestra, mike the harp about 18 inches from the front of the soundboard, or 18 inches from the player's left hand. You can mike a harp solo about 4 feet over the top.

Tape a mini omni condenser mic to the soundboard if you need more isolation. A mic on the inside of the soundboard has more isolation; a mic on the outside sounds more natural. Also try a cardioid condenser wrapped in foam, stuck into the center hole from the rear.

Horns

"Horns" in studio parlance refers to the brass instruments: trumpets, cornets, trombones, baritones, french horns, and tubas.

All the brass radiate strong highs straight out from the bell, but do not project them to the sides. A mic close to and in front of the bell picks up a bright, edgy tone. To mellow out the tone, mike the bell off-axis with a flat-response mic (Figure 8.22). The sound on-axis to the bell has a lot of spiky high harmonics that can overload a condenser mic, mixer input, or analog tape. That's another reason to mike off-axis.

Mike the trumpet with a dynamic or ribbon mic to take the edge off the sound. Use a condenser mic if you want a lot of sizzle. Mike about 1 foot away for a tight sound; mike several feet away for a fuller, more dramatic sound.

You can pick up two or more horns with one microphone. Several players can be grouped around a single omni mic, or around a stereo pair of mics. The musicians can play to a pair of boundary mics taped on the control-room window or on a large panel.

Record a classical brass quartet in a reverberant room. Use a stereo mic, or a stereo pair of mics, about 6 to 12 feet away.

Saxophone

A sax miked very near the bell sounds bright, breathy, and rather hard (Figure 8.20). Mike it there for best isolation. To get a warm, natural sound, mike the sax about 1–1/2 feet away, halfway down the wind column (Figure 8.23). Don't mike too close, or else the level varies when

MELLOW

BRIGHT

TRUMPET – TOP VIEW

Figure 8.22 Miking for trumpet tone control.

Figure 8.23 Two ways to mike a saxophone.

Figure 8.24 Typical miking setup for big-band jazz.

the player moves. A compromise position for a close-up mic is just above the bell, aiming at the holes. You can group a sax section around one mic.

Figure 8.24 shows a typical miking setup for big-band jazz. It uses the techniques already described for the drums, bass, piano, electric guitar, trumpet, and sax.

Woodwinds

With woodwinds, most of the sound radiates not from the bell, but from the holes. So aim a flat-response mic at the holes about 1 foot away (Figure 8.25).

Figure 8.25   Miking a clarinet from the side.

Figure 8.26   Two methods of miking a flute.

When miking a woodwind section within an orchestra, you need to reject nearby leakage from other instruments. To do that, try aiming a bidirectional mic down over the woodwind section. The side nulls of the mic cut down on leakage.

To pick up a flute in a pop-music group, try miking a few inches from the area between the mouthpiece and the first set of finger holes (Figure 8.26). You may need a pop filter. If you want to reduce breath noise, roll off high frequencies or mike farther away. You also can attach a mini omni mic to the flute a few inches above the body, between the mouthpiece and finger holes.

For classical music solos, try a stereo pair 4 to 12 feet away.

Harmonica, Accordion, and Bagpipe

One way to mike a harmonica (harp) is to use a cardioid dynamic mic with a ball grille. Place the mic very close to the harmonica or have the player hold it. A condenser mic about 1 foot away gives a natural sound. To get a bluesy, dirty sound, use a "bullet"-type harmonica mic or play the harmonica through a miked guitar amp.

For an accordion, try a mic about 6 to 12 inches from the sound holes near the keyboard. Some accordions have sound holes on both sides, so you'll need two mics. Follow the 3:1 rule. The distance between mics should be at least three times the mic-to-source distance. One end of the accordion is in constant motion, so you might want to attach a mini omni mic to that end. A solo accordion or concertina could be miked with a stereo pair of flat-response cardioid condenser mics about 3 to 6 feet in front.

A bagpipe has two main sound sources: the chanter, which the musician plays with the fingers, and the drone pipes, which make a steady tone. Mike the chanter about a foot away from the side, and mike the drone pipes a foot from the end. Again, follow the 3:1 rule. You could also mike the bagpipe a few feet away with one mic.

Lead Vocal

The lead vocal is the most important part of a pop song, so it's critical to record it right. First set up a comfortable environment for the singer. Put down a rug, add some flowers or candles, dim the lights. Set up a good cue mix with effects to help the singer get into the mood of the song.

You might want to turn off the reverb in the singer's headphones; this makes it easier to hear pitch. If the vocalist is singing flat, reduce their headphone volume, and vice versa.

With any vocal recording, there are some problems to overcome, but we can deal with them. Among these are proximity effect, breath pops, wide dynamic range, sibilance, and sound reflections from the music stand. Let's look at these in detail.

Miking Distance

When you sing or talk close to most directional mics, the microphone boosts the bass in your voice. This is called the proximity effect. We've come to accept this bassy sound as normal in a PA system, but the effect just sounds boomy in a recording.

To prevent boomy bass, mike the singer at a distance, about 8 inches away (Figure 8.27). A popular mic choice is a flat-response condenser mic with a large diaphragm (1–1/4-inch diameter). As always, you can use any mic that sounds good to you. If the mic has a bass rolloff switch, set it to "flat."

Singers should maintain their distance to the mic. I ask the singer to spread the fingers, touch lips with the thumb, and touch the mic with the pinky. The hand forms a spacer for keeping constant distance.

Some singers can't help but "eat" the mic. You can mike them at distance, and also give them a dummy mic to hold while singing.

If you must record the singer and the band at the same time—as in a concert—you'll have to mike close to avoid picking up the instruments with the vocal mic. Try a cardioid mic with a bass rolloff and a foam pop filter. The sound will be bassy because of proximity effect, so roll off the excess lows at your mixer. For starters, try −6dB at 100Hz. Some mics have a bass filter switch for this purpose. Aim the mic partly toward the singer's nose to prevent a nasal or closed-nose effect. This close-up method works well if you want an intimate, breathy sound.

When recording a classical-music singer who is accompanied by an orchestra, place the mic about 1 to 2 feet away. If the singer is a soloist (maybe accompanied by piano), use a stereo pair about 8 to 15 feet away to pick up room reverb.

Figure 8.27 Typical miking technique for a lead vocal.

Breath Pops

When you sing a word with "p" or "t" sounds, a turbulent puff of air shoots out of the mouth. The puff hits the mic and makes a thump or small explosion called a pop. To reduce it, put a foam-plastic pop filter on the mic. Some mics have a ball grille screen to cut pops, but foam works better. The pop filter should be made of special open-cell foam to pass high frequencies. For best pop rejection, allow a little air space between the foam and the front of the mic grille.

Foam pop filters reduce the highs a little. So they should be left off instrument mics, except for outdoor recording or dust protection. Pop filters do not reduce breathing sounds or lip noises. To get rid of these problems, mike farther away or roll off some highs.

The most effective pop filter is a hoop with a nylon stocking stretched over it (Figure 8.27) or a disk of perforated metal. You can buy those, or make one with a coat hanger and a crochet hoop. Place the filter a few inches from the mic.

Another way to get rid of pop is to put the mic at forehead height, aiming at the mouth. This way the puffs of air shoot under the mic and miss it. Make sure the vocalist sings straight ahead, not up at the mic, or the mic will pop.

Wide Dynamic Range

During a song, vocalists often sing too loud or too soft. They blast the listener or get buried in the mix. That is, many singers have a wider dynamic range than their instrumental backup. To even out these extreme level variations, ask the singer to use proper mic technique. Back away from the mic on loud notes; come in closer for soft ones. Or you can ride gain on singers: gently turn them down as they get louder, and vice versa.

Another solution is to pass the vocal signal through a compressor, which acts like an automatic volume control. Plug the compressor into the vocal channel's insert jacks. A typical compressor setting for vocals is a 2:1 ratio, −10 dB threshold, and about 3 to 6 dB of gain reduction. Of course, you should use whatever settings are needed for the particular singer. *Track 26 on the enclosed CD demonstrates vocal compression, as well as breath pops and the effects of miking distance on recorded vocals.*

If the singer moves toward and away from the mic while singing, the average level will go up and down. Try to mike the singer at least 8 inches away, so that small movements of the singer won't affect the level.

If you must mike close to prevent leakage or feedback, ask the vocalist to sing with lips touching the foam windscreen to keep the same distance to the mic. Turn down the excess bass using your mixer's low-frequency EQ (typically −6 dB at 100 Hz).

Sibilance

Sibilance is the emphasis of "s" or "sh" sounds, which are strongest around 5 to 10 kHz. They help intelligibility. In fact, many producers like sizzly "s" sounds, which add a bright splash to the vocal reverb. But the sibilance should not be piercing or strident.

If you want to reduce sibilance, use a mic with a flat response—rather than one with a presence peak—or cut the highs little around 8 kHz on your mixer. Better yet, use a de-esser signal processor or plug-in, which cuts the highs only when the singer makes sibilant sounds.

Reflections from the Music Stand and Ceiling

Suppose that a lyric sheet or music stand is near the singer's mic. Some sound waves from the singer go directly into the mic. Other sound waves reflect off the lyric sheet or music stand into the mic (Figure 8.28, top).

Figure 8.28 Preventing reflections from a music stand.

The delayed reflections will interfere with the direct sound, making a colored tone quality like mild flanging.

To prevent this, lower the music stand and tilt it almost vertically (Figure 8.28, bottom). This way, the sound reflections miss the mic.

If your studio has a low ceiling, the recorded vocal might have a colored tone quality due to phase cancellations from ceiling reflections. Try putting the mic lower and use a hoop-type pop filter. Also put a 3-foot square of acoustic foam on the ceiling over the singer and mic.

Vocal Effects

Some popular vocal effects are stereo reverb, echo, and doubling. You can record real room reverb by miking the singer at a distance in a hard-surfaced room. Slap echo provides a 1950s rock 'n' roll effect. Often a vocal is mixed dry, with no reverb. A little distortion might even be effective on some songs. You might try a vocal processor, which offers a variety of effects. Try different EQ or different effects on each section of a song.

Doubling a vocal gives a fuller sound than a single vocal track. Overdub a second take of the vocal on an empty track, in sync with the original take. During mixdown, mix the second vocal take with the original, at a slightly lower level than the original. You can double a vocal track by running it through a digital delay set at 15 to 35 msec, or through a pitch shifter that is detuned 10 to 15 cents.

Background Vocals

When you overdub background vocals (harmony vocals), you can group two or three singers in front of a mic. The farther they are from the mic, the more distant they will sound in the recording. Pan the singers left and right for a stereo effect. Because massed harmonies can sound bassy, roll off some lows in the background vocals.

If you want independent control of each background singer, give each one a close-up mic and record them with separate mixer channels or separate tracks.

Barbershop or gospel quartets with a good natural blend can be recorded with a stereo mic or stereo pair of mics about 2 to 4 feet away. If their balance is poor, close-mike each singer about 8 inches away, and balance them with your mixer. This also gives a more "commercial" sound. If you close-mike, spread the singers at least 2 feet apart to prevent phase cancellations.

Spoken Word

The tips given earlier for a lead vocalist also apply to recording the spoken word. Be sure to keep the miking distance constant and use a hoop-type pop filter. To prevent sound reflections into the mic, put the script on a padded music stand that is angled almost vertically, and put the mic in the plane of the stand near the top edge. Fold up a corner of each script page to form a handle for turning pages silently.

The engineer and announcer should both have the same script. Mark the beginning of each misread sentence. The announcer should re-read each misread sentence from the beginning to make editing easier.

Choir and Orchestra

Figure 8.29 shows three ways of miking a choir. If the mics will also be used for PA, or if the venue is noisy or sounds bad, try miking close, pan the mic as desired, and add artificial reverb (Figure 8.29A). Otherwise, try a near-coincident pair of cardioid mics (Figure 8.29B) or a pair of omni

Figure 8.29 Choir miking suggestions. (A) Close-up panned mics. (B) Near-coincident stereo pair. (C) Spaced stereo pair.

mics spaced about 2 feet apart (Figure 8.29C). Adjust the mic-to-choir distance until you hear the desired amount of hall acoustics in your monitors.

See Chapters 18 and 19 for suggestions on miking an orchestra.

Summary

We can sum up mic placement like this: If leakage or feedback are problems, place the mic near the loudest part of the instrument, and add EQ to get a natural sound. Otherwise, place the mic in various spots until you find a position that sounds good over your monitors. There is no single "correct" mic technique for any instrument. Just place the mic where you hear the desired tonal balance and amount of room reverb.

Try the techniques described here as a starting point, then explore your own ideas. Trust your ears! If you capture the power and excitement of electric guitars and drums, if you capture the beautiful timbre of acoustic instruments and vocals, you've made a successful recording.

9

DIGITAL RECORDING

In the past, everything was recorded on analog recorders. The magnetic particles on tape were oriented in patterns analogous to the audio waveform. In contrast, digital recorders convert the audio signal to a numerical code of ones and zeros.

Let's venture into the world of digital audio. We'll overview how digital recording works, explore 2-track digital recorders, and explain multitrack digital recorders. Chapter 13 covers computer recording in detail.

Analog versus Digital

Analog and digital recorders don't sound the same. Analog decks sound reasonably accurate, but they add a little warmth to the sound. It's due to slight third harmonic distortion, head bumps (bass boost), and tape compression. Analog decks also add some tape hiss, frequency response errors, wow and flutter, modulation noise, and print-through.

Digital recorders don't have these problems, so they sound very clean. Although older digital recorders sounded harsh compared to analog, they improved with each generation. In particular, digital recorders that can record at 24 bits and 96kHz can sound just as smooth as analog.

Compared to analog recorders and open-reel tape, digital recorders and their tape tend to cost less, are smaller, allow easier location of timing information, and allow easier loading of the recording medium.

Digital Recording

Like an analog tape deck, a digital recorder puts audio on a magnetic medium, but in a different way. Here's what happens in the most common digital recording method—pulse code modulation or PCM:

1. The signal from your mixer (Figure 9.1A) is run through a lowpass filter (anti-aliasing filter), which removes all frequencies above 20kHz.
2. Next, the filtered signal passes through an analog-to-digital (A/D) converter. This converter measures (samples) the voltage of the audio waveform several thousand times a second (Figure 9.1B).
3. Each time the waveform is measured, a binary number (made of 1's and 0's) is generated that represents the voltage of the waveform at the instant it is measured (Figure 9.1C). This process is called quantization. Each 1 and 0 is called a bit, which stands for binary digit. The more bits that are used to represent each measurement (the higher the bit depth), the more accurate the measurement is.
4. These binary numbers are stored on the recording medium as a modulated square wave recorded at maximum level (Figure 9.1D). For example, the numbers can be stored magnetically on a hard disk.

The playback process is the reverse:

1. The binary numbers are read from the recording medium (such as a hard disk).
2. The digital-to-analog (D/A) converter translates the numbers back into an analog signal made of voltage steps.
3. An anti-imaging filter (lowpass filter) smooths the steps in the analog signal, resulting in the original analog signal.

Digital recording reduces noise, distortion, speed variations, and data errors. Because the digital playback head reads only 1's and 0's, it is insensitive to the magnetic medium's noise and distortion. During recording and playback, numbers are read into a buffer memory and read out at a constant rate, eliminating speed variations in the rotating media.

(A) The audio waveform enters the A/D converter.

(B) The voltage is measured at regular intervals.

(C) The voltage measurements are quantized.

(D) The binary numbers are stored in memory, tape or disk.

Figure 9.1 Digital recording by pulse code modulation (PCM)

Reed-Solomon coding during recording, and decoding during playback, corrects for missing bits by using redundant data.

If a digital recording is on a defective medium such as a scratched compact disc, errors (missing samples) can occur. Usually these errors can be corrected by interpolation. This algorithm looks at data before and after the blank sample and "guesses" what its value should be. If the errors are too extended to correct, the resulting audio has a silent spot or a burst of noise.

173

Almost all digital recording devices employ the same A/D, D/A conversion process, but use different storage media: a DAT machine records on tape, a hard-disk drive records on magnetic hard disk, a compact disc and DVD recorder record on an optical disc, a memory recorder records onto a flash memory card, and a sampler records into computer memory. The sound quality of any of these devices depends mainly on its A/D and D/A converters.

Digital audio is recorded on a computer hard drive as a wave file or AIFF file. Both are standard formats for audio files. Wave (.wav) is for PC; AIFF (Audio Interchange File Format) is for Mac. Both formats use linear PCM encoding, with no data compression (explained in Chapter 20). Two wave formats are Riff and Broadcast wave, which facilitates interchange of program material between audio workstations.

Bit Depth

As we said, the audio signal is measured many thousand times a second to generate a string of binary numbers (called words). The longer each word is (the more bits it has), the greater is the accuracy of each measurement. Short words give poor resolution of the signal voltage (high distortion); long words give good resolution (low distortion). Bit depth or resolution are other terms for word length.

A word length of 16 bits is adequate (but not optimum) for hi-fi reproduction. It is the current standard for CDs. Some digital recorders offer 20- or 24-bit word lengths. More bits sound smoother and more transparent, but need more disk storage space and a faster hard drive. CDs sound better when made from 24-bit recordings.

Sampling Rate

Sampling rate is the rate at which the A/D converter samples or measures the analog signal while recording. For example, a rate of 48 kHz is 48,000 samples per second; that is, 48,000 measurements are generated for each second of sound. The higher the sampling rate, the wider the frequency response of the recording. According to the Nyquist theorem, the upper frequency limit is one-half the sampling rate. Compact discs use a 44.1-kHz sampling rate, so their frequency response extends to 22.05 kHz.

Sampling rate for high-quality audio can be 44.1 kHz, 48 kHz, 88.2 kHz, 96 kHz, or 192 kHz. Higher sampling rates sound smoother and more transparent but need more disk storage space and a faster hard drive. CD quality is 44.1 kHz/16 bits. A 96-kHz sampling rate can be used on the DVD. State-of-the-art is Super Audio CD or linear PCM at 192 kHz/24 bits (but you're more likely to see 96 kHz/24 bits).

In summary, a digital audio system samples the analog signal several thousand times a second, and quantizes (assigns a value to) each sample. Sampling rate affects the high-frequency response. Bit depth affects the dynamic range, noise, and distortion.

In a digital transmission, the two channels of a stereo program are multiplexed. That is, one word from channel 1 is followed by one word from channel 2, which is followed by one word from channel 1, and so on.

Data Rate and Storage Requirements

The data rate of digital audio (in bytes per second) is

Bit depth/8 × sampling rate × number of tracks.

Divide by 1,048,576 to get megabytes (MB) per second. For example, the data rate of a 24-bit/44.1-kHz recording of 16 tracks is

$$(24/8 \times 44,100 \times 16)/1,048,576 = 2 \text{ MB/sec.}$$

Recording digital audio on a hard drive consumes a lot of space. The storage required is

Bit depth/8 × sampling rate × number of tracks × 60
× number of minutes

Divide by 1,048,576 to get MB. Divide MB by 1024 to get gigabytes (GB). For example, suppose you record a concert at 24 bits, 44.1 kHz, 16 tracks, for 2 hours. The hard-drive space needed is

$$(24/8 \times 44,100 \times 16 \times 60 \times 120)/1,048,576 = 14,534.9 \text{ MB or } 14.2 \text{ GB.}$$

Digital Recording Level

In a digital recorder, the record-level meter is a peak-reading LED or LCD bar graph meter that reads up to 0 dBFS (FS means full scale). In a 16-bit digital recorder, 0 dBFS means all 16 bits are on. In a 24-bit digital recorder, 0 dBFS means that all 24 bits are on. The OVER indication means that the input level exceeded the voltage needed to produce 0 dBFS, and there is some short-duration clipping of the output analog waveform. This clipping can sound really nasty. Some manufacturers calibrate their meters so that 0 dBFS is less than 16 or 24 bits on; this allows a little head-room. When you set the recording level, it's a good idea to aim for −5 or −3 dB maximum so that unexpected peaks don't exceed 0 dBFS. If you're making a 24-bit recording, the recording level is not very critical because a 16-bit signal is at −48 dBFS!

The Clock

Each digital audio device has a clock that sets the timing of its signals. The clock is a series of pulses running at the sampling rate. When you transfer digital audio from one device to another, their clocks must be synchronized. One device must provide the master clock and the other must be the slave. If you send digital audio from one device, the receiver syncs to the sender's clock, which is embedded in its digital signal.

If a device (such as a digital mixer) is receiving data from many sources at once, select one device as the word clock source. Connect its word clock output to the input of a word-clock distribution unit. Connect a cable from each output of the distribution unit to the word-clock input of each of the other devices. That way all the devices will be synchronized.

Digital Audio Signal Formats

Digital audio signals come in four basic formats: AES/EBU, S/PDIF, ADAT Lightpipe, and TDIF. Let's look at each one.

- AES/EBU (also called AES3–1985): 2-channel professional format. Uses a balanced 110-ohm shielded twisted-pair cable with XLR-type connectors. The signal contains digital audio plus a word clock, or a separate word clock on another cable. The AES/EBU cable can be up to 200 m long. If the word clock cable is under 25 feet, it can be an unbalanced 75-ohm cable with BNC connectors. Word-clock cables over 25 feet should use balanced 110-ohm AES digital cable.
- S/PDIF Sony/Philips Digital Interface (also called EIAJ CP-340 Type II or IEC 958): 2-channel consumer or semi-pro format. The signal

contains digital audio plus an embedded word clock signal. Uses a 75-ohm coaxial cable with RCA or BNC connectors, or a fiber optic cable with a Toslink connector. Optical interfaces prevent ground loops and cable losses. AES/EBU signals are higher voltage than S/PDIF.

- ADAT Lightpipe: The Alesis ADAT modular digital multitracks use a Lightpipe, which sends 8 channels of digital audio in and out on a single optical cable with TOSLINK connectors. Every 8 channels of transfer requires a separate Lightpipe cable.
- Tascam TDIF (Tascam digital interface): The Tascam DA-88 and similar modular digital multitracks use a multiconductor cable with standard DB-25 connectors. TDIF sends 8 channels of digital audio in and out on a single cable, which can be up to 5m long.

Converting Signal Formats

AES and S/PDIF signals are similar but not necessarily compatible. You can convert one to the other using a format converter. Some sound cards and digital mixers do this conversion. Lightpipe and TDIF signals can be converted as well.

Some digital audio devices do not implement AES or S/PDIF correctly, so they do not interface with some other devices.

Dither

Your HD recorder, software, and digital mixer may run at 24 bits, but the result ends up on a 16-bit CD. When you save a 24-bit audio file as a 16-bit file to transfer to CD, those last eight bits are truncated or cut off. The result may be a grainy static sound at very low levels. This distortion can be prevented by adding low-level random noise (dither) to the signal. Let's explain.

A 24-bit resolution can accurately capture the quietest parts of a musical program: very low-level signals such as the end of long fades and reverb tails. But truncation of that signal to 16 bits makes those low-level signals sound grainy or fuzzy, because 16 bits is a less accurate measurement of the analog waveform than 24 bits. This fuzzy sound, called quantization distortion, doesn't exist at normal levels.

What causes this distortion? Each digital word is made of a certain number of bits. During quantization, the A/D converter assigns the closest possible digital number to represent the measured voltage of each sample. The last or rightmost bit (least significant bit or LSB) switches on or off depending on whether the converter rounds the word value up or down. If this switching occurs in the 16th bit, it may be faintly audible as a fuzzy noise during quiet passages.

Also, a 24-bit recording has 256 possible levels in the lower 8 bits. But after the signal is truncated to 16 bits, that resolution is lost.

To solve this problem, dithering adds random noise (random 1's and 0's) to the lowest 8 bits of the 24-bit signal (at about −100 dB) before they are truncated to 16 bits. That noise modulates the 16th bit with some 24-bit information (bits 17 through 24) in the form of pulse-density modulation. The average value of that modulated square wave is recovered by a low-pass filter. Then most of the 24-bit sound quality is restored, and the quantization distortion changes to a smooth hiss.

To make that added hiss less obvious, noise shaping is used. Noise shaping applies an oversampling filter to the noise, which reduces its level in the midrange where our ears are most sensitive, and increases its level in the high frequencies where it's less audible.

Compared to a truncated signal, a truncated-and-dithered signal sounds slightly more transparent. Fades and reverb tails sound smoother, and there's more sonic detail. Signals below the noise floor become audible.

For best sound quality, apply dither only once when you convert a high bit-depth source to its 16-bit CD format. For example, record at 24 bits and stay there through the entire project, then dither to 16 bits as the very last step just before you burn a CD. Do not re-dither material that has already been dithered—switch off any dithering. When doing a crossfade between two files, make sure each is non-dithered, then add dither after the crossfade during mastering.

To hear the effects of dithering, start with a clean 24-bit recording, reduce its level 50 dB in your editing software, and export it as a 16-bit file. Export it in three ways: without dither, with dither, and with noise shaping added. Next, normalize the exported recordings so that the highest peak hits 0 dBFS. Then listen to the resulting 16-bit files at high level over headphones. Compare the processing and use whatever sounds best. Generally, a signal that is truncated with no dither is accompanied by a rough, grainy noise. A signal that is truncated and dithered is accompanied by smooth, quiet hiss or silence.

Here is another application where dithering is necessary. Digital signal processing—such as level changes, EQ, or reverb—is done in a processor chip that performs mathematical calculations on each sample. These calculations create longer word lengths than existed in the original program. But the processor must output the same word length as the original signal. For example, a 16-bit audio file might result in words 32 bits long after processing. These 32-bit words must be converted back to 16 bits at the processor output. The extra bits must be truncated or cut off, but this causes distortion. So dithering is done automatically in the D/A converter. It can be set manually in some digital editing programs and outboard D/A converters. This dithering is performed only once, just before the output of the processor.

With good dithering algorithms, it's possible to preserve most of the 24-bit quality (ultra-low distortion and fine detail) when converting to 16 bits. One such system is Sony Super Bit Mapping. Another is the POW-r™ Psychoacoustically Optimized Word-Length Reduction algorithm from the Pow-r Consortium LLC. It reduces the high-resolution, higher word lengths (20 to 32 bit) to a CD-standard, 16-bit format while retaining the transparency of the high-resolution recording. In other words, the 16-bit CD sounds like the original 24-bit recording.

In addition, you could send audio through a multichannel 24-bit A/D converter, process it with POW-r, and record the result onto a 16-bit recorder. The playback would sound much like a 24-bit recording.

Jitter

Jitter is an instability in the timing of digital bits. It causes small changes in the audio waveform's shape, resulting in a slight veiling of the sound (low-level distortion). A jitter spec under 250 picoseconds is considered inaudible. Accurate A/D and D/A conversions rely on the clock precisely sampling the analog signal at equal time intervals. Any change between the sample times, even nanoseconds, causes audible amplitude errors (distortion).

Jitter occurs during A/D and D/A conversion, but not during digital-to-digital copies. One cause of jitter is analog noise and crosstalk in the recording system. They affect the switching times and switching threshold of the clock, causing frequency modulation of the clock. They also affect the analog filters and oscillators used in the clock's phase locked loops (PLLs). Jitter is also caused by inadequate digital cables. These cables pick up hum and noise, and introduce phase shift and high-frequency attenuation, which degrade the timing of the digital signal.

To reduce jitter:

- Use high-quality clock sources with low jitter specs (under 1 nanosecond). Usually the internal clock in A/D and D/A converters has less jitter than an external clock, such as AES/EBU or word clock.
- Use high-quality, well-shielded digital cables (ideally, 75-ohm RG59 video cables), and make them as short as possible.
- Keep analog and digital cabling separate.
- Use the A/D converter's internal clock as the master clock. Feed its AES or word-clock output to other slave devices in your studio to keep them in sync. If you use a separate word-clock cable, make it the same length as the digital audio cable.

Digital Transfers or Copies

When you send a digital audio signal in real time from one device to another, they must be set to the same sampling rate. For example, if the digital signal from a multitrack recorder is 48 kHz, the digital mixer it feeds must also be set to 48 kHz. In a digital transfer, the sending device usually is the master, and the receiving device is the slave. Normally you can set the slave to automatically lock onto the sampling rate of the master. Set the master device to Internal Clock, and set the slave device to External Clock. Or set the slave's clock source to whatever device the master is.

The word length (number of bits) of digital transfers is less critical. As long as the two machines handle the same digital format, transfers of any word length can be done. Formats for digital transfer are S/PDIF, AES/EBU, ADAT Lightpipe, and Tascam TDIF.

It's okay to send a lower bit signal to a higher bit device. For instance, feed a 16-bit signal from a DAT into a 24-bit digital mixer. The mixer will add more zeroes to fill out the digital word, with no effect on sound quality. It's also no problem to feed a device more bits than it can hold. The receiving device ignores the extra bits. For example, if you send a 24-bit signal to a 16-bit recorder, the last 8 bits will be truncated or cut off.

Truncation adds slight distortion, but this can be reduced by adding dither—low-level noise—to the signal before truncating. So it's best to keep the word length as long as possible as you're working on a project. Always apply dither before you reduce the word length, and only then.

What if you have a 48K recording that you want to release on CD? The CD mastering engineer can (1) convert the 48K digital signal to 44.1K, which might degrade the sound, or (2) use the analog output of his or her playback machine, and convert the analog signal to digital at 44.1K. This may or may not sound better.

To or from DAT or MDM tape and to or from CD-Rs are digital audio transfers that might or might not create perfect clones because of inadequate error correction or interpolation errors.

Does a FireWire transfer of realtime audio contribute jitter? Not really. With isochronous FireWire transfer, it's relatively easy to generate a very low-jitter D/A clock. "Isochronous" means that the data must be delivered at a certain minimum data rate. Multimedia signal streams require isochronous data flow to ensure that audio data is delivered as fast as it is played and recorded.

When you copy a wave or aiff file rather than a digital audio signal, the copy is a perfect clone of the original file. The file transfers much faster than the real-time playback of the signal. Flawless file copies can be made in these ways:

- Inside a computer from one hard drive location to another
- From one computer's hard drive to another computer's hard drive
- Via Ethernet, USB, FireWire, or the Internet
- Between computers via CD-R, Jaz, or Zip drive; floppy disks; or Flash memory cards

The following data compression algorithms don't lose any data, so they don't degrade audio quality:

- Emagic's Zap
- Waves' TrackPac
- PKWare's PKZIP
- WinZip Computing's WinZip
- Aladdin Systems' StuffIt
- Meridian Lossless Packing (MLP)

2-Track Digital Recorders

Now that we overviewed digital recording, let's look at digital recorders. They come in 2-track and multitrack formats. The 2-track formats are

- DAT (rotating-head digital audio tape); this format is obsolete and is not covered here
- Portable hard-drive recorder
- Digital Audio Workstation; this is a computer (usually a laptop) with a sound card or other audio interface, plus recording software
- CD-R recorder
- MiniDisc recorder
- Memory recorder

We'll look at each one. No matter which format you use, the sound quality depends on the A/D converter and mic preamp (if any).

Portable Hard-Drive Recorder

This device (Figure 9.2) records onto a built-in hard drive. For example, the Edirol R4 records up to 4 tracks simultaneously onto a 40-GB drive, either MP3 or WAV files up to 24-bit/96 kHz. The unit has XLR/phone inputs with phantom power, built-in stereo speakers, USB 2.0 interface to a computer, waveform editing, limiting, and a compact-flash slot. See Web site www.edirol.com.

Figure 9.2 A portable hard-drive recorder.

The Digital Audio Workstation

The Digital Audio Workstation (DAW) is a computer running recording software with a connected audio interface such as a sound card (Figure 9.3). A DAW allows you to record, edit, and mix audio programs entirely in digital form. It can store up to several hours of digital audio or MIDI data. You can edit this data with great precision on a computer monitor screen. What's more, you can add digital effects and perform automated mixdowns. We'll cover DAWs in detail in Chapter 13.

A laptop computer can make a great portable 2-track digital recorder. Just add some recording software and an audio interface connected by USB or FireWire.

CD Recordable

Another form of digital recording is the compact disc. On your own desktop, you can cut a CD by using a CD-R recorder (CD burner). It's exciting to hear one of these CDs playing your music with the purity of digital sound. The sound quality meets or exceeds CD standards.

CD-R stands for compact disc recordable. This optical medium is a write-once (nonerasable) format. CD-RW stands for compact disc re-writable; you can erase it and record a new program. Although any CD

Figure 9.3 A computer Digital Audio Workstation (DAW).

player or CD-ROM drive will play a CD-R, many CD players—especially older ones—can't play CD-RW discs. Most new CD-ROM drives support CD-RW, but not all will read CD-RW discs at full speed. CD-RW blank discs cost more than CD-R blank discs.

How can you use the CD-R format? You could make demos for your own band, or make a one-off copy of your stereo mixes for clients. Use the CD-R as a pre-master to send to a CD replicator. Another function is to compile sound libraries of production music, samples, and sound effects. If handled and stored with care, the CD format is a dependable storage medium, so it's a great way to archive your recordings.

CD-R Formats

Want to try CD-R? First you'll be faced with two basic choices of CD-R recorder:

1. A standalone CD-R recorder, sometimes called a consumer CD recorder
2. A computer peripheral CD-R recorder, also called a computer CD burner; you plug it into your computer system, or buy a computer with a CD burner built in

Both types produce discs that sound equally good. Both types of discs will play on any audio CD player.

The standalone CD recorder has everything you need in one chassis. Inside is a CD transport, laser, and microprocessor. On the back are analog and digital ins and outs. On the front are the level meters, record-level knob, display, and keypad. Because the standalone unit needs no external computer, it's user-friendly. Just connect your audio source containing an edited program, either on DAT, LP, or analog tape. Set the recording level and start recording.

The standalone unit can write audio but not data. The 63-minute blank discs it uses are the "Music CD-R" or "Digital Audio CD-R" format. Prices for standalone CD-R writers start around $400.

A computer CD recorder costs less: about $100 and up. The unit plugs into an EIDE or SCSI connector in your computer. It can write audio or computer data. Its blank discs, called "CD-R" or "Data CD-R," cost less than 35 cents in quantity. (Music CD-Rs will work in a computer CD burner as well.) Disc length is 74 (650 MB), 63 (550 MB), or 80 minutes (700 MB). Some discs permit recording at up to 52 times real time if your computer and hard drive are fast enough.

The computer CD-R recorder also requires a CD recording program, usually packaged with the recorder. You don't have to use that program; other ones are available that you might prefer. Be sure the program is compatible with your recorder. You'll also need a sound card and some software to record audio onto hard disk.

Multi-drive CD burners let you record or copy several CDs at once.

CD-R Technology

While conventional CD players follow the Sony-Philips Red Book standard, CD-Rs conform to the Orange Book part II standard. Once recorded, a CD-R disc meets the Red Book standard.

A recordable CD is the same size as a standard CD, but it is more colorful. On top is a metal reflective layer; on the bottom is a recording layer made of blue cyanine dye or yellow (gold) phthalocyanine dye. The blue layer appears green because of the gold layer behind it. Yellow dye lasts a little longer in accelerated aging tests, and it may work better with high-speed CD-R drives. A few other colors are available as well.

A blank CD-R is made of four layers:

1. Clear plastic (protects the metal layer)
2. Metal layer (gold, silver, or silver alloy, which reflects the laser light)
3. Dye (for the recording)
4. Clear plastic (protects the dye layer)

The dye fills a spiral groove which is etched in the bottom clear-plastic layer. This groove guides the laser.

To record data on a disc, the laser melts holes in the dye layer. The plastic layer flows into the holes to form pits. During playback, the same laser reads the disc at lower power. At each pit, laser light reflects off the metal layer. The reflected light enters the laser reader, which detects the varying reflectance as the pits go by.

In contrast with a standard CD, a CD-R disc has two more data areas:

- The program calibration area (PCA). The CD recorder uses this area to make a test recording, which determines the right amount of laser power to burn the disc (4 to 8 milliwatts).
- The program memory area (PMA). This area stores a temporary table of contents (TOC) as the CD-R tracks are assembled. The TOC

is a list of the tracks, their start times, and the total program time. The recorder uses the PMA for this information until it writes the final TOC.

According to CD-R manufacturers, the expected lifetime of a CD-R is about 70 to 100 years with careful handling. Avoid sunlight and high temperatures. Be sure not to damage the label side of the disc by writing on it with a ballpoint pen or pencil—use a soft felt-tip pen that is water-based, such as a CD-marking Sharpie Store CDs vertically and keep them in their cases. Avoid labels because the adhesive can attack the protective plastic layer, and the paper might warp over time. CD-RW life expectancy is claimed to be about 25 years.

CD-R Sessions, Disc-At-Once, and Track-At-Once

Before we look at the differences among CD-R recorders, we need to understand the concept of a session. A session on disc is made up of a lead-in, program area, and lead-out. Each session has its own TOC. Each lead-in and lead-out consumes 13 MB of disc space.

With the multisession feature, you can write several sessions on a disc at different times. This feature comes in handy when you need to add information to a disc a little at a time. Only the first session on disc will play on an audio CD player, so the discs are just for your own use and not for distribution.

CD-R recorders also permit Disc-At-Once recording, in which the entire disc must be recorded nonstop. You can't add new material once you write to the disc. With some software, Disc-At-Once lets you set the length of silence between tracks (down to 0 seconds), and lets you control how the tracks are laid out on disc. Use Disc-At-Once for all pro audio work.

Most CD-R recorders allow Track-At-Once recording. They can record one track (or a few tracks) at a time—up to 99 tracks. You can play a partly recorded disc on a CD-R recorder, but the disc will not play on a regular CD player until the final TOC is written. Track-At-Once is not recommended for audio because it puts 2-second spaces and clicks between audio tracks.

If you want no pauses between tracks (as on a live album), get a CD-R writer with Disc-At-Once. Also get some software that can adjust the pause length down to zero, or that can set the start ID of each song anywhere in the program. Note that a self-contained CD-R writer will copy your edited program as it is, with or without pauses.

A good CD-R writer has a buffer of at least 2MB. Some writers include SCMS copy code. Be sure that you can return the CD-R writer if it proves to be unreliable.

How to Use a Standalone CD-R Writer

Let's say that you have an edited recording of song mixes on a 2-track recorder, and you want to copy them onto a CD by using a self-contained CD-R writer.

Connect the 2-track recorder's output to the CD-R writer's input. Use a Music CD-R or Digital Audio CD-R. Set your levels and begin recording. Your program will copy to disc in real time.

Depending on the CD-R writer, the recording's start IDs may or may not convert to CD track numbers. If not, you can use a converter box, or add the track numbers manually while you record.

Using a computer CD-R burner is covered in Chapter 15. For more information on CD-R technology, see www.cdrfaq.org and www.harmony-central.com/Features/CDRecorder/.

MiniDisc Recorder

A blank MiniDisc is a rewritable, magneto-optical medium read by a laser. The disc itself is like a miniature compact disc inside a 2-1/2-inch square housing. A write-protect tab on the housing prevents accidental erasure. Estimated disc life is 30 years, but a strong magnet near the disc can erase data.

Three types of blank discs are available: the regular 74-minute MiniDisc used in 2-track recorders, the Hi-MD 1-GB disc, and the MD Data disc used in multitrack recorders.

Most MiniDisc devices record audio at 44.1kHz, 16 bits, and some at 24 bits.

To fit all this data on a small disc, MD recorders use a data compression scheme called Adaptive Transform Acoustic Coding (ATRAC). It reduces by 5:1 the storage needed for digital audio. ATRAC is a perceptual coding method that omits data deemed inaudible due to masking.

For example, if an audio signal has two sounds that are about the same frequency, and one sound is louder than the other, the quieter sound will be inaudible due to masking. So ATRAC removes the quieter sound, which would be inaudible anyway.

ATRAC has had several revision levels; the latest version (as of 2004) is ATRAC3 Plus. The higher the version number, the better the sound.

Some reviewers have claimed that ATRAC3 Plus sounds essentially the same as CDs when playing a musical program. Earlier versions are said to be near-CD in quality, and much better than MP3. All versions are compatible.

Sound quality depends not only on the ATRAC version, but also on the quality and bit depth of the A/D converter in the recorder. Although the MiniDisc format uses data compression, its stereo digital output is standard 16-bit S/PDIF.

There is a slight generation loss when ATRAC tracks are copied or bounced. The signal is ATRAC-processed with each copy. After more than five copies or so, the sound cumulatively begins to take on a mid-to-low rumble and a high-frequency squeak.

MiniDisc recorders can make ATRAC-compressed copies of CDs directly from the CD player's digital output.

Introduced in 2004, Hi-MD recorders can record uncompressed PCM audio (16-bit 44.1 kHz wave files) as well as ATRAC3 Plus audio. They record on a 1 GB magneto-optical medium, and also on the original MiniDiscs reformatted to double capacity (305 MB). A 1 GB disc can record up to 1 hour and 34 minutes of uncompressed PCM audio. Hi-MD recorders can be used as USB data drives for PCs. Recordings can be uploaded very quickly to Windows PCs via SonicStage, Sony's audio transfer software. For more information, see www.minidisc.org.

Two-track recorders come in a portable style or a component style. The portable MiniDisc Walkman™ format is popular with news reporters who need to carry a small, high-quality recorder in the field. Reporters can edit the recording using buttons on the recorder.

Portable 2-track MD recorders have been used for documenting musical groups at folk festivals. The recordist walks around the festival recording various groups of musicians (with their permission, let's hope!). MiniDisc also offers an easy way to record school concerts, sound effects, or your band's gigs.

Can you use a Hi-MD MiniDisc to record a stereo master of your mixdowns? You could, but CD-R and DAT masters are preferred. Also, many mastering houses do not accept MiniDisc masters.

Memory Recorder

This is a portable digital recorder with no moving parts and no mainte-nance. It records onto a Flash memory card, such as Compact Flash, Smart Media, Sony Memory Stick, or SD card. Here are some examples:

The Marantz PMD670 records MP3 or uncompressed CD-quality wave files (up to 48 kHz) on a Compact Flash card or IBM Microdrive. A USB port downloads data quickly to your computer for editing. Two XLR mic inputs with phantom power are included. You can record 6 hours of uncompressed 16-bit/48-kHz wave audio on a 2-GB card. The unit allows edit decision list (EDL) markings during record or playback. A smaller model, PMD660, records to Compact Flash and has built-in mics. See Web site www.d-mpro.com.

The Edirol R1 records MP3 or WAV files (up to 24-bit) onto a Compact Flash card. It has a built-in mic, effects, metronome, limiter, and tuner. Input is stereo mini phone. See Web site www.edirol.com.

The Zoom PS-02 Palmtop Studio includes three audio tracks, drum and bass machines, guitar multieffects, mixer, a built-in mic, and a tuner in a chassis that fits in your pocket. Effective as a digital sketchpad recorder, the PS-02 stores over 2 hours of program on a 128-MB Smart Media card.

The Nagra Ares-PII is a handheld audio recorder that digitizes audio with MPEG layer 2 compression or uncompressed PCM, and records to Flash RAM. The RCX220 also has a USB port for file transfer to a PC for editing. A phantom-powered mic input is included.

PocketREC is a PocketPC that works as a portable digital recorder. On Flash memory it can record 16-bit, 48-kHz linear (uncompressed) wave files. You can edit the audio and transfer data via wireless or wired connections. The unit supports the PDAudio-CD interface described below.

PDAudio is a compact memory recording system for pocket PCs (PDAs) and laptop and desktop computers (Figure 9.4). It costs $400 to $1000. The system includes:

- A stereo mic preamp/A/D converter of your choice, such as Core Sound's Mic2496
- An audio interface, such as Core Sound's PDAudio-CF (turns a PDA into a recorder by converting digital audio to Compact Flash format)
- A Personal Digital Assistant (PDA) of your choice, such as the Compaq iPAQ Pocket PC
- An expansion card adapter for Compact Flash (CF) cards or PC cards
- Recording software, such as Core Sound's PDAudio Recorder, Pocco Software's Wichita, or Gidluck Mastering's Live2496

Figure 9.4 PDAudio-CF installed in an HP iPAQ Pocket PC. A Mic2496 preamp/converter is attached to the iPAQ.

- Storage device, such as Flash memory (SD, PC Card, or CF card) or PC-card hard drive

Starting with a pocket PC (PDA), you plug in the PDAudio inter-face card, which accepts a stereo digital audio signal up to 24 bits/96 kHz. You launch the recording software, and record audio onto a flash memory card or PC card hard drive plugged into the PDA.

Multitrack Digital Recorders

So far we've covered 2-track digital recorders. Now we'll consider the multitrack digital recorders listed below:

- Modular Digital Multitrack (MDM)
- Computer DAW (covered earlier)
- Standalone hard-disk recorder
- Hard-disk recorder-mixer
- MiniDisc recorder-mixer
- DVD recorder (see Chapter 19).

Modular Digital Multitrack

Shown in Figure 9.5, a Modular Digital Multitrack (MDM) records 8 digital tracks on a videocassette, using a rotating drum like a DAT recorder. Two popular models are the Alesis ADAT-XL which records on S-VHS tape, and the Tascam DA-88 or DA-78, which record on Hi-8 video cassette tape. ADAT records up to 40 minutes on a single tape; DA-88 records up to 1 hour and 48 minutes.

With both types, you can sync several 8-track units by a cable to add more tracks, 8 at a time. Unlike SMPTE time code, MDM sync does not use up any tracks. MDM options include remote controls, remote editors, circuit boards with enhanced converters, and circuit boards that allow sync to SMPTE and MIDI.

To prevent data errors, be sure to format the MDM tape correctly. Fast-forward the tape to the end, rewind to the top, then format the tape. Clean the heads with a dry cleaning tape only when an error signal appears, because cleaning tapes are abrasive and can wear out the heads.

It's common to record music on an MDM, dump the MDM tracks to computer hard disk for editing, then dump the edited tracks back to MDM. Sound cards are available with Alesis Lightpipe or TASCAM TDIF connectors for this purpose.

MDMs have been superceded by hard-disk recorders.

Figure 9.5 An MDM recorder.

Figure 9.6 HD recorder.

Hard-Disk (HD) Recorder

This device records and plays up to 24 tracks at once on a built-in hard drive, just like the drive in your computer. The latest units record with 24-bit resolution and up to 96kHz sampling rate. Some record 24 tracks of 24-bit programs at sample rates of 44.1 and 48kHz. They record 12 tracks of 24-bit recording at 88.2 and 96kHz.

The HD recorder allows track editing, either on a built-in LCD screen or on a plug-in computer monitor. Some HD recorders have built-in removable hard drives for backing up projects. See Figure 9.6.

HD Recorder-Mixer

Shown in Figure 9.7, this device records and plays up to 36 tracks on a built-in hard drive. It's also called a standalone DAW, personal digital studio, or digital multitracker. The mixer includes faders (volume controls) for mixing, EQ or tone controls, and aux sends for effects (such as reverb). An LCD screen displays recording levels, waveforms for editing, and other functions. Some manufacturers of HD recorder-mixers are Roland, Korg, Fostex, Boss, Yamaha, Akai, and Tascam.

Listed below are some features to look for in HD recorder-mixers.

- Number of tracks: 4 to 36. The more tracks you have, the more instruments and vocals you can record on individual tracks.
- Number of tracks that can be recorded simultaneously: 8 to 24. If you want to record an entire band at once, you need to record many tracks at the same time.
- Bit depth: 16 to 24 bit. The more bits, the higher the sound quality.
- Sampling rate: 44.1 to 96kHz. The higher the sampling rate, the higher the sound quality. The latest units record with 24-bit resolution and up to 96-kHz sampling rate. Some record 24 tracks of

Figure 9.7 An HD recorder-mixer.

24-bit programs at sample rates of 44.1 and 48 kHz. They record 12 tracks of 24-bit recording at 88.2 and 96 kHz.

- Number and type of analog inputs and outputs, and digital inputs and outputs. Balanced XLR or TRS phone-jack mic connectors are preferred over unbalanced (TS) phone-jack connectors. Balanced connections let you run longer mic cables without hum pickup.
- Phantom power for condenser mics.
- Number of built-in digital effects: Up to 8 stereo or 16 mono. Examples of effects are reverberation, echo, flanging, chorus, and compression. They add sonic excitement and a professional touch to your productions. Some recorder-mixers have an expansion card that accepts downloadable plug-in effects.
- Types of EQ: 2- or 3-band, fixed or parametric. EQ means tone control. A 2-band EQ adjusts bass and treble, and a 3-band adjusts bass, midrange, and treble. A fixed-frequency EQ is less flexible than a parametric or sweepable EQ, which lets you adjust the frequency range you want to work on.
- Backup options: Removable hard drive, CD-R.

193

- Levels of undo: Up to 1000. The Undo function lets you undo an editing change you just did. If you don't like how an edit sounded, you can press Undo and go back to the way things were before you made the change.

- Number of locate points: Up to 1000. A locate point is a point in time in the recording that you can have the recorder memorize and locate later. For example, you might want to mark the time where the beginning of a song is, so you can tap a button and go there instantly.

- Expansion ports for SMPTE and MTC sync, MDM interface, and extra ins and outs. Beginners probably don't need these features, but they let you connect your multitracker to other devices.

- MIDI implementation: MIDI Machine Code (MMC), Control Change (CC). These features let you control the mixer via MIDI commands.

- Types of synchronization: SMPTE, Word Clock, MTC, MIDI Clock with Song Position Pointer, ADAT, DA-88, and RS-422. These are various formats that you might want your multitracker to sync with.

- Jog/shuttle wheel to "scrub" audio-play it slowly forward and backward to locate an edit point.

- Remote control.

- Automated mixing. A memory circuit in the multitracker remembers your song-mix settings and mix changes. The next time you play the song, the memory circuit resets the mixer to those settings automatically. Some multitrackers have motorized faders, so that the faders move up and down just as they did when you adjusted them while mixing.

- Editing (cut and paste, etc.). Editing lets you do all sorts of things with the song arrangement. You might remove parts of songs that you don't want to keep, loop a drum part so that it repeats continuously, or copy a chorus to several points in a song.

- Video output that plugs into a computer monitor screen. During editing, it's a lot easier to see details in the sound waveform if you can plug a large monitor into your multitracker.

- Mastering tools and CD-burning ability. In mastering you compile a playlist of mixed songs from which you can create an album. Some multitrackers include programs that let you record your songs on a CD-R recorder, either external or built-in.

- Number of virtual tracks (extra takes): Up to 800. A virtual track is a 1-channel recording of a single take or performance on a random-access medium. Most random-access recorders let you record several virtual tracks or takes of a single instrument, then select which take you want to hear during mixdown. For example, suppose you have a 16-track hard-disk recorder. You could record up to 255 virtual tracks of a vocal on hard disk. Then during mixdown, choose which virtual track will play as one of those 16 tracks. Some recorders can be set up to select parts of different virtual tracks during playback. For example, play vocal take 15 for the verse, play vocal take 3 for the chorus, and so on.

In other words, you can associate several virtual tracks, or takes, with a single "real" track. You might play 8 tracks at once, and you can choose which virtual track (take) plays on each track. For example, you can record several takes of a guitar solo—keeping each one—and choose the best take during mixdown. You also can create a composite track, which is made of the best parts of several takes. All HD recorder-mixers let you record virtual tracks.

MiniDisc Recorder-Mixer

Introduced by Sony in 1991, the MiniDisc (MD) is a convenient recording medium that is removable, low cost, and near-CD quality. A MiniDisc recorder-mixer is a multitrack MD recorder and mixer in a single, affordable package. It records and plays 4 or 8 audio tracks at once on a durable, removable MD data disc. About the size of a floppy disk, an MD data disc records 8 tracks for 18.5 minutes or 4 tracks for 37 minutes. A MiniDisc costs about $2.00. The built-in mixer has faders and knobs and includes a small LED screen for nongraphical editing.

Below are some MiniDisc editing features:

- Write disc title and track title on the disc.
- Song Copy duplicates a song at a new location.
- Song Erase deletes a song.
- Track Copy copies data from one track to another.
- Song Divide lets you divide a song in two at the current counter location. With this function, you can create sections—verse, chorus, etc.—that you can play or repeat in any order. You can also remove noises before and after songs.

- Song Combine joins divided parts from the same song.
- Time slipping moves individual tracks in time.
- Cue List is a list of the song sections in the order you want them played. Sections can be looped or repeated. The recorder will play down the cue list, assembling the song from its sections.
- Program Play List is a list of the songs in the order you want them played.
- Scale Factor Edit can be used to match the volume of tracks recorded at different levels, or to create fade-ins and fade-outs.

Pros and Cons of Four Multitrack Recording Systems

Let's compare a hard-drive recorder, hard-drive recorder-mixer, MiniDisc recorder-mixer, and computer DAW.

24-Track Hard-Drive Recorder

Pros:

- Portable—great for on-location recording.
- Works with your existing mixer and effects.
- Can export audio data to a computer for editing, or use its own editing software.
- Stable: Less likely to crash than a computer recording system.
- Might have higher-quality converters than a computer recording system.

   Cons (or comments)

- Requires a mixer, connecting cables, and effects or a computer with recording software.
- If used with a computer or digital mixer, requires an interface: a multichannel sound card, USB port, or FireWire port.
- Another hard drive or a DVD-R burner is needed for backup.

Hard-Drive Recorder-Mixer

Pros:

- Easy to use—hands on.
- Simple. No need for a computer, external multitrack recorder, or outboard effects.

- Portable—good for on-location recording, but bigger than an HD recorder.
- Less likely to crash than a computer recording system.
- Can export audio data to a computer for editing, or use its own editing firmware.
- All-in-one system: no external cables needed between recorder and computer or between mixer and effects.
- Low-cost CD backup.
- Automated mixing.

Cons:

- Less portable than an HD recorder for on-location recording.
- If you already have a mixer or recording software, it may be more than you need.
- Less flexible than recording software. Not as upgradable. However, some recorder-mixers have expansion cards that accept downloadable plug-in effects.
- May require copying data to a computer for sophisticated editing and plug-ins.

Mini-Disc Recorder-Mixer

Pros:

- Once you are done with a project, you simply remove the MiniDisc and put in a blank one for the next project.
- Low-cost medium.

Cons:

- ATRAC data compression reduces sound quality.
- Not for long recordings (18.5 minutes for 8 tracks on one MiniDisc).
- Can't transfer tracks to a computer for editing.

Computer Recording System (DAW)

Pros:

- Can be low cost if you already have a fast computer (except for the top-level [Pro Tools] systems).
- Flexible (due to software upgrades and plug-ins).

- All-in-one system: no external cables between recorder and computer or between mixer and effects.
- Low-cost CD backup.
- Automated mixing.
- Sophisticated editing.
- Can be used with existing analog or digital mixer.
- Saves time if used for recording: multitrack recordings and mixes do not need to be copied to computer for editing/mastering.

   Cons:

- Computer may crash.
- Harder to use than a hardware mixer, but a controller surface can help with this.
- Not easily portable except in 2-track laptop systems, but could be used with an HD recorder or recorder-mixer for on-location recordings.
- Requires a fast computer with a large hard drive optimized for digital audio.
- Requires strong computer skills (but this is a plus for some users).
- Requires an interface to get audio into and out of the computer. The interface can be one of these types:

   Sound card (2-channel or multichannel)

   Alesis ADAT Lightpipe card or Tascam TDIF card

   I/O interface (MOTU, M-Audio)

   Controller with I/O (Tascam, Edirol, etc.)

   Alesis Fireport

Backup

So far we've looked at several types of digital recording devices. With any device, it's important to back up or make a copy of your recordings' wave files. The hard drive on which you store your projects will eventually crash. DAT tapes and video tapes have a limited life span as well. So it's vital to back up your data periodically.

Some backup systems are CD-R, CD-RW, hard drive (internal or external FireWire), Iomega Zip drive (100 MB), Jaz drive (2 GB), DVD-

RAM, and DVD-RW. CD and DVD are perhaps the most long-lasting formats for archiving. Some engineers prefer to archive audio programs on analog tape as well as digital.

As we've seen, there is a wide variety of digital recording formats. Learn all you can about digital technology, then choose the formats that meet your needs.

10

EFFECTS AND SIGNAL PROCESSORS

With effects, your mix sounds more like a real "production" and less like a bland home recording. You might simulate a concert hall with reverb. Put a guitar in space with stereo chorus. Make a kick drum punchy by adding compression. Used on all pop-music records, effects can enhance plain tracks by adding spaciousness and excitement. They are essential if you want to produce a commercial sound. But many jazz, folk, and classical groups sound fine without any effects.

This chapter describes the most popular signal processors and effects, and suggests how to use them.

Effects are available both as hardware and software (called plug-ins). To add a hardware effect to a track, you feed its signal from your mixer's aux send to an effects device, or signal processor (Figure 10.1). It modifies the signal in a controlled way. Then the modified signal returns to your mixer, where it blends with the dry, unprocessed signal.

Software Effects (Plug-Ins)

Most recording programs include plug-ins: software effects that you control on your computer screen. Each effect is an algorithm (small program) that runs either in your computer's CPU or in a DSP card. Some

Figure 10.1 An effects unit.

plug-ins are already installed with the recording software; others can be downloaded or purchased on CD, then installed on your hard drive. Each plug-in becomes part of your recording program (called the host), and can be called up from within the host.

You can use plug-ins made by your recording software company or by others. Some manufacturers make plug-in bundles, which are a variety of effects in a single package.

Plug-ins are the usual way to create effects in a DAW. Some DAWs let you configure your audio interface to produce an aux-send signal, which you feed to an external hardware processor. The processed signal returns to the interface and blends with the dry signal.

All the effects described below are available as plug-ins as well as hardware.

Equalizer

Recall from Chapter 2 that an equalizer (usually in the mixer) is a sophisticated tone control, something like the bass and treble controls in a stereo system.

Equalization (EQ) lets you improve on reality: add crispness to dull cymbals or add bite to a wimpy electric guitar. EQ also can make a track sound more natural; for instance, remove tubbiness from a close-miked vocal.

To understand how EQ works, we need to know the meaning of a spectrum. Each instrument or voice produces a wide range of frequencies called its spectrum—the fundamentals and harmonics. The spectrum gives each instrument its distinctive tone quality or timbre.

If you boost or cut certain frequencies in the spectrum, you change the tone quality of the recorded instrument. EQ adjusts the bass, treble, and midrange of a sound by turning up or down certain frequency ranges. That is, it alters the frequency response. For example, a boost (a level increase) in the range centered at 10 kHz makes percussion sound bright and crisp. A cut at the same frequency dulls the sound.

Types of EQ

Equalizers range from simple to complex. The most basic type is a bass and treble control (labeled LF EQ and HF EQ). Figure 10.2 shows its effect on frequency response. Typically, this type has up to 15 dB of boost or cut at 100 Hz (for the low-frequency EQ knob) and at 10 kHz (for the high-frequency EQ knob).

With a 3-band EQ you can boost or cut the lows, mids, and highs at fixed frequencies (Figure 10.3). Sweepable EQ is more flexible because you can "tune in" the exact frequency range needing adjustment (Figure 10.4). If your mixer has sweepable EQ, one knob sets the center frequency while another sets the amount of boost or cut.

Parametric EQ lets you set the frequency, amount of boost/cut, and bandwidth—the range of frequencies affected. Figure 10.5 shows how a parametric equalizer varies the bandwidth of the boosted part of the

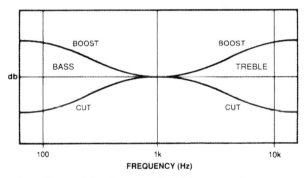

Figure 10.2 The effect of the bass and treble control.

Figure 10.3 The effect of 3-band equalization.

203

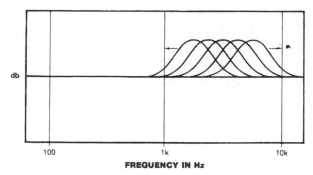

Figure 10.4 The effect of sweepable equalization.

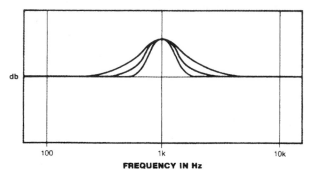

Figure 10.5 Curves that illustrate varying the bandwidth of a parametric equalizer.

spectrum. The "Q" or quality factor of an equalizer is the center frequency divided by the bandwidth. A boost or cut with a low-Q setting affects a wide range of frequencies; a high-Q setting makes a narrow peak or dip.

A graphic equalizer (Figure 10.6) is usually outside the mixing console. This type has a row of slide pots that work on 5 to 31 frequency bands. When the controls are adjusted, their positions graphically show the resulting frequency response. Usually, a graphic equalizer is used for monitor-speaker EQ, or is patched into a channel for sophisticated tonal tweaking.

Equalizers can also be classified by the shape of their frequency response. Peaking EQ shapes the response like a hill or peak when set for a boost (Figure 10.7). With shelving EQ, the shape of the frequency response resembles a shelf (Figure 10.8). *CD track 27 demonstrates various types of EQ.*

Figure 10.6 A graphic equalizer.

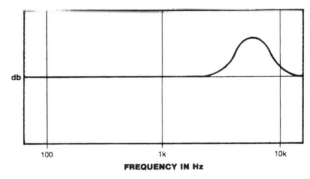

Figure 10.7 Peaking equalization at 7 kHz.

Figure 10.8 Shelving equalization at 7 kHz.

A filter causes a rolloff at the frequency extremes. It sharply rejects (attenuates) frequencies above or below a certain frequency. Figure 10.9 shows three types of filters: lowpass, highpass, and bandpass. For example, a 10-kHz lowpass filter (high-cut filter) removes frequencies above 10 kHz. Its response is down 3 dB at 10 kHz and more above that. This reduces hiss-type noise without affecting tone quality as much as a

Figure 10.9 Lowpass, highpass, and bandpass filters.

gradual treble rolloff would. A 100-Hz highpass filter (low-cut filter) attenuates frequencies below 100 Hz. Its response is down 3 dB at 100 Hz and more below that. This removes low-pitched noises such as air handler rumble or breath pops. A 1-kHz bandpass filter cuts frequencies above and below a frequency band centered at 1 kHz.

The crossover filter in some monitor speakers consists of lowpass, highpass, and bandpass filters. They send the lows to the woofer, mids to the midrange, and highs to the tweeter.

A filter is named for the steepness of its rolloff: 6 dB per octave (first order), 12 dB per octave (second order), 18 dB per octave (third order), and so on.

How to Use EQ

If your mixer has bass and treble controls, their frequencies are preset (usually at 100 Hz and 10 kHz). Set the EQ knob at 0 to have no effect (flat setting). Turn it clockwise for a boost; turn it counterclockwise for a cut. If your mixer has multiple-frequency EQ or sweepable EQ, one knob sets the frequency range and another sets the amount of boost or cut.

Table 10.1 shows the fundamentals and harmonics of musical instruments and voices. The harmonics given represent an approximate range. For each instrument, turn up the lower end of the fundamentals to get warmth and fullness. Turn down the fundamentals if the tone is too bassy or tubby. Turn up the harmonics for presence and definition; turn down the harmonics if the tone is too harsh or sizzly.

Percussion, cymbals, and muted trumpet actually have some energy up to 80 to 100 kHz.

Here are some suggested frequencies to adjust for specific instruments. If you want the effects described below, apply boost. If you don't, apply cut. Try these suggestions and accept only the sounds you like:

Table 10.1 Frequency Ranges of Musical Instruments and Voices

Instrument	Fundamentals	Harmonics
Flute	261–2349 Hz	3–8 kHz
Oboe	261–1568 Hz	2–12 kHz
Clarinet	165–1568 Hz	2–10 kHz
Bassoon	62–587 Hz	1–7 kHz
Trumpet	165–988 Hz	1–7.5 kHz
French horn	87–880 Hz	1–6 kHz
Trombone	73–587 Hz	1–7.5 kHz
Tuba	49–587 Hz	1–4 kHz
Snare drum	100–200 Hz	1–20 kHz
Kick drum	30–147 Hz	1–6 kHz
Cymbals	300–587 Hz	1–15 kHz
Violin	196–3136 Hz	4–15 kHz
Viola	131–1175 Hz	2–8.5 kHz
Cello	65–698 Hz	1–6.5 kHz
Acoustic bass	41–294 Hz	700 Hz–5 kHz
Electric bass	41–294 Hz	700 Hz–7 kHz
Acoustic guitar	82–988 Hz	1500 Hz–15 kHz
Electric guitar	82–1319 Hz	1–15 kHz (direct)
Elec. guitar amp	82–1319 Hz	1–4 kHz
Piano	28–4196 Hz	5–8 kHz
Bass (voice)	87–392 Hz	1–12 kHz
Tenor (voice)	131–494 Hz	1–12 kHz
Alto (voice)	175–698 Hz	2–12 kHz
Soprano (voice)	247–1175 Hz	2–12 kHz

- Bass: Full and deep at 60 Hz, growl at 600 Hz, presence at 2.5 kHz, string noise at 3 kHz and up.
- Electric guitar: Thumpy at 60 Hz, full at 100 Hz, puffy at 500 Hz, presence or bite at 2 to 3 kHz, sizzly and raspy above 6 kHz.
- Drums: Full at 100 Hz, wooly at 250 to 600 Hz, trashy at 1 to 3 kHz, attack at 5 kHz, sizzly and crisp at 10 kHz.
- Kick drum: Full and powerful below 60 Hz, papery at 300 to 800 Hz (cut at 400 to 600 Hz for better tone), click or attack at 2 to 6 kHz.
- Sax: Warm at 500 Hz, harsh at 3 kHz, key noise above 10 kHz.
- Acoustic guitar: Full or thumpy at 80 Hz, presence at 5 kHz, pick noise above 10 kHz.

- Voice: Full at 100 to 150 Hz (males), full at 200 to 250 Hz (females), honky or nasal at 500 Hz to 1 kHz, presence at 5 kHz, sibilance ("s" sounds) above 6 kHz.

- Example: Suppose a vocal track sounds too full or bassy. Reach for the LF EQ knob (say, 100 Hz) and turn it down until the voice sounds natural. To reduce muddiness on snare, bass, electric guitar, or vocal, cut around 300 Hz.

Set EQ to the approximate frequency range you need to work on. Then apply full boost or cut so the effect is easily audible. Finally, fine-tune the frequency and amount of boost or cut until the tonal balance is the way you like it.

What if an instrument sounds honky, tubby, or harsh, and you don't know what frequency to tweak? Set a sweepable equalizer for extreme boost. Then sweep the frequencies until you find the frequency range matching the coloration. Cut that range by the amount that sounds right. For example, a piano miked with the lid closed might have a tubby coloration—maybe too much output around 300 Hz. Set your low-frequency EQ for boost, and vary the center frequency until the tubbiness is exaggerated. Then cut at that frequency until the piano sounds natural.

In general, avoid excessive boost because it can distort the signal. Try cutting the lows instead of boosting the highs. To reduce muddiness or enhance clarity, cut 1 to 2 dB around 300 Hz—either on individual instruments or on the entire mix. Don't boost everything at the same frequency.

When to Use EQ

Before using EQ, try to get the desired tone quality by changing the mic or its placement. This gives a more natural effect than EQ. Many purists shun the use of EQ, complaining of excessive phase shift or ringing caused by the equalizer—a "strained" sound. Instead, they use carefully placed, high-quality microphones to get a natural tonal balance without EQ.

Suppose you still need some EQ. Should you EQ while recording or mixing? If you mix more than one instrument to the same track, you can't EQ them independently during mixdown unless their frequency ranges are far apart. To explain, suppose a recorded track contains lead guitar and vocals. If you add a midrange boost to the guitar, you'll hear it on

208

the vocals, too. The only solution is to EQ the lead guitar by itself when you record it.

If you assign each instrument to its own track, the usual practice is to record flat (without EQ) and then equalize the track during mixdown.

Sometimes the instruments need a lot of EQ to sound good. If so, you might want to record with EQ so that the playback for the musicians will sound good. When you play the multitrack recording through your monitor mixer, the recording may not sound right unless the tracks are already equalized. (That's assuming the monitor mixer in your board has no EQ.)

When you do a bass cut or treble boost, you'll get a better signal-to-noise (S/N) ratio by applying this EQ during recording, instead of during mixdown. But if you're doing a treble cut, apply it during mixdown to reduce any hiss.

Uses of EQ

Here are some applications for EQ:

- Improve tone quality. The main use for EQ is to make an instrument sound better tonally. For example, you might use a high-frequency rolloff on a singer to reduce sibilance, or on a direct-recorded electric guitar to take the "edge" off the sound. You could boost 100 Hz on a floor tom to get a fuller sound, or cut around 250 Hz on a bass guitar for clarity. Cut around 100 Hz to reduce bass buildup on massed harmony vocals. The frequency response and placement of each mic affect tone quality as well.

Although you can set the EQ for each track when it is soloed, a better way is to set the equalizers when the entire mix is playing. That's because one instrument can mask or hide certain frequencies in another instrument. For example, the cymbals might mask the "s" sounds in the vocal, making the vocal sound dull—even though it might sound fine when soloed.

- Create an effect. Extreme EQ reduces fidelity, but it also can make interesting sound effects. Sharply rolling off the lows and highs on a voice, for instance, gives it a "telephone" sound. A 1-kHz band-pass filter does the same thing. To make a mono keyboard track sound stereo, send it to two mixer channels. Boost lows and cut highs in one channel panned left; cut lows and boost highs in the other channel panned right.

- Reduce noise and leakage. You can reduce low-frequency noises—bass leakage, air-conditioner rumble, mic-stand thumps—by turning down the lows below the range of the instrument you're recording.

For example, a fiddle's lowest frequency is about 200 Hz, so you'd use a low-cut filter (highpass filter) set to 200 Hz (if possible). This low-cut filter won't change the fiddle's tone quality because the filtered-out frequencies are below the fiddle's lowest frequency. Similarly, a kick drum has little or no output above 9 kHz, so you can filter out highs above 9 kHz on the kick drum to reduce cymbal leakage. Filtering out frequencies below 100 Hz on most instruments reduces air-conditioning rumble and breath pops. Try rolling off the lows on audience mics to prevent muddy bass. To reduce hum, set a parametric EQ for a 24-dB cut, Q of 30, at these frequencies: 60, 120, and 180 Hz (in the United States) or 50 100 Hz, and 150 Hz (in Europe).

- Compensate for the Fletcher-Munson effect. As discovered by Fletcher and Munson, the ear is less sensitive to bass and treble at low volumes than at high volumes. So, when you record a very loud instrument and play it back at a lower level, it might lack bass and treble. To restore these, you may need to boost the lows (around 100 Hz) and the highs (around 4 kHz) when recording loud rock groups. The louder the group, the more boost you need. It also helps to use cardioid mics with proximity effect (for bass boost) and a presence peak (for treble boost).

- Make a pleasing blend. If you mix two instruments that sound alike, such as lead guitar and rhythm guitar, they tend to mush together—it's hard to tell what each is playing. You can make them more distinct by equalizing them differently. For example, make the lead guitar edgy by boosting 3 kHz, and make the rhythm guitar mellow by cutting 3 kHz. Then you'll hear a more pleasing blend and a clearer mix. The same philosophy applies to bass guitar and kick drum. Because they occupy about the same low-frequency range, they tend to mask or cover each other. To make them distinct, either fatten the bass and thin out the kick a little, or vice versa. The idea is to give each instrument its own space in the frequency spectrum; for example, the bass fills in the lows, synth chords emphasize mid-bass, lead guitar adds edge in the upper mids, and cymbals add sparkle in the highs.

- Compensate for mic placement. Sometimes you are forced to mike very close to reject background sounds and leakage. But a close mic emphasizes the part of the instrument that the mic is near. This gives a colored tone quality, but EQ can partly compensate for it. Suppose you had to record an acoustic guitar with a mic near the sound hole. The guitar track will sound bassy because the sound hole radiates strong low frequencies. But you can turn down the lows on your mixer to restore a natural tonal balance.

This use of EQ can save the day by fixing poorly recorded tracks in live concert recordings. During a concert, the stage monitors might be blaring into your recording/PA microphones, so you're forced to mike close in order to reject monitor leakage and feedback. This close placement, or the monitor leakage itself, can give the recording an unnatural tone quality. In this case, EQ is the only way to get usable tracks.

- Re-mix" a single track. If a track contains two different instruments, sometimes you can change the mix within that track by using EQ. Imagine a track that has both bass and synth. By using LF EQ, you can bring the bass up or down without affecting the synth very much. Mixing with EQ is more effective when the two instruments are far apart in their frequency ranges.

Whenever you record, the ideal situation is to use the right mic in the right position, and in a good-sounding room. Then you don't need or want equalization. Otherwise, though, your recordings will sound better with EQ than without it.

Compressor

A compressor acts like an automatic volume control, turning down the volume when the signal gets too loud. Here's why it's necessary.

Suppose you're recording a female vocalist. Sometimes she sings too softly and gets buried in the mix; other times she hits loud notes, blasting the listener and saturating the tape. Or she may move toward and away from the mic while singing, so that her average recording level changes.

To control this problem, you can ride gain—turn her down when she gets too loud; turn her up when she gets too quiet. But it's hard to anticipate these changes. You might prefer to use a compressor, which does the same thing automatically. It reduces the gain (amplification) when the input signal exceeds a preset level (called the threshold). The greater the input level, the less the gain. As a result, loud notes are made softer, so the dynamic range is reduced (Figure 10.10). *Play CD track 28.*

Compression keeps the level of vocals or instruments more constant, so they are easier to hear throughout the mix, and it prevents loud notes that might clip. Also, it can be used for special effects—to make drums sound fatter, or to increase the sustain on a bass guitar. In pro studios, compression is used almost always on vocals; often on bass guitar, kick drum, and acoustic guitar; and sometimes on other instruments.

Doesn't compression rob the music of its expressive dynamics? Yes, if overdone. But a vocal that gets too loud and soft is annoying. You need to tame it with a compressor. Even then, you can tell when the vocalist is singing loudly by the tone of the voice. It also helps to compress the bass and kick drum to ensure a uniform, driving beat.

You can avoid vocal compression if the singer uses proper mic technique. He or she should back away from the mic on loud notes, and come in close on soft notes. To tell whether you need a compressor, listen to your finished mix. If you can understand all the words, and no notes are too loud, omit the compressor.

Using a Compressor

Normally, you compress individual tracks or instruments, not the entire mix. You want to compress only the stuff that needs it. To compress a stereo mix, you need a 2-channel compressor with a stereo link, which keeps the left–right balance from changing. Multiband compression (covered later) is usually a better choice for compressing the stereo mix.

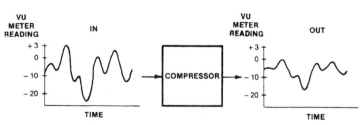

Figure 10.10 Compression.

Should you compress while tracking or mixing? If you compress while tracking, it will be difficult or impossible to change the amount of compression during mixdown. If you compress tracks during mixdown, you can change the settings at will.

Apply EQ before you compress. Often there is too much bass on a track, and that extra bass will trigger the compressor unless you EQ it out first.

Let's describe the controls on the compressor. Some compressors have few controls; most of their settings are preset at the factory.

Compression Ratio or Slope

This is the ratio of the change in input level to the change in output level. For example, a 2:1 ratio means that for every 2 dB change in input level, the output changes 1 dB. A 20 dB change in input level results in a 10 dB change in the output, and so on.

Typical ratio settings are 2:1 to 4:1. A "soft knee" or "over easy" characteristic is a low compression ratio for low-level signals and a high ratio for high-level signals. Some manufacturers say that this characteristic sounds more natural than a fixed compression ratio.

Threshold

This is the input level above which compression occurs. Set the threshold high (about –5 dB) to compress only the loudest notes; set it low (–10 or –20 dB) to compress a broader range of notes. A setting of –10 is typical. If the compressor has a fixed threshold, adjust the amount of compression with the input level control.

Many compressor plug-ins display a compression-ratio graph that shows input level on the horizontal axis and output level on the vertical axis. If the compression ratio is 1:1 (no compression), the graph is a diagonal straight line. This line bends to the right above the threshold, where the compression ratio increases to the amount that you set.

Gain Reduction

This is the number of dB that the gain is reduced by the compressor. It varies with the input level. You set the ratio and threshold controls so that the gain is reduced on loud notes by an amount that sounds right. The amount of gain reduction shows up on a meter—3 to 10 dB is typical.

Attack Time

This is how fast the compressor reduces the gain when it's hit by a musical attack. Typical attack times range from 0.25 to 10 msec. Some compressors adjust the attack time automatically to suit the music; others have a factory-set attack time. The longer the attack time, the larger the peaks that are passed before gain reduction occurs. So, a long attack time sounds punchy; a short attack time reduces punch by softening the attack.

Release Time

This is how fast the gain returns to normal after a loud passage ends. It's the time the compressor takes to reach 63% of its normal gain. You can set the release time from about 50 msec to several seconds. One-half second to 0.2 seconds is typical. For bass instruments, the release time must be longer than about 0.4 seconds to prevent harmonic distortion.

Short release times make the compressor follow rapid volume changes in the music, and keep the average level higher. But because the noise rises along with the gain, short release times can give a pumping or breathing sound. Long release times sound more natural. If the release time is too long, though, a loud passage will reduce the gain during a subsequent quiet passage. In some units, the release time varies automatically, or is factory-set to a useful value.

Some compressors disable the attack and release settings when the compressor is set to RMS or average mode. Those settings are adjusted automatically.

Output-Level Control

Also called make-up gain, this control is used to increase the output level of the compressor by the amount of gain reduction. For example, if a compressor is causing 6 dB of level reduction, increase the make-up gain by 6 dB to achieve unity gain. Some compressors keep the output level constant when other controls are varied.

Spend some free time playing with all the settings so you learn how they affect the sound. Play various instruments and vocals through a compressor, vary the settings, and take notes on what you hear.

Some compressors have a side chain. This is a pair of in/out jacks for connecting an equalizer. To compress only the sibilant sounds on a vocal track, boost the side-chain EQ around 10 kHz. To compress only the breath pops on a vocal track, boost the side-chain EQ around 20 Hz.

A multiband or split-frequency compressor divides the audio band into three or four bands (bass, mids, treble) and compresses each band separately. That way, the compressor can squash a loud bass note, or soften "s" sounds, without bringing down the overall level. Multiband compression is often applied to the final mix of each song during mixdown or mastering.

Connecting a Compressor

Connect a compressor in line with the signal you want to compress, in one of the following ways:

- To compress one instrument or voice while recording: Locate the input module of the instrument you want to compress. On the back of that module, connect the insert send jack to compressor in; connect compressor out to the insert return jack. (Chapter 11 explains these terms.) Or, take a signal from the input module's direct out. Feed that into the compressor, and feed the compressor output to the recorder track input.

- To compress a group of instruments while recording: Locate the bus output of the instruments you want to compress. Go from bus out to compressor in, and go from compressor out to recorder-track in. If the bus has insert jacks, you could connect to them instead.

- To compress one track during mixdown: Go from track out to compressor in, and go from compressor out to mixer channel in. Or locate the mixer input module for that track, then find the insert send and return jacks in that module. (There might be a single Insert jack with send and return terminals.) Connect the insert send to compressor in; connect compressor out to insert return.

- To compress one track in DAW, select a compressor plug-in for that track. Do not use an aux send for compression.

Suggested "Ballpark" Compressor Settings

- **Vocals**: Ratio 2:1 to 3:1, fast attack, 1/2 second release, set threshold for 3 to 6 dB of gain reduction. Singers with extreme dynamic range might need 12 dB of gain reduction and a ratio of 4:1.

215

- **Bass and drums**: Ratio 4:1, slow attack, slow release, set threshold for 3 to 6 dB of gain reduction on loud "pops." Adjust attack time depending on how much you want to soften the attack. Short attack time = soft attack, long attack time = loud attack.

- **Electric guitar**: 4:1 to 8:1 ratio, 10-dB gain reduction, 400 msec release.

- **To reduce breath pops**: Use a multiband compressor. Enable only the lowest frequency band. Try these settings: ratio 30:1, upper frequency 600 Hz, make-up gain 0 dB, attack 1 msec, release 100 msec, threshold –18 dB. Experiment with the threshold setting.

- **To reduce sibilance**: Use a multiband compressor. Disable all bands except the upper-midrange. Try these settings: ratio 20:1, make-up gain 2 dB, attack 1 msec, release 100 msec, hi-mid 5 kHz, high 5 kHz, threshold –32 dB. Experiment with the threshold setting.

- **To compress the stereo mix**: Try these settings: 2:1, soft knee, attack 20 msec, release 200 msec. Set the threshold to get 5 to 10 dB of gain reduction.

Limiter

A limiter keeps signal peaks from exceeding a preset level. While a compressor reduces the overall dynamic range of the music, a limiter affects only the highest peaks (Figure 10.11). To act on these rapid peaks, limiters have a very fast attack time—1 microsecond to 1 millisecond. The compression ratio in a limiter is very high—10:1 or greater—and the threshold is set high, say at 0 dB. For input levels up to 0 dB, the output level matches the input. For input levels above 0 dB, the output level stays at 0 dB. This prevents overload in the device following the limiter

A compressor/limiter carries out both of the functions in its name. It compresses the average signal levels over a wide range, and limits

Figure 10.11 Limiting.

216

peaks to prevent overload. It has two thresholds: one low for the compressor and one high for the limiter.

Limiters can be used to prevent recorder overload during field recording, or to prevent PA power amps from clipping. When you master a program of several mixed songs in a DAW, you might use limiting to reduce the level of signal peaks in the program. Set the threshold about 6 dB below the highest peak level. Then apply normalization, which raises the level of the entire program until the highest peak in the program reaches maximum level. Limiting and normalization create a louder program on your finished CD without compressing the music's dynamics.

Noise Gate

A noise gate (expander) acts like an on-off switch that removes noises during pauses in an audio signal. It reduces the gain when the input level falls below a preset threshold. That is, when an instrument stops playing for a moment, the noise gate drops the volume, which removes any noise and leakage during the pause (Figure 10.12).

Note: The gate does not remove noise while the instrument is playing.

Where is it used? The noise gate helps to clean up drum tracks by removing leakage between beats. It can shorten the decay time of the drums, giving a very tight sound. If you're recording a noisy guitar amp, try a gate to cut out the buzz and hiss between phrases.

How do you use a noise gate? Patch it between a recorder-track output and a mixer line input, or use a gate plug-in in a DAW. Solo the track that you want to gate. Set the gate's threshold so that noise and leakage go away during pauses. If the gate chops off each note, the threshold is set too high—turn it down. To fix a boomy kick drum, adjust the threshold until the kick sounds as "tight" as you want. That is, use the

Figure 10.12 Gating.

gate to shorten the decay portion of the kick-drum's envelope. Set the release time short for drums and longer for instruments that have a long sustain.

Excellent recordings can be made without gating. But if you want a tighter sound, gates come in handy. Some signal processors have compression, limiting, and noise gating in a single package.

Some gates have a side-chain input or key input. It's an input for an external signal that controls the gating action. The control signal triggers the output of the gate's main audio path. For example, you could feed a bass guitar through the noise gate, and gate the bass with a kick-drum signal fed into the side chain. Then the bass will follow the kick drum's envelope.

Delay: Echo, Doubling, Chorus, and Flanging

A digital delay (or a delay plug-in) takes an input signal, holds it in a memory chip, then plays it back after a short delay—about 1 msec to 1 second (Figure 10.13). Delay is the time interval between the input signal and its repetition at the output of the delay device.

If you listen to the delayed signal by itself, it sounds the same as the undelayed (dry) signal. But if you combine the delayed and dry signals, you may hear two distinct sounds: the signal and its repetition. By delaying a signal, a processor can create several effects such as echo, repeating echo, doubling, chorus, and flanging.

Echo

If the delay is about 50 msec to 1 second, the delayed repetition of a sound is called an echo. This is shown in Figure 10.14 by the two pulses. Echoes occur naturally when sound waves travel to a distant room surface, bounce off, and return later to the listener—repeating the original sound.

Figure 10.13 Delaying the signal.

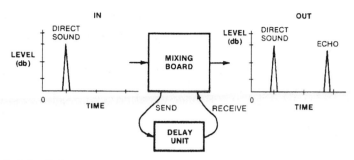

Figure 10.14 Echo

A delay unit can mimic this effect. Many people use the term delay to mean echo.

In setting up a mix with echo, you want to hear both the dry sound and its echo. You do this by creating an effects loop: from the mixer, to the effects box, back to the mixer. Here's how:

1. On the delay unit, set the dry/wet mix control all the way to "wet" or "100% mix." Then the output of the delay unit will be only the delayed signal.
2. Suppose you want to use aux1 as the echo control. Connect aux1 send to delay unit IN. Connect delay unit OUT to Bus 1 and 2 IN (or to the effects-return jacks).
3. Find the mixer module for the instrument you want to add echo to.
4. Assign the instrument to busses 1 and 2. Monitor busses 1 and 2.
5. Find the knobs labeled Bus 1 IN and Bus 2 IN. They might be called "Aux Return" or "Effects Return." Turn them up to 0, about three-fourths of the way up.
6. Turn up the aux1 send knob, and there's your echo.

The delayed sound mixes with the dry sound in busses 1 and 2. You hear both sounds, which together make an echo. Each aux knob controls the amount of echo on each track, while the effect-return knobs control the overall amount of echo on all tracks that are feeding the echo unit.

To set up echo in a DAW, follow this procedure:

1. Create or use a stereo aux bus that has an Echo or Delay plug-in enabled.
2. On a track that you want to have echo, enable and turn up the virtual aux-send knob.

3. Open the Echo or Delay plug-in. Set its dry/wet mix control all the way to wet or 100% mix. Adjust the parameters for the desired effect.

Slap Echo

A delay from 50 to 200 msec is called a slap echo or slapback echo. It was often used in 1950s rock 'n' roll tunes, and still is used today.

Repeating Echo

Most delay units can be made to feed the output signal back into the input, internally. Then the signal is re-delayed many times. This creates a repeating echo—several echoes that are evenly spaced in time (Figure 10.15, and *CD tracks 29 and 30*). The regeneration (feedback) control sets the number of repeats.

Repeating echo is most musical if you set the delay time to create an echo rhythm that fits the tempo of the song. The formula is

```
Delay in seconds 60/tempo
```

So if the tempo is 120 bpm, the delay is 0.5 sec (500 msec). That's one echo per quarter note. Use half that delay to get one echo per eighth note. Use one-third that delay for triplets. A slow repeating echo—0.5 second between repeats, for example—gives an outer-space or haunted-house effect.

Doubling

If you set the delay around 30 to 60 msec, the effect is called doubling or automatic double tracking (ADT). It gives an instrument or voice a fuller sound, especially if the dry and delayed signals are panned to opposite sides. The short delays used in doubling sound like early sound reflections in a studio, so they add some "air" or ambience.

Figure 10.15 Repeating echo.

Doubling a vocal can be done without a delay unit. Record a vocal part, then overdub another performance of the same vocal part. Mix the parts, pan them both to center, or pan them left and right.

Chorus

This is a wavy or shimmering effect. The delay is 15 to 35 msec, and the delay varies at a slow rate. Sweeping the delay time causes the delayed signal to bend up and down in pitch, or to detune. When you combine the detuned signal with the original signal, you get chorusing.

Stereo Chorus

This is a beautiful effect. In one channel, the delayed signal is combined with the dry signal in the same polarity. In the other channel, the delayed signal is inverted in polarity, then combined with the dry signal. Thus, the right channel has a series of peaks in the frequency response where the left channel has dips, and vice versa. The delay is slowly varied or modulated. *Hear a demonstration on CD track 33.*

Bass Chorus

This is chorus with a high-pass filter so that low frequencies are not chorused, but higher harmonics are. It gives an ethereal quality to the bass guitar.

Flanging

If you set the delay around 0 to 20 msec, you usually can't resolve the direct and delayed signals into two separate sounds. Instead, you hear a single sound with a strange frequency response. The direct and delayed signals combine and have phase interference, which puts a series of peaks and dips in the frequency response. This is called a comb-filter effect (Figure 10.16). It gives a very colored, filtered tone quality. The shorter the delay, the farther apart the peaks and dips are spaced in frequency.

The flanging effect varies or sweeps the delay between about 0 and 20 msec. This makes the comb-filter nulls sweep up and down the spectrum. As a result, the sound is hollow, swishing, and ethereal, as if the music were playing through a pipe. Flanging is easiest to hear with

221

broadband signals such as cymbals but can be used on any instrument, even voices. *Hear a demonstration on CD track 34.*

Some examples of flanging are on many Jimi Hendrix records, and on the oldies "Itchycoo Park" by the Small Faces and "Listen to the Music" by the Doobie Brothers. The first use of flanging was on "The Big Hurt" sung by Toni Fisher.

Positive flanging refers to flanging in which the delayed signal is the same polarity as the direct signal (Figure 10.16). With negative flanging, the delayed signal is opposite in polarity to the direct signal, which makes a stronger effect. The low frequencies are canceled (the bass rolls off), and the "knee" of the bass rolloff moves up and down the spectrum as the delay is varied. The high frequencies are still comb-filtered (Figure 10.17). Negative flanging makes the music sound like it's turning inside out.

When the flanger feeds some of the output signal back into the input, the peaks and dips get bigger. It's a powerful "science fiction" effect called resonant flanging.

Reverberation

This effect adds a sense of room acoustics, ambience, or space to instruments and voices. To know how it works, we need to understand how

Figure 10.16 Flanging (or positive flanging).

Figure 10.17 Negative flanging.

reverb happens in a real room. Natural reverberation in a room is a series of multiple sound reflections that make the original sound persist and gradually die away or decay. These reflections tell the ear that you're listening in a large or hard-surfaced room. For example, reverberation is the sound you hear just after you shout in an empty gymnasium.

A reverb effect simulates the sound of a room—a club, auditorium, or concert hall—by generating random multiple echoes that are too numerous and rapid for the ear to resolve (Figure 10.18). Digital reverb is available either in a dedicated reverb unit, as part of a multi-effects processor, or as a plug-in.

The most natural sounding digital reverb is a sampling reverb or convolution reverb, which creates the reverb from impulse-response samples (wave files) of real acoustic spaces, rather than from algorithms. One convolution reverb plug-in is SIR at www.knufinke.de/sir/index_en.html. Free impulse-response samples are at www.noisevault.com and www.echochamber.ch.

Reverb and echo are not the same thing. Echo is a repetition of a sound (HELLO hello hello); reverb is a smooth decay of sound (HELLO-OO-oo-oo).

Multichannel digital reverbs are available for surround sound, both as hardware and software. Some examples are Eventide's Orville, Sony's DRE-S777, TC Electronics' System 6000, Lexicon's 960L, and Kind of Loud's RealVerb 5.1 Pro Tools plug-in. Surround reverb plug-ins for Steinberg's Nuendo platform include Steinberg's Surround Edition plug-in bundle and TC Works SurroundVerb plug-in.

Reverb Parameters

Here are some controls in a reverb unit or plug-in:

- Reverb Time (RT60): The time it takes for reverberation to decay 60 dB below its original level. Set it long (1 1/2 to 2 seconds) to

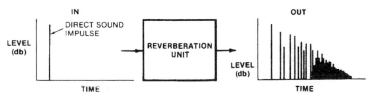

Figure 10.18 Reverberation.

simulate a large room; set it short (under 1 second) to simulate a small room. Generally you use short reverbs (or no reverb) for fast songs, and long reverbs for slow songs.

- Pre-delay (pre-reverb delay): A short delay (30 to 100 msec) before the onset of reverb to simulate the delay that happens in real rooms before reverb starts. The longer the pre-delay, the bigger the room sounds. Using pre-delay on an instrument's reverb often helps to clarify the sound by removing the onset of reverb from the direct sound of the instrument. If your reverb unit does not have pre-delay built in, you can create it by connecting a delay unit between your mixer's aux-send and the reverb input. *CD track 31 demonstrates reverb: short reverb time, long reverb time, and pre-delay.*

- Density: A high density setting produces many echoes spaced close together. It gives a smooth decay but increases the load on the CPU. A low-density setting produces fewer echoes spaced farther apart, and may be adequate for vocals.

- Damping: Adjusts the reverb time or decay at high frequencies. Set the damping frequency high (say, 10 kHz) to simulate a hard-surfaced room; set it low (say, 2 kHz) to simulate a soft-surfaced room. The latter is also called a "warm room" reverb.

- Presets: Algorithms or small programs that simulate the reverb patterns of small rooms, auditoriums, halls, and so on. A plate reverb setting duplicates the bright sound of a metal-foil plate, which used to be the most popular type of reverb in pro studios. Unnatural effects are available, such as nonlinear decay, reverse reverb that builds up before decaying, or gated reverb. With gated reverb, the reverb cuts off suddenly shortly after a note is hit. It's often used on a snare drum. A good example is the oldie "You Can Call Me Al" on Paul Simon's album *Graceland*.

Reverb Connections

To connect a reverb unit to your mixer, connect a cable from the mixer aux-send to the reverb input. Connect a cable (two for stereo) from the reverb outputs to the mixer aux returns (effects returns or bus inputs). Set the mix control on the reverb unit all the way to "wet" or "reverb." Turn the mixer's aux-return or bus-in knobs (if any) about two-thirds of the way up, and adjust the amount of reverb on each track with the aux-send knobs. Try to get an overall reverb-send level near 0 on the meter; then fine-tune the aux return level for the desired amount of reverb.

To enable reverb in a DAW, use the same procedure as for setting up echo, but choose a reverb plug-in.

Preverb

Preverb is reverb that precedes a note rather than follows it. The reverb starts from silence and builds up to a note's attack (*CD track 32*). When used on a snare drum, preverb gives a whip-cracking kind of sound, like "shSHK!"

Here's how to add preverb to a snare drum track in a DAW:

1. Set up an aux1 bus with reverb. Set the reverb all the way to wet or 100% mix.

2. Mute all tracks except the drum track.

3. On the drum track, turn up the aux1 send and set it to pre-fader.

4. Turn down the drum-track fader and play the track. You should hear only the reverb from the aux1 bus.

5. Export or save the mix as "Drum reverb.wav".

6. Select a blank track and call it Drum Reverb. Import "Drum reverb.wav" into that track.

7. In the drum-reverb track you just imported, find a good snare hit with reverb. Make it a clip or region (a selected area) and delete the rest of the track.

8. Select the drum-reverb clip, then select the Reverse processing in your DAW. This reverses the drum-reverb clip.

9. On the snare-drum track, disable or turn down the aux1 send. Turn up the track fader.

10. Slide in time the reversed drum-reverb clip so that it ends just as a snare-drum hit starts (check the waveforms).

11. Play the reversed drum-reverb track along with the snare-drum track. You should hear preverb.

Some signal processors have a reverse reverb effect in which the reverb comes after the note that produced it, but builds up before it fades out. This is not quite the same as preverb. Reverse reverb can upset the musical timing; preverb doesn't.

Enhancer

If a track or a mix sounds dull and muffled, you can run it through an enhancer to add brilliance and clarity. An enhancer works either by adding slight distortion (as in the Aphex Aural Exciter) or by boosting the treble when the signal has high-frequency content (as in the Alesis Micro Enhancer and the Barcus Berry Sonic Maximizer).

The latter device also divides the frequency range into three bands. The lows are delayed about 1.5 msec; the midrange is delayed about 0.5 msec; and the highs are delayed only a few microseconds. In this way, the Maximizer aligns the harmonics and fundamentals in time for added clarity.

Octave Divider

This unit takes a signal from a bass guitar and provides deep, growling bass notes one or two octaves below the pitch of the bass guitar. It does this by dividing the incoming frequency by 2 or 4: If you put 82 Hz in, you get 41 Hz out. Some MIDI sound modules have bass patches with extra-deep sound, and some bass guitars have an extra string tuned especially low.

Harmonizer

Basically a delay unit with delay modulation, a harmonizer makes a variety of pitch-shifting effects. It can create harmonies, change pitch without changing the duration of the program, change duration without changing pitch, and many other oddities. You've heard harmonizers on radio-station spots when the announcer's voice sounds like a Munchkin or Darth Vader. *Play CD track 35.*

Vocal Processor

This device or plug-in can affect the vocal's inflection, add growls or whispers, correct the pitch, add vibrato, make the voice more-or-less nasal or chesty, and so on. The latest vocal formant-corrected pitch-shifters maintain the voice formant structure when they shift pitch; this prevents the "chipmunk" effect. Examples are TC-Helicon VoicePrismPlus.

Automatic Pitch Correction

Auto-Tune by Antares provides automatic or manual pitch correction. It corrects flat or sharp notes by changing their pitch to match a musical scale of your choice. You also can use Auto-Tune as a "robotic" effect where the sung notes change pitch in a step-wise, jerky way rather than smoothly. Pitch-correction plug-ins are available from other companies as well.

Tube Processor

This device uses a vacuum tube or a transistor simulation of one. Tubes have euphonic even-order harmonic distortion, which is claimed to add "richness" or "warmth" when the tube distorts (*CD track 36*). There are tube mics, tube mic preamps, tube compressors, and standalone tube processors.

Rotary Speaker Simulator

This effect simulates the sound of a Leslie organ speaker, which plays music through rotating horns. It's a complex sound effect of pitch shifting, tremolo, and phase shifting. The speed and depth of the effect are adjustable.

Analog Tape Simulator

Analog tape saturation is mainly third-harmonic distortion and compression. An analog tape simulator adds this distortion to digital recordings in an attempt to smear or warm up the sound in a pleasant way (*CD track 37*).

Spatial Processor

Spatial processors enhance the stereo imaging or spatial aspects of a mix heard over two speakers. Some units have joystick-type pan pots, which move the image of each track anywhere around the listener. Other units make the stereo stage wider, so that images can be placed to the left of the left speaker, and to the right of the right speaker. The listener might hear images toward the sides of the listening room. In 5.1 surround systems, this spatial processing is done by surround panning and surround reverbs.

Microphone Modeler

Antares and Roland offer a microphone modeler or simulator. You tell it which mic you are using and which mic you want it to sound like. A wide variety of vintage and current mic simulations are available. Mic modeling comes in three forms: hardware device, plug-in, and firmware (programmed into a chip) in a recorder-mixer.

Guitar Amplifier Modeler

Another simulator takes the sound of a direct-recorded guitar, and makes the guitar sound like it is played through a guitar amp (*CD track 38*). Several amp models can be simulated, as well as effects, tone, drive, the mic used to pick up the amp, and the mic's position.

Two hardware examples of amp modelers are the Line 6 Pod and the Johnson J Station. Amp Farm is a guitar modeling plug-in for Pro Tools. Roland's digital workstations offer COSM mic modeling and guitar-amp modeling.

Guitar processors or guitar stomp boxes can be used on any instrument or vocal to add distortion, or to generally "shred" the sound for a low-fi effect.

De-Click, and De-Noise

Also called "Audio restoration progams", these are plug-ins—or stand-alone programs—that can remove the clicks and pops from LP records, or remove hiss and hum from noisy recordings.

Surround Sound

Recent plug-ins for surround sound are surround panning, surround reverb, and surround encoding/decoding.

Multieffects Processor

This provides several effects in a single device or plug-in. Some units let you combine up to four effects in any order. Others have several channels, so you can put a different effect on different instruments. With most processors, you can edit the sounds and save them in memory as new programs. *On the enclosed CD, tracks 29–38 demonstrate various effects on voice.*

An extension of the multieffects processor is the vocal processor. It includes a high-quality mic preamp or two, plus EQ, compression, gating, de-esser, and perhaps some tube saturation distortion.

A multieffects processor uses a digital signal processing (DSP) chip and RAM memory. The amount of memory is limited, so the more memory that one effect uses, the less is available for other effects. For example, suppose you're combining reverb and echo. If you use a reverb with a long decay time (which takes a lot of memory), you may have to settle for an echo with a short delay.

Most units have a frequency response up to 20 kHz and at least 16-bit resolution. They offer 100 or more programmable presets with MIDI control over any parameter. For example, with some units you can place an instrument in a simulated room, and use a MIDI controller to continuously change the size of that room.

Many signal processors can be controlled by MIDI program-change commands. You can quickly change the type of effect, or effect parameters, by entering certain program changes into a sequencer.

Suppose you want each tom-tom hit in a drum fill to have a different size room added to it. For example, put the high-rack tom in a small room; put the low-rack tom in a concert hall; and put the floor tom in a cave. To do this, first assign a different program number (patch or preset number) to each effects parameter. You do this with the effects device. Then, using the sequencer, punch in the appropriate program number for each note.

A MIDI program-change footswitch lets guitarists call up different effects on MIDI signal processors. By tapping a footswitch, they can get fuzz, flanging, wah-wah, spring reverb, and so on.

A MIDI mapper lets you control some effects parameters with any controller. For example, vary reverb decay time with a pitch wheel, or vary a filter with key velocity.

Looking Back

We've come a long way with effects. Looking back over the past few decades, each era had its own "sound" related to the effects used at the time. The '50s had tube distortion and slap echo; the '60s used fuzz, wah-wah, and flanging. Much of the early '70s sounded dry, and the early '90s emphasized synth, drum machines, and gated reverb. Now vacuum tubes and acoustic instruments are back, along with occasional low-fi (tinny, distorted, or noisy) sounds and dry vocals. Whatever effects you choose, they can enhance your music if used with taste.

Sound-Quality Glossary

The sound of effects and EQ can be hard to translate into engineering terms. For example, what EQ should you use to get a "fat" sound or a "thin" sound? The glossary below may help. It's based on conversations with producers, musicians, and reviewers over many years. Not everyone agrees on these definitions, but they are common.

AIRY Spacious. The instruments sound like they are surrounded by a large reflective space full of air. A pleasant amount of reverb. High-frequency response that extends to 15 or 20 kHz.

BALLSY OR BASSY Emphasized low frequencies below about 200 Hz.

BLOATED Excessive mid-bass around 250 Hz. Poorly damped low frequencies, low-frequency resonances.

BLOOM Adequate low frequencies. Spacious. Good reproduction of dynamics and reverberation. Early reflections or a sense of "air" around each instrument in an orchestra.

BOOMY Excessive bass around 125 Hz. Poorly damped low frequencies or low-frequency resonances.

BOXY Having resonances as if the music were enclosed in a box. Speaker cabinet diffraction or vibration. Sometimes an emphasis around 250 to 500 Hz.

BREATHY Audible breath sounds in vocals, flute, or sax. Good high-frequency response.

BRIGHT High-frequency emphasis. Harmonics are strong relative to fundamentals.

BRITTLE High-frequency peaks, or weak fundamentals. Slightly distorted or harsh highs. Opposite of round or mellow. See **Thin**. Objects that are physically thin and brittle emphasize highs over lows when you crack them.

CHESTY A vocal signal with a bump in the low-frequency response around 125 to 250 Hz.

CLEAN Free of noise, distortion, and leakage.

CLEAR See **Transparent**.

CLINICAL Too clean or analytical. Emphasized high-frequency response, sharp transient response. Not warm.

COLORED Having timbres that are not true to life. Non-flat response, peaks, or dips.

CONSTRICTED Poor reproduction of dynamics. Dynamic compression. Distortion at high levels. Also see **Pinched**.

CRISP Extended high-frequency response. Like a crispy potato chip, or crisp bacon frying. Often referring to cymbals.

CRUNCH Pleasant guitar-amp distortion.

DARK Opposite of bright. Weak high frequencies.

DELICATE High frequencies extending to 15 or 20 kHz without peaks. A sweet, airy, open sound with strings or acoustic guitar.

DEPTH A sense of closeness and farness of instruments, caused by miking them at different distances. Good transient response that reveals the direct/reflected sound ratio in the recording.

DETAILED Easy-to-hear tiny details in the music; articulate. Adequate high-frequency response, sharp transient response.

DRY Without effects. Not spacious. Reverb tends towards mono instead of spreading out. Overdamped transient response.

DULL See **Dark**.

EDGY Too-strong high frequencies. Trebly. Harmonics are too strong relative to the fundamentals. When you view the waveform on an oscilloscope, it even looks edgy or jagged, because of excessive high frequencies. Distorted, having unwanted harmonics that add an edge or raspiness to the sound.

EFFORTLESS Low distortion, usually coupled with flat response.

ETCHED Clear but verging on edgy. Emphasis around 10 kHz or higher.

231

FAT   See **Full** and **Warm**. Also, a diffuse spatial effect. Also, smeared out in time, with some reverberant decay.

FOCUSED   Referring to the image of a musical instrument which is easy to localize, pinpointed, having a small spatial spread.

FORWARD   Sounding close to the listener, projected. Emphasis around 2 to 5 kHz.

FULL   Opposite of **Thin**. Strong fundamentals relative to harmonics. Good low-frequency response, not necessarily extended, but with adequate level around 100 to 300 Hz.

GENTLE   Opposite of edgy. The harmonics—highs and upper mids— are not exaggerated, or may even be weak.

GLARE, GLASSY   A little less extreme than edgy. A little too bright or trebly.

GRAINY   The music sounds like it's segmented into little grains, rather than flowing in one continuous piece. Not liquid or fluid. Suffering from harmonic or IM distortion. Some early A/D converters sounded grainy, as do current ones of inferior design. "Powdery" is finer than "grainy"!

GRUNGY   Lots of harmonic or IM distortion.

HARD   Too much upper midrange, usually around 3 kHz. Or, good transient response, as if the sound is hitting you hard.

HARSH   Too much upper midrange. Peaks in the frequency response from 2 to 6 kHz. Or, excessive phase shift.

HEAVY   Good low-frequency response below about 50 Hz. Suggesting great weight or power, like a diesel locomotive or thunder.

HOLLOW   See **Honky**. Or, too much reverberation. Or, a mid-frequency dip.

HONKY   The music sounds the way your voice sounds when you cup your hands around your mouth. A bump in the response around 500 to 700 Hz.

LIQUID   Opposite of grainy. A sense of seamless flowing of the music. Flat response and low distortion. High frequencies are flat or reduced relative to mids and lows.

LOW-FI (low fidelity)   "Trashy" sounding. Tinny, distorted, noisy, or muddy.

MELLOW   Reduced high frequencies, not edgy.

MUDDY   Not clear. Weak harmonics, smeared time response, IM distortion. Too much reverb at low frequencies. Too much emphasis around 200 to 350 Hz.

MUFFLED   The music sounds covered up. Weak highs or weak upper mids.

MUSICAL   Conveying emotion. Flat response, low distortion, no edginess.

NASAL   The vocalist sounds as if he or she is singing with the nose closed. Also applies to strings. Bump in the response around 300 to 1000 Hz. See **Honky**.

NEUTRAL   Accurate tonal reproduction. No obvious colorations. No serious peaks or dips in the frequency response.

PAPERY   Referring to a kick drum that has too much output around 400 to 600 Hz.

PINCHED   Narrowband. Midrange or upper-midrange peak in the frequency response. Pinched dynamics are overly compressed.

PIERCING   Strident, hard on the ears, screechy. Having sharp, narrow peaks in the response around 3 to 10 kHz.

PRESENT, PRESENCE   Adequate or emphasized response around 5 kHz for most instruments, or around 2 to 5 kHz for kick drum and bass. Having some edge, punch, detail, closeness, and clarity.

PUFFY   Bump in the response around 500 to 700 Hz.

PUNCHY   Good reproduction of dynamics. Good transient response. Sometimes a bump around 5 kHz or 200 Hz.

RASPY   Harsh, like a rasp. Peaks in the response around 6 kHz which make vocals sound too sibilant or piercing.

RICH   See **Full**. Also, having euphonic distortion made of even-order harmonics.

ROUND   High-frequency rolloff or dip. Not edgy.

SHARP See **Strident** and **Tight**.

SIBILANT, ESSY Exaggerated "s" and "sh" sounds in singing, caused by a rise in the response around 5 to 10 kHz.

SIZZLY See **Sibilant**. Also, too much highs on cymbals.

SMEARED Lacking detail. Poor transient response. This may be a desirable effect in large-diameter mics. Also, poorly focused images.

SMOOTH Easy on the ears, not harsh. Flat frequency response, especially in the midrange. Lack of peaks and dips in the response. Low distortion.

SPACIOUS Conveying a sense of space, ambience, or room around the instruments. To get this effect, mike farther back, mix in an ambience mic, add reverb, or record in stereo. Components that have out-of-phase crosstalk between channels may add false spaciousness.

SQUASHED Overly compressed.

STEELY Emphasized upper mids around 3 to 6 kHz. Peaky, nonflat high-frequency response. See **Glassy**, **Harsh**, **Edgy**.

STRAINED The component sounds like it's working too hard. Distorted. Inadequate headroom or insufficient power. Opposite of effortless.

STRIDENT See **Harsh** and **Edgy**.

SWEET Not strident or piercing. Flat high-frequency response, low distortion. Lack of peaks in the response. Highs are extended to 15 or 20 kHz, but they are not bumped up. Often used when referring to cymbals, percussion, strings, and sibilant sounds.

THIN Fundamentals are weak relative to harmonics. Note that the fundamental frequencies of many instruments are not very low. For example, violin fundamentals are around 200 to 1000 Hz. So if the 300-Hz area is weak, the violin may sound thin—even if the mic's response goes down to 40 Hz.

TIGHT Good low-frequency transient response. Absence of ringing or resonance when reproducing the kick drum or bass. Good low-frequency detail. Absence of leakage.

TINNY, TELEPHONE-LIKE Narrowband, weak lows, peaky mids. The music sounds like it's coming through a telephone or tin can.

TRANSPARENT   Easy to hear into the music, detailed, clear, not muddy. Wide, flat frequency response, sharp time response, very low distortion and noise.

TUBBY   See **Bloated**. Having low-frequency resonances as if you're singing in a bathtub.

VEILED   The music sounds like you put a silk veil over the speakers. Slight noise or distortion, or slightly weak high frequencies.

WARM   Good bass, adequate low frequencies, adequate fundamentals relative to harmonics. Not thin. Or, excessive bass or mid-bass. Or, pleasantly spacious, with adequate reverberation at low frequencies. Or, gentle highs, like from a tube amplifier. See **Rich**.

WOOLY, BLANKETED   The music sounds like there's a wool blanket over the speakers. Weak high frequencies or boomy low frequencies. Sometimes, an emphasis around 250 to 600 Hz.

MIXERS AND MIXING CONSOLES

The heart of your recording studio is the mixer. It's a control center where you plug in all sorts of signals; mix or blend them; add effects, EQ, and stereo positioning; and route the signals to recorders and monitor speakers. A mixing console (also called board or desk) is a large mixer with many controls. This chapter covers three types of mixers: analog hardware, digital hardware, and digital software mixers.

A recorder-mixer combines a mixer and a multitrack recorder in a single portable chassis. This convenient unit is also called a ministudio or portable studio. Low-end recorder-mixers record 4 tracks, and high-end units record 8 to 32 tracks.

This chapter covers both mixers and mixing consoles. First we'll look at typical features of both, then at features found only in large mixing consoles.

Stages of Recording

This chapter will refer to the three stages in making a multitrack recording: recording, overdubbing, and mixdown.

1. Recording (tracking) The mixer accepts mic-level signals and amplifies them up to line level. You send the line-level signal from each

mic to a separate track in a multitrack recorder. The multitrack recorder records several tracks on tape, MiniDisc, or hard disk. One track might be a lead vocal, another track might be a saxophone, and so on.

2. Overdubbing While listening to prerecorded tracks over headphones, the musician records new parts on open (unused) tracks.

3. Mixdown: After all the tracks are recorded, mix or combine them into 2-track stereo or 6-channel surround. Add effects. Record the stereo mix with a 2-track recorder, such as a DAT recorder, CD-R burner, or hard drive. This recording can be duplicated on a CD-R (recordable compact disc).

Mixer Functions and Formats

Although the knobs and meters on a mixer may appear intimidating, you can understand them if you read the manual and practice with the equipment. A mixer is complicated because it lets you control many aspects of sound:

- The loudness of each instrument (to control the balance among instruments in the mix)
- The tone quality of each instrument (bass, treble, midrange)
- The room that the instruments are in (reverberation)
- The left-to-right position of each instrument (panning)
- Effects (flanging, echo, chorus, etc.)
- Track assignments (put one or more instruments on each recorder track)
- Recording level (to prevent distortion and noise in the recorder)
- Monitor selection (what you want to listen to)

Mixers come in many formats:

- An analog mixer is a control device that works on analog signals, and sends them to an external multitrack recorder and 2-track recorder.
- A digital mixer is the same, but works internally with digital signals. It accepts analog or digital signals.

238

- A software mixer exists only in your computer as part of digital recording software. You control it with a mouse or a controller surface (described next). The recording is done on your hard drive.
- A controller surface is a device that looks like a mixer with faders and knobs. It plugs into your computer's USB or FireWire port and controls the software mixer.

A mixer can be specified by the number of inputs and outputs it has. For example, an 8-in, 2-out mixer (8 × 2 mixer) has 8 signal inputs that can be mixed into 2 output channels (buses) for stereo recording. Similarly, a 16-in, 8-out (16 × 8) mixing board has 16 signal inputs and 8 output channels for multitrack recording. A 16 × 4 × 2 mixing board has 16 inputs, 4 submixes or groups (explained under the heading "Output Section") and 2 master outputs. There also are connectors for external equipment, such as effects devices and a monitor power amplifier. The more inputs your mixer has, the more instruments you can record at the same time. If you're recording only yourself, you may need only two inputs.

Let's look at the analog mixer in more detail. Knowing how it works will help you understand the other types.

Analog Mixer

A mixer can be divided into three sections: input, output and monitor. Here are the main parts of each section and what they do:

Input section

- Inputs connect to your mics, electric instruments, and recorder outputs.
- Faders are sliding volume controls that affect the loudness of each instrument. This lets you control the balance among instruments in the mix.
- Equalization (EQ) knobs adjust the tone quality of each instrument (bass, treble, midrange).
- Aux knobs set the amount of reverb or other effects, and also can be used to set up a monitor mix or headphone mix.
- Pan pots place the monitored sound of each track where desired between your stereo speakers—left, center, right, or anywhere between. In some mixers, the pan pot is also used with the assign switch during recording to send signals to the desired tracks.

- Channel assign buttons route each input signal to the desired recorder track.

Output section

- Master faders set the overall level.
- Outputs connect to your recorder inputs and the monitor power amplifier.
- Meters help you set the correct recording level (to prevent distortion and noise).

Monitor section

- Monitor controls select what you want to listen to.
- Aux knobs or channel faders set up the monitor mix.

Let's look at each part in more detail.

Input Section

A mixer is made of groups of controls called modules. An input module (Figure 11.1) affects a single input signal—from a microphone, for instance. The module is a narrow vertical strip, one per input. Several modules are lined up side-by-side. Each input module is the same, so if you know one, you know them all.

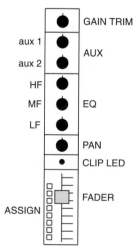

Figure 11.1 A typical input module.

Let's follow the signal flow from input to output through a typical input module (Figure 11.2). Every mixer is a little different, but you are likely to find features like those described here.

Input Connectors

On the back of each input module are input connectors with these labels:

MIC: Accepts signals from a microphone or direct box

LINE: Accepts an electric musical instrument, or a track output of a multitrack recorder

TRACK: Accepts a track output of a multitrack recorder; not included in all mixers

Some units use a single jack for both mic and line inputs; others have separate jacks for each. The mic input is either an unbalanced 1/4-inch phone jack (a 1/4-inch hole) or an XLR-type connector (with three small holes). The line input is either a 1/4-inch phone jack, an RCA phono jack (like you see on a stereo system), or an XLR-type connector.

You can plug a synthesizer directly into a phone-jack line input without using a direct box if the cable is under 10 feet; a longer cable may pick up hum. In that case, use a direct box plugged into a mic input.

In some mixers, a phone-jack input is switchable between low impedance (for microphones) or high impedance (for electric guitar pickups). Other mixers might have a separate low-impedance input and a high-impedance input.

Figure 11.2 Signal flow through a typical input module.

Phantom Power (P48, +48)

This switch (not shown) turns on phantom power (48V DC) for condenser microphones. In the mic input connector, the 48V appears on pins 2 and 3 relative to pin 1. The microphone receives phantom power and sends audio along the same two cable leads.

Mic Preamp

After entering the mic connector, the microphone signal goes to a mic pre-amplifier inside the mixer. This preamp boosts or amplifies the weak microphone signal up to a higher voltage, making it a line-level signal.

Trim (Gain)

The TRIM or GAIN control adjusts the amount of amplification in the mic preamp. If the gain control is turned up full, and the incoming mic signal is very strong due to a loud instrument or vocal, this signal can overload the mic preamp. This causes distortion—a gritty sound. To get a good signal-to-noise ratio (S/N) in your mixer, set the gain control as high as possible, but not so high that the preamp distorts.

Here's how: Start with the trim turned up all the way down (counterclockwise). In most mixers, each input module has a tiny light (LED) labeled "clip," "peak," or "OL" (overload). It flashes when the mic preamp is distorting. When an instrument is playing its loudest signal through an input module, gradually turn up the trim control until the clip LED starts flashing. Then turn down the trim control just to the point where the light stays off, and turn it down another 10 dB for extra headroom.

Some low-cost mixers do not have a trim control. The input fader serves this function.

Input Selector Switch

This switch lets you select the type of signal you want to work with. Some common switch labels are below:

MIC (Mic or direct box)

LINE (For line-level signals: a synthesizer, drum machine, electric guitar, or multitrack recorder's track output)

INPUT (Mic or line)

TRACK (Multitrack recorder's track output)

242

MUTE No signal is processed; during mixdown, it's a good idea to mute tracks that have nothing playing at the moment to reduce tape hiss

Using the input selector is simple. If you plugged in a microphone or direct box to record its signal, set the input selector to MIC or INPUT. If you plugged in a synth, drum machine, or electric guitar, set the selector to LINE or INPUT. When you're ready to mix, select TRACK (if available) or LINE (if the recorder track outputs are plugged into mixer line inputs).

Some mixers have no input selector. The mixer processes whatever signal is plugged in.

In some recorder-mixers, a single 1/4-inch phone jack is used both for mic-level and line-level signals. A mic-level signal is typically about 1 to 2 millivolts. A line-level signal is about 0.3 to 1.23 volts. The two levels are handled either by a MIC/LINE switch or a TRIM control.

Insert Jacks

Following the input selector switch are the insert-send jack and insert-return jack (on the back of each input module). Or the mixer might have a single insert jack, a TRS type that has the send on the tip terminal and the return on the ring terminal. Inside the mixer, the send is connected to the return so that the signal passes through to the rest of the mixer.

If you insert a plug into the insert jack(s), you break the signal path so you can insert an external device there in series with the input module's signal. You might insert a compressor into the signal path of one module for automatic volume control, or insert any other signal processor (reverb/delay, for instance). That way, if all your aux sends are tied up, you can add another signal processor. On the reverb/delay unit, set the dry/wet mix control for the desired amount of effect.

Another use for the insert jack is to send the mic-preamp output signal to a multitrack recorder track. The output of each track returns to the insert jack and continues though your mixer. In this case, you use the trim controls to set recording levels, and use the faders, EQ, and aux sends to set up a monitor mix.

Low-cost mixers omit the insert jacks. Some units have insert jacks on only two inputs.

243

Input Fader (Channel Fader)

Next, the input signal that you selected goes to a fader. This is a sliding volume control for each input signal. During recording, you can use the fader in two ways. If you are recording one instrument per track, record off the insert send and use the fader to adjust the instrument's level in the monitor mix. If you are recording two or more instruments on one track, record off a group output, and use each instrument's fader to set up a mix within that group. For example, if you are recording several drum mics onto one track, set up a drum mix with the drum-mic faders.

During mixdown, you use the faders to set the loudness balance among instruments.

EQ

The signal from the input fader goes to an equalizer, which is a tone control. With EQ you can make an instrument sound more or less bassy, and more or less trebly, by boosting or cutting certain frequencies. See Chapter 10 for details.

Direct Out

The direct out is an output connector following each input fader and equalizer. The signal at the direct-out jack is an amplified, equalized version of the input signal. The fader controls the level at the direct-output jack. You can use the direct-out jack when you want to record one instrument per track (with EQ) on an external multitrack recorder. Connect the direct-out jack to a multitrack recorder track input. Because the direct output bypasses the mixing circuits farther down the chain, the result is a cleaner signal.

Suppose your mixer has 8 inputs and 2 outputs. You can use this mixer with an 8-track recorder. Just connect the direct-out jack in each input module to a separate track input. Or do the same with the insert-send jacks.

Channel Assign Switch

The equalized signal also goes through a pan pot (explained in the section below) to the channel assign switch or track selector switch. It lets you send the signal of each instrument to the recorder track you want to record that instrument on.

A mixer with four groups or buses would have an assign switch labeled 1, 2, 3, 4. If you want to record bass on track 1, for instance, find the assign switch for the input module the bass is plugged into, and push assign switch 1. If you want to record four drum mics on track 2, push assign switch 2 for all those input modules.

Some mixers assign tracks with two controls: a selector switch and a pan pot.

Pan Pot

This knob sends a signal to two channels in adjustable amounts. By rotating the pan-pot knob, you control how much signal goes to each channel. Set the knob all the way left and the signal goes to one channel. Set it all the way right and the signal goes to the other channel. Set it in the middle and the signal goes to both channels equally.

Here's how you might use a pan pot to assign an instrument to a track. The channel-assign switch might have two positions labeled 1–2 and 3–4. If you turn the pan pot left, the signal goes to odd-numbered tracks (either 1 or 3, depending on how you set the assign switch). If you turn the pan pot right, the signal goes to even-numbered tracks (2 or 4). Suppose you want to assign the bass to track 1. Set the assign switch to 1–2, and turn the pan pot far left to choose the odd-numbered track (track 1).

During mixdown, the pan pot has a different function: It places images between your speakers. An image is an apparent source of sound, a point between your speakers where you hear each instrument or vocal. Set the pan pot to locate each instrument at the left speaker, right speaker, or anywhere in between. If you set the pan pot to center, the signal goes equally to both channels, and you hear an image in the center.

Aux

The aux or aux-send function (Figures 11.2 and 11.3) sends some of the input module's signal to equipment outside the mixer. A pre-fader aux send is before the fader, and usually goes to a power amp that feeds monitor speakers or headphones. You can use the pre-fader aux knobs to set up a monitor mix—a balanced blend of input signals you hear over speakers or headphones. A post-fader aux send is after the fader and EQ, and usually goes to an effects unit. You use the post-fader aux knobs to set the amount of effects (reverb, echo) heard on each instrument in a mix.

Some mixers have no aux sends, some have one aux-send control per module, and some have two or more (labeled aux 1, aux 2, etc.). The more aux sends you have, the more you can play with effects, but the

245

Figure 11.3 Aux sends.

greater the cost and complexity. The aux number (1 or 2) is not necessarily assigned a specific function; you decide what you want aux 1 and aux 2 to do.

During recording and overdubbing, the aux knobs of all the input modules can be used to create a monitor mix. The monitor mix that you create with the aux knobs is independent of the levels going to the multitrack recorder. You use the gain-trims during recording to set recording levels, and you use the aux knobs to create an independent mix that is heard over your monitor system.

In Figure 11.3, the aux-2 send control is just before the fader. In this mixer, the signals from all the aux-2 knobs in the mixer combine at a connector jack labeled "aux-2 send." You can connect that jack to your power amplifier, which drives monitor speakers and headphones.

In Figure 11.3, the aux-1 send control is just after the fader. During mixdown, each aux-1 knob controls how much effects (reverb, echo) you hear on each track. In each input module, the aux knob adjusts how much of that input signal is sent to an external effects unit. The effected signal returns to the mixer's aux-return or bus-in jacks, where it blends with the original signal.

For example, suppose the aux-1 send is connected to a digital reverb. The more you turn up the aux-1 send knob, the more signal goes to the reverb. The output of the reverb unit returns to the mixer's bus-in jack, and blends with the original dry signal, adding a spacious effect.

A few mixers have an aux-return control (also called effects-return or bus-in) that sets the overall effects level returning to the mixer.

Follow these steps to use the aux controls to adjust the amount of effects heard on each track:

1. Patch an effects unit between your mixer's aux-send and aux-return (bus-in) jacks.
2. On the effects unit, set the dry/wet mix control all the way to "wet" or "effect."
3. If your mixer has aux-return (bus-in) knobs, turn them about half way up and pan their signals hard left and right.
4. Turn up the aux-send knob for each input, according to how much effect you want to hear on that input signal. Suppose you're using reverb as an effect. You might turn it up by different amounts for the vocals, drums, and lead guitar, and leave it turned down for the bass and kick drum.

As you're setting the aux levels, check the overload indicator on the effects unit. If it's flashing, turn down the input level on the effects unit just to the point where the overload light stops flashing. Then turn up the output level on the effects unit (or turn up the aux return on the mixer) to achieve the same amount of effect you heard before.

There might be a pre/post switch next to the aux-send knob. When an aux knob is set to pre (pre-fader), its level is not affected by the fader setting. You use the pre setting for a headphone mix during recording or overdubbing because you don't want the fader settings to affect the monitor mix.

The post setting (post-fader) is used for effects during mixdown. In this case, the aux level follows the setting of the fader. The higher you set the track volume with the fader, the higher the effects level. But the dry/wet mix stays the same.

Output Section

The output section is the final part of the signal path; the section that feeds mixed signals to the recorder tracks. It includes mixing circuits, submaster or group faders (sometimes), master faders, and meters (Figure 11.4).

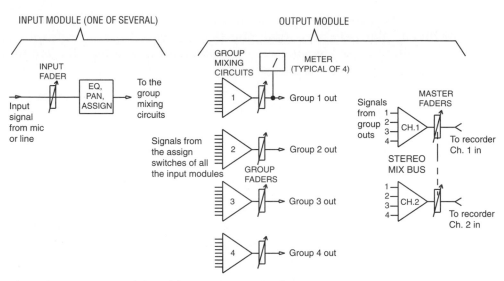

Figure 11.4   Input module and output section of a mixer.

Mixing Circuits (Active Combining Networks), Group Faders, and Bus Output Connectors

The group mixing circuits are in the center of Figure 11.4. Recall that you use the assign switches to send each input signal to the desired channel or bus, and each bus feeds a different recorder track. A bus is a channel in a mixer containing an independent mix of signals. The bus 1 or group 1 mixing circuit accepts the signals from all the inputs you assigned to bus 1 and mixes them together to feed track 1 of the recorder. The bus 2 mixing circuit mixes all the bus 2 assignments, and so on.

Following each group mixing circuit is a group fader. If you assigned all the drum mics to group 1, the group 1 fader adjusts the overall level of the drum mix. The signal from each group fader goes to a group or bus output connector in your mixer. You can connect each bus output to a recorder track input.

Stereo Mix Bus, Master Faders, Main Output Connectors

The stereo mix bus are two group mixing circuits: one for channel 1 and one for channel 2. Three types of signals feed into the stereo mix bus:

248

1. The group output signals.

2. Signals from input modules. You can assign an input module's signal directly to the stereo mix bus, bypassing the groups. This results in lower noise.

3. Effects-return signals, such as the reverberated signal from an external digital reverb.

Located on the right side of your mixer, the stereo master faders are one or two sliding volume controls that affect the overall level of the stereo mix bus. Usually, you set the master fader(s) within design center, the shaded area about three-quarters of the way up on the scale. This setting minimizes mixer noise and distortion. You can fade out the end of a mix by turning down the master faders gradually.

After the master faders, the signal goes to a pair of main output connectors, which feed a 2-track recorder of your choice.

You feed the multitrack recorder either from group outputs, direct outs, or insert sends. If you're mixing several instruments to track 5, for example, assign those instruments' signals to Group 5. Connect the Group 5 output to recorder track 5 in. If you're recording one instrument on track 5, however, connect that instrument's direct out or insert send to track 5 in. The signal is cleaner at the direct out or insert send than at the group output.

Some mixing consoles have voltage controlled amplifier (VCA) group faders. A VCA group fader acts like a remote control for groups of channel faders. For example, you could assign each drum mic's fader to a group, which is an audio path, and also assign each drum mic's fader to a VCA group, which is a fader that controls all the drum mic channels at once.

Meters

Meters are an important part of the output section. They measure the voltage level of various signals. Usually, each group or bus output has a meter to measure its signal level. If these buses feed the recorder tracks, you use the meters to set the recording level for each track.

A mixer has either VU meters or LED bar graph meters.

- A VU meter (now rarely used) is a voltmeter that shows approximately the relative loudness of various audio signals. Set the record level so that the meter needle reaches +3 VU maximum for most signals, and about −6 VU maximum for drums, percussion, and piano. That's necessary because the VU meter responds too slowly to show the true level of percussive sounds.

- An LED bar graph meter has a column of lights (LEDs) that shows peak recording level. Usually you set the recording level to peak near 0 dB maximum.

Monitor Section

The monitor section is used to control what you're listening to. It lets you select what you want to hear, and lets you create a mix over headphones or speakers to approximate the final product. The monitor mix has no effect on the levels going to your recorder.

During recording, you want to monitor a mix of the input signals. During playback or mixdown, you want to hear a mix of the recorded tracks. During overdubs, you want to hear a mix of the recorded tracks and the instrument that you're overdubbing. The monitor section lets you do this.

Monitor Select Buttons

These buttons let you choose what signal you want to monitor or listen to. Because the configuration of these buttons varies widely among different mixers, they are not shown in Figure 11.4.

If you want to use aux 2 as the monitor bus, select the aux-2 bus as the monitor source. Some mixers have no monitor-select switches. Instead, you always monitor the stereo mix bus.

Monitor Mix Controls and Connectors

One way to set up a monitor mix is with the aux knobs. Suppose you want aux 2 to be the monitor mix. Connect the aux-2 send jack to your power amp and speakers. Or if you're using headphones, switch them to monitor the aux-2 bus. Turn up all the aux-2 knobs about halfway, then turn each knob up or down to set a good loudness balance. You do this during recording or overdubbing.

Here's another way to set up a monitor mix using the faders. Connect your multitrack recorder ins and outs to the insert jacks (Figure 11.5). Connect the insert-jack 1 tip (send) to track 1 in; connect track 1 out to the insert-jack ring (return). Make similar connections for the other tracks. Also connect the mixer's monitor-out jacks to your power amp

250

Figure 11.5 Using insert jacks to send each input signal to a recorder track. The track signal returns to the mixer, where you adjust level, panning, EQ, and effects.

and speakers. Monitor the stereo mix bus. With this setup, use the trim controls to set recording levels. Use the faders to set up a monitor mix, cue mix, or mixdown with EQ and effects. It's a convenient way to work.

In a split console (side-by-side console), a separate monitor mixer is built into the control surface. This monitor mixer has level, pan, and aux controls for each track. In an in-line console (I/O console), the monitor mix is done with the aux knobs or fader in each input module. This is the most common type of console.

During mixdown, monitor the stereo mix bus.

SOLO

The SOLO button in an input module lets you monitor one instrument or vocal at a time so you can hear it better. By pressing two or more SOLO buttons, you can monitor more than one input signal at a time.

Suppose you hear a buzz in the audio and suspect it may be in the bass guitar signal. If you push the SOLO button in the bass guitar's input module, you'll monitor only the bass guitar. Then you can easily hear whether the buzz is in that input.

On British consoles, the SOLO function is called pre-fader listen (PFL) or after-fader listen (AFL).

Additional Features in Large Mixing Consoles

Large mixing consoles have more features than small mixers do. If you're working only with a small mixer or a recorder-mixer, you might want to skip this section.

251

FOLD BACK (FB): Another name for cue, or headphone mix.

PHASE (POLARITY INVERT): Used only with balanced lines, this switch inverts the polarity of the input signal. That is, it switches pins 2 and 3 to flip the phase 180 degrees at all frequencies. You might use it to correct a miswired mic cable whose polarity is reversed. If you mic a snare drum top and bottom, you need to invert the polarity of the bottom mic.

AUTOMATED MIXING CONTROLS: These controls (Read, Write, Update, Record Automation, Play automation) set up the console for various automation functions. With automation, a memory circuit in the mixer remembers your console settings and mix moves. You can recall the settings with the push of a button, then continue working on the mix. More on automated mixing is at the end of Chapter 12.

EFFECTS PANNING: This feature places the images of the effects signals wherever desired between the monitor speakers. Some consoles let you pan effects in the monitor mix as well as in the final program mix.

EFFECTS RETURN TO CUE: This is an effects-return level control that adjusts the amount of effects heard in the studio headphone mix. These monitored effects are independent of any effects being recorded.

EFFECTS RETURN TO MONITOR: This effects-return control adjusts the amount of effects heard in the monitor mix. These monitored effects are independent of any effects being recorded.

BUS/MONITOR/CUE switch for effects return: A switch that feeds the effects-return signal to your choice of three destinations: program bus (for mixdown), monitor mix, cue mix, or any combination of the three.

METER SWITCHES: In many consoles, the meters can measure signal levels other than console output levels. Switches near the meters can be set so that the meters indicate bus level, aux-send level, aux-return level, monitor-mix level, etc.

Those readings help you set optimum levels for the outboard devices receiving those signals. Too low a level results in noise; too high a level causes distortion in the outboard unit. For example, if the aux-return signal sounds garbled or distorted, the cause may be an excessive aux-send level. Verify that condition by checking the meters switched to read the aux or effects bus.

DIM: A switch that reduces the monitor level by a preset amount so you can talk (as in "dim the lights").

TALKBACK: An intercom between the control room and studio. A mic built into the console lets you talk to the musicians in the studio when you push the talkback button.

SLATE: This function routes the control-room microphone signal to all the buses so you can record the name of the tune and take number.

OSCILLATOR or TONE GENERATOR: This is used to put a level-calibration tone on a DAT tape, or to reference the recorder's meters to those on the console. You also can use it to check signal path, levels, and channel balance.

Digital Mixer

So far we've looked at the analog mixing console, which works entirely with analog signals. A digital mixing console accepts analog or digital signals. It converts the analog signals to digital, and processes all signals internally in digital format. The signal stays in the digital domain for all mixer processing. Level changes, EQ, and so on are done by digital signal processing (computer calculations) rather than by analog circuits.

In a digital console, each analog input signal goes through an analog-to-digital (A/D) converter so that the mixer can process the signal digitally. Some of the digital output signals from the mixer go through digital-to-analog (D/A) converters. The resulting analog signals feed a power amp, effects unit, and so on.

Analog and digital consoles operate differently. To use EQ in an analog console, you find the channel you want to EQ, and adjust its EQ knobs. To use EQ in a digital console, you press a button to select the channel you want to equalize, and press an EQ button. Then the EQ settings for that channel show up on an LCD screen. You press buttons and turn a knob to adjust the EQ frequency and boost/cut for that channel.

Because one knob controls the EQ for all the channels, digital consoles have fewer controls than analog consoles. One knob or switch can have several functions. This makes digital consoles harder to operate because you can't just reach for an EQ knob for a particular channel. You have to press a few buttons to set the EQ parameters.

253

On the other hand, digital consoles have built-in effects and automated mixing. A sequencer circuit in the mixer remembers your mix so you can recall it at a later time. One type of automated mixing is called scene or snapshot automation. When you press the snapshot button, a memory circuit in the mixer takes a "snapshot" of all the mixer settings for later recall. Another type of automation is dynamic. Your mix moves are stored in memory for later recall.

When you recall a mix, some mixers make the faders move into the positions you set up. This feature is called "flying faders" or "motorized faders." Other mixers do not move the faders when you recall a mix. You have to set them manually by looking at a display, which is a disadvantage. Motorized faders are more expensive but easier to use.

Digital Mixer Features

Look for the following features in digital mixers when making a buying decision:

- Number of mic inputs: 2 to 16 or more.
- Number and type of digital inputs and outputs: S/PDIF, AES/EBU, TDIF, ADAT
- Number and type of option card slots: extra I/O, DSP, sync, effects, plug-ins
- Number of effects processors
- Ease of use
- Snapshot or dynamic automation
- Motorized or non-motorized faders
- Surround-sound monitoring

Software Mixer

Here's another type of mixer. A software mixer (virtual mixer) is a simulated mixer you see on a computer screen. It exists only in your computer as part of digital recording software. You control it with a mouse or with a controller surface. Recordings made with this mixer are stored on your hard drive.

In a software mixer, the input and output connectors are in an audio interface (such as a sound card) that connects to the computer. This interface is 2-channel or multichannel. The software mixer works with effects

254

that are either plug-ins (software) or external (hardware). Automation is a standard feature.

For more on software mixers, see Chapter 13.

Controller Surface

A controller surface resembles a standalone mixer with faders and knobs. It plugs into your computer's USB or FireWire port and controls the software mixer. Some people prefer a controller because it is easier to use than a mouse.

Now that you understand the typical features of mixers and mixing consoles, you are ready to learn how to use them.

12

OPERATING THE MULTITRACK RECORDER AND MIXER

Get your hands on those knobs. You're going to operate a mixer as part of a recording session. This will be a basic run-through. Detailed session procedures are described in Chapter 15. Most of these procedures also apply to operating a virtual mixer in a Digital Audio Workstation (DAW).

First recall the stages in making a recording:

1. Session preparation
2. Recording
3. Punching in
4. Overdubbing
5. Bouncing tracks
6. Mixdown

This chapter considers each stage in turn.

Session Preparation

If you're recording on hard drive, make sure you have enough drive space for the project. Chapter 13 has simple equations and a table that show how much space you need.

If you're recording on a modular digital multitrack, fast-forward the tape to the end and rewind to the top. This loosens the tape pack, distributes the tape lubricants more evenly, and aligns the tape with the tape guides. Then format the tape from beginning to end. For efficiency, you might want to format an entire box of videocassettes at once.

Plan your track assignments. Write a track sheet that tells what instrument goes on which track. **Note**: If you assign multiple instruments to the same track, you can't separate their images in the stereo stage. That is, you can't pan them to different positions; all the instruments on one track sound as if they're occupying the same point in space. If you're recording 4-track, you may want to do a stereo mix of the rhythm section on tracks 1 and 2; then overdub vocals and solos on tracks 3 and 4.

Studio setup for the musicians is covered in Chapter 15.

Recording

To start the process, first zero or neutralize the mixer by setting all the controls to "off," "flat," or "zero." This establishes a point of reference and avoids surprises later on. Set all faders down.

If you have a separate mixer and multitrack recorder, you need to make their meter readings match. To do this, play a steady tone into the mixer to get a 0 reading on the meters for all channels. Then set the multitrack recorder's record level (if any) to get 0 readings on all the tracks. This works only if the mixer meters and recorder meters have the same reaction speed (ballistics). If you are feeding the multitrack recorder from your mixer's insert jacks or direct-out jacks, you'll need to watch the recorder meters.

Some audio interfaces use on-screen volume controls to set recording levels. Some interfaces have level knobs for this purpose.

Suppose you're ready to record a vocal or an acoustic instrument. Place the microphone and plug it into a mic input. If you want to record the audio output of a hardware synth or drum machine, connect a cable between the instrument's output and a line input on the mixer.

Attach a "scribble strip" of masking tape or white removable tape along the bottom of the faders, and label each fader according to what instrument you plugged into that input. Some consoles and software mixers have scribble strips that you can type on.

Set the input selector (if any) to Mic or Line depending on what is plugged into each input.

Plug in headphones to hear what you're recording. Turn up the headphone volume control. Or, if you're in a control room and the musicians are in a studio, turn up the monitor level to listen over the monitor loudspeakers. On a hardware mixer, set the MONITOR SELECT switch to hear the signal you're recording, and turn up the musicians' cue mix.

Set the master faders about three-quarters of the way up, at 0, or within the shaded portion of fader travel. This is called design center. Do the same for the input fader(s) in use (Figure 12.1). These settings give the best compromise between noise and distortion.

Assign Inputs to Tracks

If your equipment wiring is mics > mixer > insert sends > track inputs, your inputs are already assigned. Input 1 goes to track 1; input 2 goes to tracks 2, and so on.

If you're using a computer DAW, select the input source for each track. For example, you might set track 5's input to channel 5 in your audio interface. Then, whatever is plugged into interface channel 5 will go to track 5. If your sound card has just 2 channels, set track 5's input to channel 1 or 2 in your audio interface, and likewise for the other tracks.

If your equipment wiring is mics > mixer > busses > track inputs, assign each input signal to the desired output channel (bus) as specified on your track sheet. Each bus is connected to the corresponding numbered track on the multitrack recorder. **Note**: If only one instrument is assigned to a track, you can eliminate the noise of the console's combining amplifier by patching the instrument's signal directly to the recorder track. To do that, locate the direct output jack (or the insert jack) of the input module for that instrument and patch it to the desired track. Some mixing boards also require pressing a Direct button on the input module.

Set Recording Levels

Now you're ready to "get a level." Have each instrument play the loudest part of the music, one at a time or all at once. For each input signal, set

(A) Mixer.

DESIGN CENTER ─

MASTER FADERS

INPUT FADERS

(B) Recorder–
mixer.

DESIGN CENTER ─

Figure 12.1 Setting mixer faders at design center.

the TRIM control so the recording level is as high as possible without causing distortion. While setting levels for a digital multitrack, peak each track around −5 dBFS (−5 decibels Full Scale). This allows some headroom for surprises. Also, musicians generally play louder during a performance than during a level check. Generally keep the maximum level around −2 or −3 dBFS. If you exceed 0 dBFS you'll hear digital clipping which makes a loud click.

If you are mixing several instruments to one or two groups (as in a drum submix), follow this procedure:

1. Monitor the group(s).
2. Set its submaster fader (group fader) to design center.
3. Set the submix balances, panning, and recording level with the input faders.
4. Fine-tune each submix level with the submaster fader.

Set EQ

Although it's common to record flat (without EQ), you may want to apply equalization at this point to each instrument heard individually. Filter out frequencies above and below the range of the instrument. However, don't spend too much time on EQ until all the instruments are mixed together. The EQ that sounds right on a soloed instrument may not sound right when all the instruments are heard together. In creating the desired tonal balance, use EQ as a last resort after trying different mics and mic placements. You also can apply EQ during playback or mixdown. This may be preferable because EQ applied when recording cannot always be undone if you're unhappy with it.

Recording

Next, set the track(s) you want to record to "record ready" mode. Now start recording. Write down the counter time for this take. "Slate" the recording: record the name of the tune and the take number. Then record a two-measure count-off. This is done to set the tempo for overdubs. For example, if the time signature is 4/4, you say, "1, 2, 3, 4, 1, 2, (rest) (rest)." The rests are silent beats. You need some silence before the song starts to make editing easier later on. A DAW has a built-in metronome that can be used for count-offs.

Playback

After recording the track, go to the beginning of the song using the return-to-zero or locate function. If necessary, set the monitor selector to "track," or "mix," and play back the recording to check the performance and sound quality. You can set a rough mix with the monitor mix (aux) knobs. If your multitrack is patched to the insert jacks, use the faders, pan pots, EQ, and aux knobs to set a rough mix.

261

Get or write a sheet of the song's arrangement. It shows the lyrics, verses, choruses, bridge, and so on. Play the song and set a locate point at the start of each verse and chorus. That way, when the musician says "Let's fix that flat note at the end of the second chorus," you can go instantly to that part of the song.

Overdubbing

After your first track is recorded with a good performance, you might want to add more musical parts. This procedure is called overdubbing. When you overdub, the musician listens to tracks already recorded, and records a new part on an unused track.

Ready to overdub? Here's what to do:

1. Turn off the control room speakers and listen on headphones.

2. Set up a headphone monitor mix of the recorded tracks. Some mixers use the aux knobs for this. If your multitrack recorder is connected to the mixer's insert jacks, monitor the stereo mix, and set up a mix with the channel faders.

3. Plug in the mic or direct box for the instrument or vocal you want to record. Assign it to an unused track. Turn up its fader to design center (the 0 point about three-quarters of the way up).

4. Set up your multitrack to monitor the playback of recorded tracks and the input signal of the instrument or vocal you want to record. In some DAWs this function is called "echo input monitor." If your mics and DAW output are plugged into a mixer, disable "echo input monitor" because you will be monitoring all signals through the mixer.

5. Have the musician play or sing. Can you hear the signal in the phones? Set the recording level using the trim (gain) control for that mic's channel. If you are using a DAW and you hear a lot of latency (delay) in the musician's signal as heard over headphones, decrease the latency setting in software. To prevent latency, you might want to monitor the output of your hardware mixer instead. Feed your audio interface output into the mixer, and listen to the recorded tracks that way.

6. Play the recording. As the musician plays or sings along, set up a good mix of the recorded tracks and the live instrument you're going to record. Ask the musician if he is hearing what he needs to hear.

Change the mix if needed. Some musicians want to hear effects when they overdub; some want it dry. If you're overdubbing backup vocals one at a time, often it helps to remove certain other vocals from the headphone mix.

7. When you're ready to record the new part, go to a point about 10 seconds before the part of the song where the musician plays. Set the recorded tracks to SAFE and set the track you're recording on to RECORD READY.

8. Before you hit that RECORD button, stop! Are you recording on the correct track(s)? Are you going to accidentally erase any tracks? Double-check your track sheet, and make sure the record-enable buttons are on only for the tracks that are safe to record over.

9. Start recording and have the musician play along with the tracks. If the musician makes a mistake, you can re-record or punch-in the new part without affecting the parts already on tape.

Punching-In

Punching-in is used to fix mistakes in a recorded performance, or to record a musical part in segments. A punch is also called an insert. You enable record mode on a track, play the multitrack recording, then "punch" or press the record button at the right spot, record a new part, then punch-out of record mode.

Some musicians record the same musical part over and over until the part is perfect. This process is tedious, but you have to pay attention. You need to be aware of where you are in the song at all times, and not erase anything you want to keep.

Some musicians like to record a performance a phrase at a time, perfecting each phrase as they go. Others record a complete take, then go back and fix the weak parts.

To do a punch-in, grab your song-arrangement sheet and follow these steps:

1. Go to a point about 10 seconds before the point where you want to start recording. Note the counter time and write it down. Also enter a memory location point there if your recorder can do that.

2. Play the song to the musician over headphones. The musician plays along to practice the part. Write down the tape counter times where you want to punch-in and punch-out. Otherwise you might erase a good take.

3. Finally you're ready to record. Before you hit that RECORD button, stop! Tell the musician what you're going to do so there's no chance of a mistake. Be very clear. For example, "I'm recording your keyboard part on two new tracks." Or, "I'm punching-in over your old performance—is that what you wanted?"

4. Okay, ready to go. Play the recording. During a rest or pause in the music just before the part needing correction, punch-in the record button (or use a footswitch). Have the musician record the new part, and punch-out right away. You don't want to erase the rest of the track!

5. Press LOCATE to go to the location point you set, about 10 seconds before the punch. Play the recording to see if the punch was okay. If necessary, you can re-record the punch. After you go to the locate point, notice whether the musician wants to practice the part. Don't redo the punch until he or she is ready.

Some multitrack recorders have an autopunch function. You set the punch-in and punch-out points into the machine's memory. As the recording plays, the recorder automatically goes into and out of record mode at those points. Some recorders can loop repeatedly between those two points.

Composite Tracks

If several open tracks are available, you can record a solo performance in several takes, each on a separate track or virtual track. Then combine the best parts of each track into a single track. Use only that track in the final mix, and you'll hear the best parts of all the takes in succession. This is called "recording composite tracks" or "comping." Here's an example of comping vocal tracks with a multitrack recorder and mixer:

1. Suppose you have four takes of a vocal recorded on tracks 10, 11, 12, and 13. Solo each track, and mark on the lyric sheet which tracks have the best performance of each section. For example, you might prefer track 13 on Verse 1, track 10 on Chorus 2, and so on.

2. Assign all the vocal tracks to an open track, which we'll call the comp track. Match their levels.

3. Start recording on the comp track.

4. As the song plays, mute and unmute the tracks to copy the best performances to the comp track. To do that, you can stop the recorder

between sections, change the mute settings, and record a section at a time.

5. Once you're happy with the comp track, you can erase or archive the original tracks.

Comping with a DAW is even easier. Pick the best overall track, then copy and paste good sections from other tracks into the best track.

Some recorders have virtual tracks (explained in Chapter 9). They let you comp a performance with virtual tracks rather than real tracks.

Getting More Tracks

What if you want to overdub more parts, but all the tracks are full? With care, you can punch-in more instruments by recording them in the pauses on recorded tracks. For example, suppose all the tracks are full but you want to add a cymbal crash at the beginning of the chorus. Find a track that has a pause at that moment, and punch in the cymbal crash there.

Here's another way to free up tracks. Suppose that one track is mostly blank except for a short riff at certain points in the song. Using a patch cord (or using cut-and-paste in a DAW), copy that riff to a blank area in another recorded track at exactly the same point in time. Then you can erase the first track and record a new part on it.

As an alternative, you can bounce tracks—mix several tracks to one or two open tracks, and record the mix on that track. Then you can erase the original tracks, freeing them for more overdubs. Bouncing procedures are in the instruction manual for your recorder or software.

Flying In

Suppose you have a multitrack recording of a pop song. One of the tracks is vocals. During the first chorus, the vocals sound great, but during the second chorus, the vocals are out of tune. So you want to copy the vocal track from the first chorus to the second. If you're using a DAW or hard-disk recorder, you can use the cut-and-paste editing function.

If you're using a Modular Digital Multitrack (MDM), copying parts can be done by a process called flying in. You copy the first chorus on the vocal track to an external DAW or sampler, then fly-in (copy) the external recording back to the multitrack where the second chorus would be. The fly-in part must stay in sync with the other tracks.

Here's another use for a fly-in. You give a keyboard player a CD-R of a song you recorded. The keyboardist takes the song home, figures out a keyboard part, and records it on another CD-R while listening to the song. He or she sends the part to you, and you fly it into the multitrack recording, or rip the CD track to a wave file and import it.

Here are the steps to do a fly-in:

1. On the multitrack recorder, find the track you want to copy. Connect its output to a DAW audio interface input or sampler input.

2. Play the multitrack recording and copy the part to the DAW or sampler. If the part you want to fly-in is already on CD-R, DAT, or MiniDisc, copy it to your DAW or sampler.

3. Connect the DAW or sampler output to the input of the track on which you want to record the fly-in.

4. Locate the multitrack recorder to a point a few seconds before where you want to fly-in the part.

5. Start recording on the correct track. While listening to the song play, press PLAY on the DAW, or trigger the sample, so that the flown-in part starts at the right time. Practice this until the timing is correct. The tempos should match if you recorded to a click track.

Drum Replacement

Suppose you've recorded the drum tracks, but you don't like the sound of them. Maybe the kick drum is too flabby, and no amount of EQ or gating seems to help. You might try a technique called drum replacement. You take recorded drum tracks and replace them with drum samples generated by a drum machine, sampler, or sound module. Here are the steps:

1. Select the desired drum sound in the sampler.

2. Feed the audio from the drum track into the sampler's trigger input.

3. Connect the sampler's audio output to an unused track input on your recorder.

4. Play the original drum track and record the sampler's output signal on the unused track. When each drum hit triggers the sampler, it instantly plays its internal drum sample.

In this way you can replace individual drum tracks or an entire kit. You can also mix the samples with the original tracks to get a bigger drum sound.

In a MIDI/Digital Audio recording program, you can replace an audio drum track with a MIDI track that plays drum samples. First, set up a MIDI track with a drum sample ready to play (for details, see Chapter 16). Then use one of these methods:

Method 1: In the MIDI drum track you want to create, open the MIDI sequencer editing screen. Draw a note for each drum hit on the correct beats.

Method 2: Start recording on the MIDI drum track, and tap your MIDI controller key in sync with the playback of the audio drum track.

Method 3: Select the audio drum track, then enable the "Extract Rhythm" or "Extract Beat" feature, if any. Copy and paste the extracted MIDI beats to the MIDI drum-sample track.

Mute the original audio drum track and press PLAY. You should hear the replacement drum sample playing.

Mixdown

After all your tracks are recorded (maybe with some bouncing), it's time to mix or combine them to 2-track stereo. You will use the mixer faders to control the relative volumes of the instruments, use panning to set their stereo position, use EQ to adjust their tone quality, and use the aux knobs to control effects.

Set Up the Mixer and Recorders

To prepare for a mixdown, first locate the mixer jacks for output channels 1 and 2 (they might be called bus 1 and 2, or stereo mix bus). Plug these outputs into the line inputs of your 2-track recorder (Figure 12.2). If you're using a DAW, omit this step.

What if you're synching an external MIDI sequencer to your audio recorder? Use a separate line mixer to combine your audio mixer's output with the sequencer's audio output. Or if your audio mixer has enough input channels, plug the sequencer audio outputs into the audio mixer. Mix the audio tracks with the MIDI tracks.

Once connections are made, you can begin. Set all the mixer controls to "off," "zero," or "flat." You should start from ground zero in building a mix.

Figure 12.2 Connections for mixdown.

Tape a strip of paper or masking tape along the front of the mixer to write which instrument(s) each fader affects. Keep this strip with the multitrack tape for use each time you play it. Or type in this information in your DAW or mixing console.

If necessary, set the input-selector switches on the mixer to "track" because you'll be mixing down the multitrack recording. Monitor the 2-track stereo mix bus. These steps are unnecessary if you are using a DAW.

For starters, put the master faders at design center (about three-quarters of the way up, at the shaded portion of fader travel). This sets the mixer gain structure for the best compromise between noise and headroom.

Erase Unwanted Material

Mixing will be a lot easier if you first erase noises before and after each song, and within each track.

Play the multitrack recording and listen to each track alone. Erase unwanted sounds, outtakes, and entire segments that don't add to the song. To avoid mistakes, it's best to do this while the musicians are around.

What if a noise occurs just before the musician starts playing? When you erase the noise, you might erase the beginning of the performance. It might be safer to mute the track during mixdown and then unmute it just before the musician plays. Or you could set up an automatic punch-in/-out at the correct times to erase the noise.

Removing noises with a DAW is easier. Look at a recorded track's waveform to see where the music and silences are. Divide the track into music clips and silent clips (segments), and delete the silent ones. This removes the noises that are in the silent (unplayed) portions of the track. If you recorded a track in separate segments or clips, trim or slip-edit the start and end points of each clip to remove noises.

Panning

You need to pan the tracks before doing the mix, because the loudness of a track depends on where it's panned. Assign each track to busses 1 and 2 (or the stereo mix bus), and use the pan pots to place each track where desired between your stereo speakers. Typically the bass, snare, kick drum, and vocals go to center; guitars can be panned left or right, and stereo keyboards and drum overheads go left and right.

Pan tracks to many points between the monitors: left, half-left, center, half-right, right. Try to achieve a stereo stage that is well balanced either side of center. For clarity, pan to opposite sides any instruments that cover the same frequency range.

You may want some tracks to be unlocalized. Harmony singers and strings should be spread out rather than appearing as point sources. Stereo keyboard sounds can wander between speakers. You could fatten a lead-guitar solo by panning it left, and panning the solo delayed to the right. (This delay might come from a distant room mic you used while recording, or from copying and sliding a track a few milliseconds in a DAW.) Pan doubled vocals left and right for a spacious effect.

Consider creating some front-to-back depth. Leave some instruments dry so they sound close; add reverb to others so they sound farther away.

If you want the stereo imaging to be realistic (say, for a jazz combo), then pan the instruments to simulate a band as viewed from the audience. If you're sitting in an audience listening to a jazz quartet, you might hear drums on the left, piano on the right, bass in the middle, and sax slightly right. The drums and piano are not point sources, but are somewhat spread out. If spatial realism is the goal, you should hear the same ensemble layout between your speakers. In most rock recordings, the piano and drums are spread all the way between speakers—interesting but unrealistic.

Pan-potted mono tracks often sound artificial; each instrument sounds isolated in its own little space. It helps to add some stereo reverb. It surrounds the instruments and "glues" them together.

When you monitor the mix in mono, you'll likely hear center channel buildup. Instruments in the center of the stereo stage will sound louder in mono than they did in stereo, so the mix balance will change in mono. To prevent this, note which tracks are panned hard left or right, and bring them a little toward the center: 9 and 3 o'clock on the pan knobs.

Compression

Sometimes the lead vocal track might be too loud or too quiet relative to the instruments because vocals have a wider dynamic range than instruments. You can control this by running the vocal track through a compressor. It will keep the loudness of the vocal more constant, making it easier to hear throughout the mix. Patch the compressor into the insert jack(s) of the vocal input module, or between the vocal track output and the mixer input. Set the desired amount of compression (typically 2:1 ratio, −10-dB threshold). It's also common to compress the kick drum and bass. (For more information, see Chapter 10.)

Set a Balance

Now comes the fun part. The mixdown is one of the most creative parts of recording. Here are some tips to help your mixes sound terrific.

Before doing a mix, tune up your ears. Play over your monitors some CDs whose sound you admire. This helps you get used to a commercial balance of the highs, mids, and lows.

Choose a CD with tunes like those you're recording. Check out the production. How is the balance set? How about EQ, effects, and sonic surprises? Try to figure out what techniques were used to create those sounds, and duplicate them. Of course, you might prefer to break new ground.

Using the input faders, adjust the volume of each track for a pleasing balance among instruments and vocals. You should be able to hear each instrument clearly. Some mixing consoles have trim knobs that set the playback gain of the multitrack recorder tracks. In that case, set all faders in use to design center, and adjust the trims to get a rough mix.

Here's one way to build the mix. Make all the instruments and vocals equally loud. Then turn up the most important tracks and turn down background instruments. Or, bring up one track at a time and blend it with the other tracks. For example, first bring up the kick drum to about −10 dB, then add bass and balance the two together. Next add drums and set a balance. Then add guitars, keyboards, and finally vocals.

In a ballad, the lead vocal is usually on top. You might set the soloed lead-vocal level to peak at −5 dB. Bring up the monitor level so that the vocal is as loud as you like to hear it, then leave the monitor level alone. Bring in the other tracks one at a time and mix them relative to the vocal track.

When the mix is right, everything can be heard clearly, yet nothing sticks out too much. The most important instruments or voices are loudest; less important parts are in the background. In a typical rock mix, the snare is loudest, and the kick is nearly as loud. The lead vocal is next in level. Note that there's a wide latitude for musical interpretation and personal taste in making a mix.

Sometimes you don't want everything to be clearly heard. Once in a while, you might mix in certain tracks very subtly for a subconscious effect.

It's a good idea to monitor around 85 dBSPL. If you monitor louder, the bass and treble will be weak when the mix is played softly.

To test your mix, occasionally play the monitors very quietly and see if you can hear everything. Switch from large monitors to small, and make sure nothing is missing.

Set EQ

Next, set EQ for the tonal balance you want on each track. If a track sounds too dull, turn up the highs or add an enhancer. If a track sounds too bassy, turn down the lows, and so on. Cymbals should sound crisp and distinct, but not sizzly or harsh; kick drum and bass should sound deep, but not overwhelming or muddy. Be sure the bass is recorded with enough edge or harmonics to be audible on small speakers.

You'll need to readjust the mix balances after adding EQ. The EQ that sounds right on a soloed track seldom sounds right when all the tracks are mixed together. So make EQ decisions when you have the complete mix happening.

In pop-music recordings, the tone quality or timbre of instruments does not have to be natural. Still, many listeners want to hear a realistic timbre from acoustic instruments, such as the guitar, flute, sax, or piano.

The overall tonal balance of the mix shouldn't be bassy or trebly. That is, the perceived spectrum should not emphasize lows or highs. You should hear the low bass, mid-bass, midrange, upper midrange, and highs roughly in equal proportions. Frequency bands that are too loud can tire your ears.

When your mix is almost done, switch between your mix and a commercial CD to see whether you're competitive. If the tonal balance of your mix matches a commercial CD, you know your mix will translate to the real world. This works regardless of what monitors you use. An effective tool for this purpose is Harmonic Balances (www.har-bal.com).

Add Effects

With the balances and EQ roughed in, it's time to add effects. You might want to plug in a reverb or delay to add spaciousness to the sound (see Chapter 10). This device connects between your mixer's aux-send and aux-return jacks (or aux-send and bus-in jacks).

Find the AUX RETURN or BUS IN controls (if any), set them half of the way up, and pan them hard left and right. Using the AUX knobs on the mixer, adjust the amount of delay or reverb for each track as desired.

Too much effects and reverb can muddy the mix. You might turn up the reverb only on a few instruments or vocals. Once you have the reverb set, try turning it down gradually and see how little you can get by with.

The producer of a recording is the musical director and decides how the mix should sound. The producer might be the band members or yourself. Ask to hear recordings having the kind of sounds the producer desires. Try to figure out what techniques were used to create those sounds.

Also try to translate the producer's sound-quality descriptions into control settings. If the producer asks for a "warmer" sound on a particular instrument, turn up the low frequencies. If the lead guitar needs to be "fatter," try a stereo chorus on the guitar track. If the producer wants the vocal to be more "spacious," try adding reverb, and so on. *CD track 39 demonstrates a mixdown.*

Set Levels

Set the overall recording level as you're mixing. To maintain the correct gain staging, keep the master faders at design center. Then adjust all the input faders by the same amount so your stereo output level peaks around −5dB. You can touch up the master faders a few decibels if necessary. Don't exceed a 0-dB recording level if you're recording to a digital medium.

Judging the Mix

When you mix, your attention scans the inputs. Listen briefly to each instrument in turn and to the mix as a whole. If you hear something you don't like, fix it. Is the vocal too tubby? Roll off the bass on the vocal track. Is the kick drum too quiet? Turn it up. Is the lead-guitar solo too dead? Turn up its effects send.

Check the mix while listening from another room, where the lows and highs are weakened. Is the balance still good?

The mix must be appropriate for the style of music. For example, a mix that's right for rock music usually won't work for folk music or acoustic jazz. Rock mixes typically have lots of production EQ, compression, and effects; and the drums are way up front. In contrast, folk or acoustic jazz is usually mixed with no effects other than slight reverb, and the instruments and vocals sound natural. A rock guitar typically sounds bright and distorted; a straight-ahead jazz guitar usually sounds mellow and clean.

Suppose you are mixing a pop song, and you're aiming for a realistic, natural sound. Listen to the reproduced instruments and try to make them sound as if they're really playing in front of you. That is, instead of trying to make a pleasant mix or a sonically interesting recording, try to control the sound you hear to simulate real instruments—to make them believable. To do this you must be familiar with the sound of real instruments.

It's like an artist trying to draw a still-life as realistically as possible. The artist compares the drawing to the real object, notes the difference, and then modifies the drawing to reduce the difference.

When you're striving for a natural sound, compare the recorded instrument with your memory of the real thing. How does it sound different? Turn the appropriate knob on the console that reduces the difference.

Alternatively, when you're mixing, imagine that you're creating a sonic experience between the monitor speakers, rather than just reproducing instruments. Sometimes you don't want a recording to sound too realistic. If a recording is very accurate, it sounds like musical instruments, rather than just music itself.

This approach contradicts the basic edict of high fidelity—to reproduce the original performance as it sounded in the original environment. Some songs seem to require unreal sounds. That way, you don't connect the sounds you hear with physical instruments, but with the music behind the instruments—the composer's dream or vision.

Here's one way to reproduce pure music rather than reproducing instruments playing in a room: Mike closely or record direct to avoid picking up studio ambience. Then add reverb. Also add EQ, double-tracking, and effects to make the instrument or voice slightly unreal. The idea is to make a production, rather than a documentation—a record, rather than a recording.

Try to convey the musician's intentions through the recorded sound quality. If the musician has a loving, soft message, translate that into a warm, smooth tone quality. Add a little mid-bass or slightly reduce the highs. If the musical composition suggests grandeur or space, add reverberation with a long decay time. Ask the musicians what they are trying to express through the music, and try to express that through the sound production as well.

Try to keep the mix clean and clear. A clean mix is uncluttered; not too many parts play at once. It helps to arrange the music so that similar parts don't overlap. Usually, the fewer the instruments, the clearer the sound. Mix selectively, so that not too many instruments are heard at the same time. Have guitar licks fill in the holes between vocal phrases, rather than playing on top of the vocals.

In a clear-sounding recording, instruments do not "crowd" or mask each other's sound. They are separate and distinct. Clarity arises when instruments occupy different areas of the frequency spectrum. For example, the bass provides lows; keyboards might emphasize mid-bass; lead guitar may provide upper mids, and cymbals fill in the highs.

Often the rhythm guitar occupies the same frequency range as the piano, so they tend to mask each other's sound. You can aid clarity by equalizing them differently. Boost the guitar at, say, 3 kHz, and boost the piano around 10 kHz. Or pan them to opposite sides.

More on judging sound quality is seen in Chapter 14.

Changes During the Mix

It's rare to do a mix in which you set the faders and leave them there. Often you need to mute tracks, change fader levels, or change EQ during a mix.

To reduce background noise, mute all tracks that have nothing playing at the moment. That is, if there is a long silence during a track, mute that silent portion. Unmute these tracks just before their instruments start playing. Mute unrecorded tracks as well.

Level changes during the mix should be subtle, or else instruments will "jump out" for a solo and "fall back in" afterwards. Set faders to preset positions during pauses in the music. Nothing sounds more amateurish than a solo that starts too quietly then comes up as it plays. You can hear the engineer working the fader. If you need to reduce the level of a loud passage, do so at the end of the preceding soft passage before the loud one begins.

If you need to change fader levels during the mix, and you don't have automation, you might mark these levels next to each fader on a thin piece of tape. Make a cue sheet that notes the mixer changes at various tape-counter times. For example:

0:15 Unmute vocal

1:10 Lead solo −5

1:49 Lead −10

2:42 Synth EQ +6 at 12K

3:05 Start fade, out by 3:15.

What if you want the sound of an instrument to change drastically during a song, but there are too many mixer changes to handle at once? You can do this by multing and muting. Here's what to do:

Suppose you want a track to have a radically different level, EQ, and effects during the chorus. Using a Y-cord or a mult on a patch bay, connect that track to two mixer channels—say, channels 5 and 6. If you're using a DAW, copy track 5 to track 6. On channel 5, set the level, EQ, and aux sends as you want them to be for most of the song. On channel 6, set those controls as you want them to be for the chorus.

When you're ready to mix, mute channel 6. Play the song and mix it. When the chorus comes up, unmute channel 6 and mute channel 5. The sound will change during the chorus. In a DAW you make this change part of an automated mix.

Record the Mix

When you're happy with the mix and recording levels, record the mix on your 2-track recorder: DAT, CD-R, or hard drive. If you will record the mixes on DAT, put in a blank DAT tape and exercise it: fast-forward to the end and rewind to the top. You might want to clean the tape path with a cleaning cassette, but usually only if the DAT is producing errors. Keep a log noting the start and stop times for each song. You'll use these times when you're ready to edit. If you're using a DAW, export or save the mix as a new sound file.

If the mix is very difficult, you can record it a section at a time, and then edit the sections together.

To fade out the end of the tune, pull down the master faders slowly and smoothly. Try to have the music faded out by the end of a musical phrase. The slower the song, the slower the fade should be. The musical meaning of a fade is something like, "This song is continuing to groove, but the band is leaving on a slow train." You might want to postpone fades until mastering, which is done in a DAW.

The mixdown is complete. If you're using a standalone 2-track recorder, leave it running for a few seconds—make a blank space so you don't accidentally erase the end of the mix you just did with subsequent recordings. Play back the mix to listen for dropouts and errors. You might record several different mixes of one song, then choose the best mix. It's common to record a mix with the lead vocal up 1 dB, and another with the lead vocal down 1 dB.

Repeat these mixdown procedures for the rest of the good takes, leaving about 20 seconds of silence between each mix recording. Give your ears a rest every few hours! Otherwise, your hearing loses highs and you can't make correct judgments.

After a few days, listen to the mix on a variety of systems—car speakers, a boom box, a home system. The time lapse between mixdown and listening will allow you to hear with fresh ears. Do you want to change anything? If so, make it right. You'll end up with a mix to be proud of.

If you lack good multitrack recordings with which to practice mixing, go to www.raw-tracks.com. There you can download individual tracks in wav or mp3 format, or purchase a CD of raw tracks.

Summary

The following are summaries of the procedures for recording, overdubbing, and mixdown. Use these steps for easy reference.

Recording

1. Turn up the headphone or monitor volume control. Monitor the aux bus that you're using for the monitor mix. If your multitrack is wired to the insert jacks, monitor the stereo bus instead. If you're using a DAW, monitoring is automatic.
2. Assign instruments to tracks. To record one instrument per track, connect its direct-out (or insert send) to a track input. If you're using a DAW, select each track and assign it an input signal from your audio interface.
3. Turn up the input faders, submaster, and master faders to design center (the shaded portion of fader travel, about three-quarters of the way up).
4. Adjust the input attenuators (trim) to set submixes and recording levels. If you're recording with a DAW, use the volume controls of the audio interface (they might be a software application).
5. Set the monitor/cue mix.
6. Record onto the multitrack recorder.

Overdubbing

1. Assign the instruments or vocals to be recorded to open tracks. An open track is blank or has already been bounced.
2. Turn up the monitor/cue system.
3. Turn up the submasters and master to design center.
4. Play the multitrack recording and set up a cue mix of the already-recorded tracks.
5. While a musician is playing, adjust the input attenuation and recording level.
6. Set the monitor/cue mix to include the sound of the instrument or vocal being added.
7. Record the new parts on open tracks.
8. Punch in and comp tracks as needed.

Mixdown

1. If necessary, set the input selectors to accept the multitrack recorder output signals.

2. Monitor busses 1 and 2 (or the stereo mix bus). Monitoring is automatic with a DAW.

3. Assign tracks to busses 1 and 2 (or the stereo mix bus).

4. Turn up the master fader to design center. In some mixers, the submasters should also be up.

5. Set preliminary panning.

6. Set a rough mix with the input faders. Maybe start with all faders at −12 dB, then adjust from there.

7. Set equalization and effects.

8. Perfect the mix and set recording levels. Set up automation if your system has it (described next).

9. Record onto the 2-track recorder. If you have a DAW, export or save the mix as a new wave file.

Automated Mixing

A multitrack mixdown is often a complicated procedure. It can be difficult to change the mixer settings correctly at all the right times. So you might want to use automated mixing—have a MIDI sequencer remember and set the changes for you.

As you mix a song, you might adjust the mixer controls several times. For example, raise the piano's volume during a solo, then drop it back down. Mute a track to reduce noise during pauses in the performance. An automated mixing system can remember your mix moves, and later recall and reset them accordingly each time you play back the mix. You can even overdub mix moves; for example, do the vocal-fader moves on the first pass, drum moves on the second pass, and so on. You also can punch-in fader moves to correct them. Effects changes can be automated as well in some units. Automated mixing is a feature in digital mixers, software DAWs, and many hard-disk recorder-mixers.

Automated mixing has many advantages. With it you can:

- Perform complicated mixes without errors
- Fine-tune the mix moves
- Recall mixes weeks or months after storing them, without having to reset the mixer manually each time
- Listen to the mix without the distraction of having to adjust faders

Types of Automation Systems

Three types of automation systems are

1. Automated mixer with non-motorized faders
2. Automated mixer with motorized faders (Flying Faders)
3. Recording software with automated mixing

Each of these is worth a closer look.

Automated Mixer with Non-Motorized Faders

Suppose you have mixer or controller surface with non-motorized faders. The mix is playing and you are adjusting the faders. As you set the position of each fader, this action produces a MIDI signal that is recorded by a sequencer. This sequencer is either built into the mixer or is external.

When the mix plays back, so does the sequence of mix moves. MIDI signals from the sequencer set the volume level for each channel. This is done by varying the gain of a voltage controlled amplifier (VCA) or digitally controlled amplifier (DCA) in each channel. Because the faders do not move as the mix plays back, the fader positions do not represent the mix you are hearing.

Automated Mixer with Motorized Faders

This system works as follows. First, you adjust the faders to set up a mix. This generates MIDI signals that are recorded by a sequencer. When you play back the mix, the sequencer makes the motorized faders move up and down as if controlled by a ghost, matching your mix moves. The position of the mixer faders show the mix levels. VCAs are not needed, which is a benefit because VCAs can degrade the audio slightly.

Recording Software with Automated Mixing

Chapter 13 defines a DAW as a personal computer running audio recording software. A sound card or I/O box plugged into the computer has connectors to get audio into and out of the computer.

Once the musical tracks are recorded and edited, you're ready to mix. The monitor screen shows virtual faders that you adjust with a mouse or with a controller surface. A software sequencer remembers your mix moves, EQ settings, and so on, and does an automated mix during playback.

In most DAWs, a fader envelope appears on each track. This envelope is a graph of a track's fader setting versus time. You might prefer to automate the mix by tweaking the fader envelope instead of moving the faders.

As we said earlier, a MIDI/audio recording program includes both digital audio tracks and sequencer tracks. The audio track levels are automated by adjusting their gain; the MIDI tracks are automated by adjusting their MIDI volume or key-velocity scaling.

Lower cost programs let you control volume only; higher cost programs let you control all parameters (EQ, panning, and effects). Some or all of these parameters can be automated.

Snapshot versus Continuous Automation

Two types of automation are snapshot and continuous (dynamic). With snapshot automation, you push a button to take a "snapshot" or "scene" of the mixer settings. The sequencer stores the snapshot as the MIDI program changes. To reset the mixer to any of these stored settings, you punch-up the appropriate number (MIDI program change). Alternatively, your sequencer/audio recording software can reset the mixer's virtual controls at the correct times as a song plays.

With continuous or dynamic automation, the sequencer records the motion of the mixer controls—not just their position. Continuous automation costs more than snapshot and consumes more memory, but permits finer resolution of mix moves. Some digital mixers do both snapshot and continuous automation with their internal sequencer. Other mixers require an external MIDI sequencer to record continuous automation moves.

Some automated mixers let you specify a fade time between snapshots. When the sequencer changes from one snapshot to the next, the sequencer fades or adjusts the control settings gradually between the two snapshots. This acts like continuous or dynamic automation.

Automated Mixing Procedure

Here is a suggested order of steps in doing an automated mix:

1. Set up a rough mix manually. You might prefer to start with a song section where all the instruments are playing.
2. Record the mutes. Mute tracks that aren't being used, or that don't add to the arrangement. For example, a chorus might sound better without the horns. Unmute each muted track just before you want it to be heard.

3. Record the fader moves. For example, turn up the guitar during the solo, turn up the toms during the fills, turn down the background vocals when they get too loud. Save your changes often.

4. Record the effects changes. You might bring up the chorusing on the acoustic guitar during the bridge, and so on.

5. Sit back and listen to the mix. Is there anything you want to change?

Another way to do an automated mix is with snapshots. At several points in the song where you need mutes or fader changes, set the mute or fader level on the appropriate track and take a snapshot. This can be done in a hardware mixer or a software DAW.

Suppose you want a track's sound to change radically during the song. Maybe you want the lead guitar to change level, EQ, and reverb during the chorus. Here's a slick way to do it: Using a Y-cable or patch-bay mult, split the guitar track's signal to two adjacent channels on your mixer—say, channels 5 and 6. Channel 5 is set up the way you want it for most of the song. Channel 6 is set up with the EQ, level, and reverb you want during the chorus. Mute channel 6 and start playing the song. When the chorus comes up in the song, mute channel 5 and unmute channel 6. Save this mute change. This technique also works with nonautomated mixers. In a DAW, you can cut and paste the chorus from track 5 onto track 6, which is set up with different EQ and effects than track 5.

Here is a "poor man's" automation system suggested by producer/engineer Dave Aron. Suppose you are using a standalone digital multitrack recorder and you have two empty tracks.

1. Connect your mixer's stereo outputs to the inputs of those two tracks.

2. Set up the mix at the beginning of the tune, and start recording the mix on the empty tracks, as if you were recording to a 2-track recorder.

3. When you come to a point in the song where the mix needs to be changed, stop recording.

4. Rewind a few seconds, hit PLAY, and reset the faders as desired.

5. Again, rewind the tape a few seconds and hit PLAY. On the two mix tracks, punch into record mode at the point where the mix changes.

6. Continue this process for all the changes in the mix, updating the mix as you go.

13

COMPUTER RECORDING

With recording software and a sound card, you can turn your computer into a powerful digital recording studio. This Digital Audio Workstation (DAW) lets you record dozens of audio tracks, edit them, add effects, do a mixdown with automation, and burn a professional-quality CD—all in your computer. The cost is only a few hundred dollars.

In addition to recording audio, most DAW software can act as a sequencer by recording MIDI data. You can record, edit, and play both audio and MIDI tracks in the same program.

Another function of a DAW is to edit tracks that were originally recorded on a standalone multitrack recorder. You can transfer eight or more tracks at once to your computer by using a sound card with Alesis Lightpipe or Tascam TDIF connectors, or with an Alesis FirePort that works with an Alesis HD24 hard-disk recorder.

A DAW has three parts (Figure 13.1):

1. A fast computer with lots of memory and a large hard drive.
2. An audio interface to get audio and MIDI into and out of your computer.
3. Recording software.

You also need some powered monitor speakers, at least one mic, and a mic preamp or mixer. Although you can use the mic input in a sound card, it is likely to be low quality and noisy. You'll get better sound with

Figure 13.1 A computer DAW.

a separate mic preamp or mixer. Some optional extras are a control surface and DSP cards. These are explained later under those headings.

Basic Operation

When you launch the recording software, you see simulated tracks and recorder transport buttons such as fast-forward, rewind, record and play. You also see a mixer with virtual controls: simulated faders, knobs, buttons, and meters.

DAW software has several windows or views (Figure 13.2). You view and manipulate the tracks in the Track window, do edits in the Track or Edit window, and adjust mixer controls in the Mixer window. Each window can be opened to fill the entire screen.

Recording and Playback

Here's how the system works. Audio from your mixer or mic preamp goes to the inputs on the audio interface, which converts the audio to computer data and sends it to the computer. The software lets you record this data on the computer's hard drive. During playback, the recorded data streams from the hard drive into the interface, which converts the

Figure 13.2 An example of DAW software windows.

data back into audio at the interface outputs. Monitor speakers connected to those outputs play the audio.

Audio data on the hard drive is read by an electromagnetic head. Because the head can be controlled to jump to any location on disk, it has random access: it can instantly locate any part of the audio program. As the head is jumping around, the data pieces it picks up are read into a buffer memory, then read out at a constant rate.

The waveform of the recorded audio appears on your monitor screen (Figure 13.3). You can zoom out to see the entire program, or zoom in to see individual samples.

Like a multitrack tape recorder or hard-disk recorder, a DAW has several tracks to record on. You might record a MIDI drum pattern on track 1, a bass audio signal on track 2, guitar on 3, keys on 4 and 5, vocals on 6, and so on. During playback those tracks mix or combine into a stereo signal that plays through your audio interface.

Figure 13.3 A waveform editing screen (Source: Adobe Audition).

Editing

Editing is a major feature of all DAW programs. Listed below are some editing functions you'll find in most DAW recording software:

- Cut or Trim: Remove or truncate unwanted portions of the waveform or program. For example, cut out silent areas in each track to reduce leakage and background noise. Delete audio glitches and clicks.

- Cut and paste: Remove a section of a song and put it somewhere else in the song. If you think the bridge section of a song should come earlier, you can remove it from its current location and put it where you want it.

- Copy and paste: For example, copy a chorus in a song and put it at the measures where the chorus is repeated. Copy a track and paste it into a blank track with different EQ or processing. Or copy an in-tune note and paste it over an out-of-tune note (zoom-in to do this with precision).

- Fade-in, fade-out: Do an automated fade (a gradual change in volume).

- Crossfade: Fade out of one song while fading into another. Or cross-fade across an edit point to make it sound smoother. A logarithmic crossfade gives a more even volume level than a linear crossfade.

- Slip: You can slip tracks forward or backward in time, independent of other tracks. You also can move single notes in time. If a note in a bass track comes in too late, select the note and slide it or nudge it to the left until it is on the beat. Or adjust the start times of sound effects so they align with events in video clips.

- When mastering a program for CD, you can select entire songs and move them around to change the order they will play in.

When you edit audio, you select a portion of the audio program, creating a **clip** or **region**. Examples of clips are an entire song, the chorus of a song, a guitar track, a drum riff, or a single note. Using a mouse, you can create a clip by marking its beginning and end points in the waveform. In Figure 13.3, a clip is selected and highlighted.

A clip is actually a pair of pointers to part of an audio file on your hard drive. One pointer is the data address for the beginning of the clip, and the other pointer is the address for the end of the clip. Rather than containing audio data, clips tell the software which section(s) of the audio file to play.

When you set up the sequence of clips in a mix, you're telling the hard-drive head which pointers to play in what order. Or when you delete a clip, you're telling the hard-drive head to skip the clip's pointers during playback. If you copy an audio clip, the software does not make copies of the audio file it points to—instead, it plays the same audio file each time it sees a clip pointing to that file. Or when you split a clip into parts and put them in a different order, the audio file itself is not split. Instead, the software plays the sections of the audio file that the clips point to, in the desired order.

These types of edits are called **nondestructive edits**. Only the pointers change; the data on disk is not changed or destroyed. Nondestructive edits are not permanent. If you don't like an edit, you can undo it and try it again.

Some types of edits or processing are **destructive**: they write over the data on disk. However, some recording programs save the data before you edit it. Then you can undo the change by reverting to the saved data.

Editing can create unusual effects. For instance, copy a syllable in a vocal track and paste it several times to create stuttering. Suppose you want to double a guitar that is in the left channel to make it stereo. Copy the guitar track, paste it to another track, slide the pasted guitar track to the right about 20 to 30 msec (which delays the guitar signal), and pan the delayed guitar track to the right.

Mixdown

Once all your tracks are recorded (and maybe edited), it's time for mixdown. Here is the general procedure:

1. Adjust each on-screen fader with your mouse to set the level of each track until you create a good balance among tracks.
2. Make selections with the mouse to add EQ, compression, and effects to various tracks.
3. Set up automation so that the computer remembers your mix settings and resets them the next time you play the mix.
4. Once your mix is perfected, export it to a stereo wave file or aiff file on your hard drive.
5. Repeat steps 1 to 4 for all the songs in a demo or album.
6. Open a new project, and import the mixes into a stereo track. Put the mixes in order with some silence between them.
7. Burn a CD of the finished mixes.

With this basic understanding, let's take a closer look at DAW components: the computer, audio interface, and recording software.

The Computer

Either Mac or PC will give great results with audio software. Just be sure that the software you want to use is compatible with your computer platform.

In order to play a lot of tracks and effects in real time, you need a computer with a fast central processing unit (CPU); lots of RAM; and a large, fast hard drive. A minimum system would be a 2 GHz CPU,

Table 13.1 Hard-Drive Storage Required for a 1-Hour Recording

No. of tracks	Bit depth	Sampling rate	Storage needed
2	16	44.1 kHz	606 MB
2	24	44.1 kHz	909 MB
2	24	96 kHz	1.9 GB
8	16	44.1 kHz	2.4 GB
8	24	44.1 kHz	3.6 GB
8	24	96 kHz	7.7 GB
16	16	44.1 kHz	4.8 GB
16	24	44.1 kHz	7.1 GB
16	24	96 kHz	15.4 GB
24	16	44.1 kHz	7.1 GB
24	24	44.1 kHz	10.7 GB
24	24	96 kHz	23.3 GB

512 MB RAM, and an 80-GB hard drive. Two hard drives are faster: one for system files and programs, and another for audio data. The drive should be capable of high sustained transfer rate (thruput). Current ATA-100 or ATA-133 drives can sustain up to about 40 MB per second, which is fast enough for multitrack recording.

Multitrack audio consumes a lot of disk storage space. Table 13.1 shows the amount of hard-drive space needed for a 1-hour recording with various recording formats.

If you plan to record a 1-hour album with about four takes per song, multiply the "Storage Needed" by four. In addition to the disk storage for each song's tracks, you will need about 30 to 200 MB per song mix, and up to 750 MB for a CD album of the song mixes.

Audio Interfaces

Once you have a fast computer with a large hard drive, you need a way to get audio signals into and out of the computer. An audio interface does the job. Four types of interface are listed below, and we'll look at each one (see Figure 13.4).

- Sound card (2-channel or multichannel).
- I/O interface or breakout box (2 to 16 channels).
- Controller surface with I/O (Input/Output connectors).

Figure 13.4 Four types of audio interface used with a DAW.

- Alesis FirePort. This converts wave files from an Alesis HD24 FST-formatted hard drive to FireWire, and sends the data to your computer. FireWire (IEEE-1394) is a high-speed data link for connecting digital devices.

Sound Card

The simplest form of interface is a 2-channel sound card, which plugs into a PCI user slot in your computer's motherboard. Low-cost sound cards have unbalanced 1/8-inch (mini) phone jack connectors, which include a mic input, stereo line input, and stereo line output. Generally, the sound quality and connectors of low-cost cards are not up to professional standards. Current high-quality sound cards can record with 24-bit resolution. Many sound cards have MIDI connectors and an onboard synthesizer. Some have a FireWire port.

The next step up is a sound card with 1/4-inch TRS (Tip-Ring-Sleeve) connectors (Figure 13.5) or XLR connectors on cables. It offers 2 to 8 balanced inputs. Examples are sound cards by Digital Audio Labs, Frontier Design, SEK'D, Lynx Studio Technology, Echo Digital Audio, Midiman, Turtle Beach, and RME Hammerfall. A digital-only sound card is a low-cost choice if you work only with digital signals. A sound-card comparison is at www.pcavtech.com/soundcards/compare/index.htm.

Figure 13.5 A sound card.

Choose either a 2-channel or multichannel card. A 2-channel sound card is adequate if you are recording one instrument at a time (such as a vocal, sax, keyboard, guitar, or a mix of several drum mics). You can overdub more parts using the same stereo input on different tracks. Also, a 2-channel sound card is sufficient if you are using an Alesis FirePort, which copies several Alesis HD24 recorder tracks to your hard drive, bypassing the sound card. In that case, you'd use the sound card just for monitoring.

You'll need a multichannel sound card (or a multitrack hard-disk recorder) if you want to record several instruments at once, such as a band or individual mics on a drum set. This type of card has several connectors on short cables, one per channel. Each instrument's signal goes to a separate channel in the sound card, and each channel's signal is recorded on a separate track. You can install two or more sound cards side-by-side to get more channels.

Check the sound card's Web site to make sure that it is compatible with your computer and recording software.

Here are some desirable features to look for when shopping for a sound card:

- 16-bit, 44.1-kHz minimum, full duplex recording. A full duplex card can record and play back simultaneously. The card works on two DMA channels.

- 85-dB or greater signal-to-noise ratio, 20 Hz to 20 kHz frequency response (+/−0.5 dB or less).
- XLR connectors, 1/4-inch TRS phone jacks, RCA phono jacks, MIDI connectors.
- Operates at +4 dBu as well as −10 dBV.

If the sound card includes an onboard synthesizer chip, look for these features:

- General MIDI (GM) compatible.
- MPU-401 MIDI interface compatible.
- Wavetable synthesis (better sound than FM synthesis)
- Programmable synth patches.
- 24-note polyphony minimum.
- 2 MB of wavetable ROM or RAM minimum.

I/O Breakout Box

A more convenient setup than a sound card, but more expensive, is an outboard audio interface (I/O breakout box) in which all the connectors are in a common chassis. See Figure 13.6. Because the interface's analog circuits are outside of the computer, they tend to pick up less computer electrical noise than analog sound cards do. This interface accepts analog audio or MIDI signals, converts them computer data, and sends the data to the computer via a PCI, USB, or FireWire connection (explained later under the heading "Data Transfer Format").

All breakout boxes accept line-level signals from a mic preamp or mixer. Some have mic preamps built in. A few have a high-impedance (hi-Z) 1/4-inch input jack for an electric guitar.

Figure 13.6 A small outboard audio interface, which is a stereo mic preamp and A/D converter.

I/O breakout boxes are made by Aardvark, Digidesign, M-Audio, Echo, Metric Halo, PreSonus, Creative Labs, Tascam, and MOTU (Figure 13.7), among others. Typical prices are $200 to $2495.

When deciding which interface to purchase, some features to look for are listed below.

Data Transfer Format

The interface has a connector to send digital audio to the computer. This data can be transferred by PCI, USB, or FireWire formats. PCI is a common format of slots on a computer motherboard, used for sound cards. USB and FireWire are protocols for transferring digital data quickly from one device to another. For example, they transfer digital audio from an audio interface to your computer.

Outboard interfaces with the USB format are convenient: they plug into a USB port on the outside of your computer, so you don't need to open up the computer to install the interface. Figure 13.8 shows a USB port and icon.

FireWire comes in two speeds: FireWire 400 (IEEE 1394) which runs at 400 Mbps (megabits per second), and FireWire 800 (IEEE 1394b) which runs at 800 Mbps (100 MB per second). In contrast, USB 2.0 high-speed is 480 Mbps.

Figure 13.7 A 16-channel outboard audio interface or I/O breakout box.

Figure 13.8 USB port and cable connector.

You can add a FireWire port to your computer with a $35 FireWire card, and some computers have FireWire built in. Figure 13.9 shows a FireWire port and icon.

USB and FireWire devices are hot-swappable: you can insert or remove the connector while the computer is on. Both formats are compatible with Mac or PC.

Both ports are also available in PCMCIA cards and CardBus cards, which fit into laptops. A PCMCIA card is a credit-card-size memory card or I/O device that connects to a slot in a computer. CardBus is an advanced PCMCIA card with faster speed due to its direct memory access (DMA) and 32-bit data transfer.

Digital I/O

Some interfaces have digital inputs and outputs as well as analog. If you have a digital mixer or an outboard A/D converter, a digital-only card may be all you need. Four digital formats are available:

1. S/PDIF—a coaxial cable with RCA plugs, or an optical connection. Used with DAT recorders and digital mixers.
2. AES/EBU—an XLR connection used with DAT recorders and digital mixers.
3. ADAT Lightpipe—an optical connection to Alesis ADAT recorders. Transfers eight channels of audio at once.
4. Tascam TDIF—a D-sub connection to Tascam multitrack recorders. Transfers eight channels of audio at once.

Analog I/O

Check whether the analog I/O on the interface is balanced or unbalanced. Balanced connections reduce electrical interference picked up by cables. However, unbalanced connections cost less, and usually are adequate for

Figure 13.9 FireWire port and cable connector.

cable runs under 10 feet. Balanced connectors are XLR or 1/4-inch TRS phone jacks. Unbalanced connectors are RCA phono jacks or 1/4-inch TS (Tip-Sleeve) phone jacks.

Sampling Rate and Bit Depth

Available sampling rates in audio interfaces are 32 to 192 kHz. Recordings made with higher sampling rates give better sound quality but consume more hard-drive space. If you are making CDs, a sampling rate of 44.1 kHz does the job.

Some interfaces can handle 24-bit as well as 16-bit signals. Recordings made with 24-bit resolution sound a little smoother and cleaner (less distorted) than 16-bit recordings. Even if your final product is a 16-bit CD, it will sound better with a 24-bit recording that is dithered down to 16 bits during mastering. Dithering is explained in Chapter 9.

MIDI Ports

Many interfaces have MIDI ports, which accept MIDI signals from a MIDI controller and send them to your computer's sequencing software. If your audio interface has MIDI ports, you don't need a separate MIDI card in your computer.

Word Clock

Some interfaces offer word clock connectors, which send and receive timing signals for digital audio. It's best to use a single word clock to drive all the digital devices in your studio with a common clock signal; this reduces jitter (see details in Chapter 9).

Driver Support

An audio driver is a program that allows recording software to transfer audio to and from an audio interface. Most interfaces are sold with several types of drivers. Be sure that your interface has the drivers that your recording software requires.

A good driver has a minimum latency spec under about 5 msec. Latency is the signal delay through the driver and interface to the monitor output. This can be a problem in overdubbing, in which the monitored signal of the sound source you're overdubbing is heard later than the pre-recorded tracks.

The most popular audio driver formats are ASIO, DAE, Direct I/O, GSIF, MAS, SoundManager, Wave, WDM, and MME. Here's a brief description of each driver format:

- ASIO (Audio Streaming Input and Output) (Mac, PC): A very popular driver developed by Steinberg. Allows multiple channels of simultaneous input and output, and low latency with software synthesizers.
- DAE (Mac, PC): Used only with Digidesign audio interfaces. It's a multichannel driver that runs with a compatible host such as Pro Tools, Logic, and Digital Performer. DAE lets you use RTAS and/or TDM plug-ins (explained later under the heading "Digidesign Pro Tools").
- Direct I/O (Mac, PC): Works with Digidesign interfaces as a multichannel driver only. Does not let you run RTAS or TDM effects.
- GSIF (PC): Permits very low latency when playing samples from hard disk with Tascam's GigaStudio software sampler.
- MAS (Mac): Developed by Mark of the Unicorn. Offers resolutions up to 24/96 and multiple simultaneous input and output channels. It's also a format for plug-ins (software audio effects).
- SoundManager (Mac): Macintosh's standard audio driver. It lets you record and play mono and stereo files up to 16-bit and 44.1 kHz. Has a moderate amount of latency.
- Wave (PC): The PC standard audio driver. Wave can be used with a variety of audio interfaces (like Sound Blaster-type sound cards) to record and play mono or stereo audio. Has a moderate amount of latency.
- WDM (PC): Win32 Driver Model, a multichannel driver. Allows low latency with WDM-compatible audio hardware and DXi software instruments. DXi stands for DirectX Instruments, Cakewalk's virtual instrument integration standard. DirectX audio effects can be used live on input signals, not just during play-back. This lets you monitor and record effects in real time as you're recording.
- MME is an older driver that offers lower performance than newer ones.

CAUTION: If you have multiple drivers installed, they may conflict. Then the computer might crash or the recording software might not access the audio interface. Delete unused drivers.

Other Options

Listed below are several features that some interfaces offer.

- Zero-latency monitoring: The input signal to the interface is copied to its output, so you hear the input signal without any signal-processing delay (latency).
- Pro Tools compatible: The interface works with Pro Tools recording software and hardware.
- Surround sound: Provides 5.1 or 7.1 surround sound monitoring.
- Powered by FireWire bus or USB bus: The FireWire or USB connection powers the interface; you don't need another power supply.
- Battery powering: This makes the interface portable for on-location recording with a laptop.
- Supplied recording software: The interface is packaged with recording software, so you might not need to buy other software.
- A/D/A converter mode: The interface can act as a real-time analog-to-digital (A/D) and digital-to-analog (D/A) converter.
- SMPTE sync: The interface will synchronize with a SMPTE time code signal (see details in Appendix C).

Control Surface

So far we've covered the computer and audio interface. Let's look at other hardware for a computer studio.

Using a mouse to adjust the controls on-screen is slow and can lead to repetitive stress syndrome. An alternative is a DAW control surface or controller (Figure 13.10). It's a chassis with physical controls for software functions. Resembling a mixer with real faders, knobs, and transport buttons, it lets you adjust the software's virtual controls that you see on the monitor screen. The controller attaches to the computer through a MIDI connector, Ethernet connector, USB, or FireWire port.

Many control surfaces also act as audio interfaces: they include mic/line inputs and outputs (I/O) and MIDI connectors.

Some control surfaces are dedicated to one DAW recording program, such as Pro Tools. Others are universal: they work with several different DAW programs. Be sure to check whether the controller you want to buy will work with your recording software.

Figure 13.10 A control surface.

What if your controller has 8 faders, but your project has 16 virtual faders on screen? You can make the controller affect groups (banks) of virtual faders, 8 at a time, by pressing a bank switch. Normally, your controller faders will affect virtual faders 1 through 8. Press the bank switch to make the controller faders affect virtual faders 9 through 16. Press it again to access virtual faders 17 through 24, and so on. Controllers that have 4 faders can access banks of virtual faders 4 at a time.

Advanced controllers offer these features:

- Motorized faders: The faders in the control surface are motorized, so they move like the virtual faders on screen.
- Standalone mixer mode: You can use the control surface as a regular mixer.
- Footswitch jack: Accepts a footswitch for punch-in/out. The footswitch works only if your recording software supports this function.
- Meters.
- Monitoring section.
- Aux send/return: Allows you to connect an external analog signal processor.
- Insert jacks: An insert jack in series with each channel's signal allows you to plug in an analog compressor.
- Expandability: You can add more controllers, side-by-side, to control more virtual faders at once.

Note that a MIDI controller or a MIDI fader box can control a sequencer's MIDI tracks, but not audio virtual controls. A control surface can control both.

The Frontier Design TranzPort is a wireless remote control for a DAW, letting you control your DAW from the studio. It includes transport controls, metering, a footswitch jack, and more. See the Web site www.frontierdesign.com for more information.

Alesis FirePort

This is a special type of audio interface that transfers audio tracks from an Alesis HD24 multitrack recorder to a computer. The HD24 records tracks as wave files on an Alesis FST-formatted hard drive that is plugged into the recorder. After recording tracks, you can remove the hard drive and plug it into an Alesis FirePort (Figure 13.4), a small device that converts those tracks' wave files to FireWire format and sends them to your computer hard drive for editing and mixing.

As we've seen, there are a wide range of features and connectors among different interface models. A good interface is well worth the price because of its top-quality sound, convenient connections, and easy control of audio signals.

DSP Card

Another DAW option is a digital signal processing (DSP) card that you plug into a user slot in your computer's motherboard. Why is it helpful? The more software effects (plug-ins) you use, the fewer tracks you can mix, because effects put a big load on the CPU. Normally this is not a problem except in mixes with more than 24 tracks and loads of plug-ins in use. You can resolve this issue by installing a DSP card, which handles the processing for the plug-ins. The plug-ins access the card rather than the CPU. You can install multiple cards to expand the system. Examples are TC Works Powercore, Digidesign TDM-based Pro Tools cards, Mackie UAD-1 PCI card, and CreamWare Pulsar PCI cards.

Analog Summing Amplifier

This device is a mixer without faders, used to mix tracks or submixes from a DAW through analog circuitry. Two examples are the Dangerous 2BUS 16 × 2 mixer and the Manley Labs 16 × 2. By using a summing amplifier, you bypass the DAW's internal mix bus which might create slight distortion due to rounding errors, numeric overload, and bit truncation. (Some DAW software claims not to produce those errors, such as Pro Tools TDM systems.) You connect each DAW track's analog output to a summing amplifier input, and use the DAW's faders and automation to set mix levels. Some engineers claim to avoid the distortion of mixing inside the computer by mixing at lower levels.

Recording Software

Now let's turn from hardware to software. Some popular programs for recording, editing, and mixing are Adobe Audition, MOTU Digital Performer (Mac only), Steinberg Cubase SX and Nuendo, Digidesign Pro Tools, Apple GarageBand, Cakewalk Music Creator, Home Studio and Sonar; Emagic Logic (for Mac only), Sony Pictures Digital Vegas and Sound Forge, BIAS Deck, Pro Tracks Plus, Magix Samplitude, Mackie Tracktion, Magix Sequoia, and RML Labs SAW Studio. They share a lot of the same features but implement them differently. Also, each product has unique features and its own GUI.

Check out the products online to see how you might like working with them. Some Web sites have interactive demos; others have free trial versions that you can download.

Free multitrack recording programs are available for download. Although they lack extensive features, they offer a chance to practice your skills at no cost. Examples are Pro Tools Free, N Track Studio, and Kristal (from Kreative.org).

When you first install recording software or hardware, expect some frustration. Your software may not be compatible with your hardware, or may not perform as expected. You might need to modify certain settings in your computer system or recording software. Read the manual and any readme files that came with your product. Also, check out the products' Web sites for knowledge bases, FAQs, etc. Each DAW program has online discussion groups. If you're still having problems, call or e-mail the tech support for your product.

For each recording program, books are available that provide operating details and power-user tips. To find them, search www.amazon.com or the references in Appendix D. Type the name of the program and manufacturer in the search field.

This chapter is not meant to duplicate the information in those specialized books. Instead, it provides an overview of recording software.

Features

Here are some features you'll find in most recording programs:

- Multitrack digital recording from 2 tracks on up: Some programs can record and play back an unlimited number of tracks depending on the speed of your computer, RAM, and hard drive.

- MIDI sequencer: This lets you import or record several tracks of MIDI performance data, edit it, and play it. MIDI sequencer tracks can be mixed with audio tracks.

- Customizable GUI: Change the graphical user interface to suit the way you work. Create configurations to use as templates for similar projects.

- Keyboard shortcuts: Instead of constantly using a mouse, tap certain keys on your computer keyboard. This speeds operation and gives your mouse hand a rest.

- Virtual tracks: Extra takes of a musical performance that are recorded on the hard drive. During mixdown you can choose the best takes to use, or the best parts of each take. Some DAWs have 256 virtual tracks, but 16 real tracks. In that case you can mix 16 tracks during mixdown, but some of those 16 tracks might have several virtual tracks or alternate takes that can be accessed during mixdown.

- Automated mixing: All settings for a song project can be saved and automated. The computer remembers your mix moves, and sets the mixer faders and pan pots accordingly during subsequent mix-downs. Some programs let you automate effects settings as well. For more information, see the section on Automated Mixing in Chapter 12.

- Locate and marker points: Mark several points in a song (intro, verse, chorus, solos) so that you can go to them instantly.

- Routing or virtual patchbay: Assign any input to any track.

- Video display: An on-screen window that shows video clips. Found in high-end software, this feature lets you synchronize music and sound effects to a video program.

- CD recording: Record your mixes on a CD burner.

- PQ editing: Set up a list of song start times before burning a CD of your mixes.

301

- Spectral analysis: A display of level versus frequency of the audio program as it progresses in time.
- Notation application: Converts a MIDI file to musical notes on a treble and bass clef.
- Sync: Synchronization to SMPTE time code or MIDI time code (see details in Appendix C).
- Plug-ins: Described below.

Plug-Ins

A plug-in is a software module that adds digital signal processing or effects to a DAW recording program. For example, you can insert a reverb, compressor, or equalizer plug-in into an audio track.

To create effects, each plug-in runs an algorithm (short program) in your computer's CPU, or in DSP cards that plug into your computer. Because plug-ins work inside the computer, no external processors or patch cords are needed. Any effect available in hardware is also available as a software plug-in. Unlike hardware effects, plug-ins are instantly upgradeable—just download the latest version.

Recording software comes with several plug-ins already installed. You also can purchase and download plug-ins from the Web. When you install the plug-in software, it becomes part of the DAW program, and you can call it up when you need it. Your DAW becomes the "host," and the plug-in provides extra effects that did not come with the host program.

When you click on the name of a plug-in, a screen pops up that looks like a hardware processor with knobs, faders, lights, and meters. Parameters (such as reverb time, chorus depth, or compression ratio) can be adjusted, and most parameters can be saved and automated.

A number of plug-ins can simulate or model the sound of specific devices. For example, Amp Farm by Line 6 is a guitar-amp modeling plug-in, while Antares Microphone Modeler can simulate the sound of popular studio mics. Vintage compressors, tube mic preamps, concert halls, and analog tape recorders have been modeled as well.

Some plug-ins run in real time as the audio is playing. Others process a sound file and create a new file as with normalization or time compression/expansion.

Plug-in effects can be applied in three ways:

1. Master effect: The effect is on the master output bus, and processes the entire mix. Multiband compression is a typical master effect.
2. Aux or send effect: The effect is on an aux bus. Set the plug-in's dry/wet mix control all the way to "wet" or "100% mix." Adjust the amount of effect on each track by turning up the track's aux send control. This option reduces the number of plug-ins needed for a mix, and so reduces the load on your CPU. Delay effects—reverb, echo, chorus and flanging—are aux effects.
3. Insert effect: This effect is on a specific track and affects only that track. EQ and compression are insert effects. If you insert a delay effect into a track, you adjust the dry/wet mix control in the effect to control the amount of effect that you hear.

Plug-ins come in several formats; the most popular for Windows are Steinberg's Virtual Studio Technology (VST) and DirectX. DirectX is a group of application-program interfaces that enhance multimedia (video and audio) on Windows systems. Some plug-in manufacturers make plug-in bundles, which are a variety of effects in a single package. An example is the WAVE audio bundle.

You can find plug-ins on vendors' Web sites or on the Web sites of recording-software companies. For example, Digidesign's Web site lists dozens of plug-in partners.

Examples of DAW Software

We've looked at DAW software in general. Now let's examine a few representative applications: Adobe Audition, Cakewalk Sonar, and Digidesign Pro Tools.

Adobe Audition

This low-cost but powerful program uses the computer's CPU for digital signal processing (Figures 13.11 and 13.12). It does not support MIDI sequencing or VSTi/DXi virtual instruments. Instead, it focuses on audio work. Listed below are some of its unique features.

- Up to 128 stereo tracks.
- Up to 32-bit resolution, up to 10 MHz sampling rate, including standard rates.

Figure 13.11 Adobe Audition multitrack view.

- All edits are sample-accurate. They can be automatically snapped to zero crossings (points where the waveform crosses the zero-volts line), which results in click-free edits. Short crossfades can be added for smooth, pop-free cuts.
- More than 45 DSP tools and effects, mastering and analysis tools, and audio restoration features are included. Third-party DirectX and VST effects plug-ins are supported.
- Precise sample rate conversions from 44.1 to 48 kHz for video or 96 kHz for DVD.
- Track-volume envelopes (fader-setting graphs) can be adjusted to produce gradual, nonlinear changes, and can be scaled.
- Correct pitch. Remove vocals from a stereo program, or reduce the vocal level. Change a clip's length without changing its pitch. Select and edit sounds in frequency or time views.

Figure 13.12 Adobe Audition edit view.

- Import audio from Adobe Premiere Pro or After Effects software for use in Adobe Audition. Use the Edit Original command to make changes to the original wave file and import the changes into Adobe Audition.

- Organizer window lets you access currently open audio, MIDI, and video files; effects; and favorites. Create keyboard shortcuts.

- Multichannel encoder transforms a multitrack mix into 5.1 surround sound.

- Audio for video. Open video files in the multitrack view (AV, native DV, MPEG, WMV) and edit the soundtrack in Adobe Audition. Create new soundtracks, sweeten existing recordings, reduce noise, etc.

- Looping tools let you create music for songs or movie soundtracks. Loops (repetitive drum or music patterns) automatically match the recording's tempo and key. Loop-based recording is covered in Chapter 16 on MIDI.

- Includes 5000 original, performance-based loops in a wide range of musical genres. Supports more than 20 file formats and variations, including Windows PCM (WAV), AIFF, MP3, and WMA 9. Twenty sample sessions included.

- Digitally extract audio CDs to your hard drive for use in projects. Integrated CD burning.

- Transfer audio quickly to other Rewire-supporting applications.

The main page about Adobe Audition is at http://www.adobe.com/products/audition/main.html. From there, you can download a trial version, check system requirements, see a video, read a tutorial, and read an overview or in-depth information. Details about Adobe Audition are at http://www.adobe.com/products/audition/pdfs/audition_nph.pdf

Cakewalk Sonar

Like Adobe Audition, Cakewalk software is cost-effective but powerful, and uses the computer's CPU for all processing. Products range from Home Studio ($125 street) to Sonar Producer ($600 street) (Figure 13.13). Below are some unique features of Sonar Producer:

- Graphical envelope automation control of audio, MIDI, synths, and effects.

- Professional metering (peak, rms, peak and rms, adjustable ballistics, pre-fader, post-fader, and pre-fader post effects).

- Loop-based composition, construction, and editing tools.

- ACID-loop and MIDI Groove Clip support.

- Support for DirectX and VST audio effects. (DirectX is a standard audio driver.)

- 29 included audio effects.

- Support for DXi and VSTi soft synths. (VSTi stands for Virtual Studio Technology for Instruments, Steinberg's virtual instrument integration standard.)

- Several included DXi soft synths and samplers.

Figure 13.13 Sonar screen view.

- Support for ReWire 1.0 and 2.0 clients (Project5, Reason, etc.).
- 13 included MFX (MIDI effects) plug-ins. Support for real-time, non-destructive MIDI FX plug-ins. Full plug-in delay compensation.
- Works with many MIDI-compatible control surfaces.
- Surround mixing (5.1 and 7.1)
- Custom screen layouts and colorization. User-definable keyboard shortcuts and templates for other DAWs.
- Multitrack Piano Roll and Drum Editor. Custom, multi-port drum mapping.
- Full notation of MIDI with lyrics, chord symbols, guitar tablature, etc.
- Event List view with display filtering, Tempo view with list and graph displays, Markers view.

307

- Track Folders let you file multiple tracks into a single folder and edit them simultaneously.
- Track Layers displays multiple takes on one track for easy comping.
- Freeze function mixes down effects and edits on a track to free up CPU resources. Unfreeze returns to the original setup.
- Advanced project and file management tools.
- OMFI and Broadcast Wave import/export for cross-platform collaboration with Digital Performer, Logic, Nuendo, and Pro Tools studios.
- Import formats: AIF, ASF, AU, AVI (with stereo or 5.1 audio), Broadcast Wave, MIDI, MP2, MP3, MPEG, MPG, OMFI, QuickTime, SND, WAV, Windows Media Audio 9 (WMA), WMA9 Pro 5.1, WMA9 lossless, Windows Media Video, and proprietary Cakewalk formats (.bun, .cwb, .cwp, .wrk).
- Export: WAV, ACID-format WAV, Broadcast Wave, AVI (with stereo or 5.1 audio), OMFI, MIDI, MP3 (30-day trial encoder), QuickTime, Real Audio G2, Windows Media Audio 9 (WMA), WMA9 Pro 5.1, WMA9 lossless, Windows Media Video (with stereo or 5.1 audio), and proprietary Cakewalk formats (.cwb, .cwp); Other audio formats supported via external command-line encoders.
- Open support for external encoders: surround, LAME, Ogg Vorbis, Monkey's Audio, etc.
- 32-bit floating-point resolution of the DSP.
- Audio bit depths up to 24 bit, any sample rate.
- Reliable synchronization of audio, MIDI, video, external hardware.
- Sample-accurate timing for audio, soft synths and automation; 960 PPQN for external MIDI devices. Frame-accurate SMPTE sync with auto-detection of timecode.
- Yamaha OPT Level 1 MIDI Hardware Link support.
- Multiple video monitor support. Multi-processor support.
- ASIO and WDM compatibility.
- Surround panning and effects.

Sonar Studio Edition is a special version of Sonar designed for project studios and aspiring professionals. Sonar Studio and Sonar Producer are the same, except that Sonar Producer has these extra features: an enhanced virtual mixing console (with per channel EQ, assignable effects controls), VSampler 3.0 DXi, Ultrafunk Sonitus Effects Suite, a professional sample library, and Lexicon Pantheon Reverb (the Studio version has Lexicon LE), surround mixing and editing, Sonitus Surround Compressor, video thumbnail track, POW-r Dithering, and MPEX time scaling.

Digidesign Pro Tools

Pro Tools is the industry standard DAW system. However, other excellent programs are available. People often use the term "Pro Tools" generically to mean "digital audio workstation."

There are two basic categories: Pro Tools LE and Pro Tools HD.

Pro Tools LE is a 32-track recording system that uses your computer's CPU for all processing. Your CPU speed, hard-drive speed, and amount of RAM determine how many plug-ins and soft synths you can use simultaneously.

LE systems are intended for singer/songwriters, project and personal studios, remote recording engineers, and bands. Pros can use it at home because it is compatible with Pro Tools HD. Cost is $500 to $3000.

Pro Tools HD is a high-resolution professional system with DSP cards, up to 192-kHz sample rate, expandable I/O, and many options. It comes in three Core systems (each with more DSP power and more I/O) and one or more audio interfaces. Each system requires the Pro Tools 192 I/O or the 96 I/O audio interface, and supports legacy Pro Tools audio interfaces for additional inputs and outputs. Cost is $10,000 to $100,000 or more.

Pro Tools LE components include Pro Tools LE software and a choice of Mbox, Digi 002 Rack, or Digi 002 Rack interfaces.

- Mbox is a USB audio interface with two Focusrite mic preamps. It includes Digitial S/PDIF I/O, line and insert I/O and 2-channel stereo outputs.

- Digi 002 is a control surface and audio interface. It communicates with your computer via FireWire. It includes 8 analog ins and outs, 4 mic preamps, 8 channels of ADAT I/O, stereo S/PDIF and MIDI I/O, 8 touch-sensitive faders, and 8 motion-sensitive rotary encoders. It includes transport buttons and automation. Any changes you make in your Pro Tools software are duplicated on the control surface and vice versa.

- Digi 002 Rack is Digi 002 in a 2U rack-mount chassis.

Pro Tools HD Systems offer dedicated DSP cards that free up your CPU for other tasks. The result is you can run more tracks, more plug-ins, and more soft synths at the same time. HD offers a wide range of sample rates, hi-res audio, and up to 192 tracks. An HD system can have 8 to 96 ins and outs. The three basic Core systems have one, two, or three DSP cards, and a system can use up to seven DSP cards.

Listed below are the three Core systems. Each can be expanded by adding more HD Accel cards.

- Pro Tools HD 1 includes the HD Core DSP card, which handles up to 32 channels of I/O and up to 96 tracks.
- Pro Tools HD 2 Accel includes the HD Core card and an HD Accel card, which provide more than four times the mixing and processing power of HD 1 systems and 64 channels of I/O and 192 tracks.
- Pro Tools HD 3 Accel, the most powerful system, includes the HD Core card and two HD Accel cards, which provide up to 96 channels of I/O and 192 tracks.

The Digidesign 96 I/O Audio Interface converts audio at up to 24-bit/96 kHz. It includes 8 channels of analog I/O, 8 channels of ADAT optical I/O, 2 channels of AES/EBU and S/PDIF I/O, and word clock. It can link to other 96 I/O interfaces and to Digidesign 888/24, 882/20, 1622, or 24-bit ADAT Bridge.

The Digidesign 192 I/O Audio Interface converts audio at up to 24-bit/192 kHz. It includes 8 channels of analog I/O, 8 channels of AES/EBU, 8 channels of TDIF, 16 channels of ADAT, and 2 additional channels of ADAT or S/PDIF digital I/O. A 192 Digital expansion card provides 8 more channels of AES/EBU, TDIF, and ADAT I/O connections and a Legacy port to link to older Digidesign interfaces.

New features of the latest Pro Tools software are on the www.protools.com Web site.

Pro Tools FREE is free software that records up to 8 tracks and uses the computer's CPU for processing (Figure 13.14).

Plug-ins for Pro Tools come in four formats: TDM, HTDM, RTAS (Real Time Audio Suite), and AudioSuite. TDM plug-ins work only with Pro Tools' professional TDM systems, and use dedicated cards that plug into your computer. HTDM is the same for the Pro Tools HD (high definition) system. Many RTAS plug-ins are the same as TDM plug-ins, except that they use your computer's processor instead of specialized TDM DSP cards. All the above plug-ins work in real time.

Figure 13.14 Pro Tools 6.4 screen view.

AudioSuite plug-ins don't work in real time. Instead, they process a track's signal and create a new track in which the effect is part of the recorded sound. That is, the effects are embedded in the track. Playing a track with embedded effects is less work for the CPU than playing a track with real-time effects. Less load on the CPU means you can play more tracks without drop-outs.

Loads of plug-ins are available for Pro Tools. Some are included, and some can be downloaded from third-party vendors. Use the plug-in finder on the Pro Tools Web site (www.protools.com).

Listed below are some unique features of Pro Tools LE:

- Up to 32 audio tracks at 24-bits/96 kHz.
- Accepts music created in ReWire-compatible audio applications such as Propellerhead's Reason soft synth and Ableton Live looping application. (ReWire is Propellerhead/Steinberg's technology for transferring audio data between software applications.)

311

- Version 6.1 or later works with Windows XP or Mac OS X.
- MIDI Time Stamping, Groove Quantize, Restore Performance, and other enhancements.
- In 6.1 and later versions, Time Trimmer can change the tempo of audio files and loops.
- All data is compatible with other Pro Tools systems, and can be moved back and forth between them.
- Has nearly the same user interface as professional Pro Tools TDM systems.

Included with every Pro Tools system is a Music Production Enhancement Suite, which includes Reason soft synth, Live looping, AmpliTube guitar-amp, and stomp-box modeling, SampleTank sample-playback module, and T-Rack's tube-based parametric EQ.

ProControl and Control|24 are powerful control surface options for Pro Tools. Listed below are the features of each:

ProControl Control Surface with I/O:

- High-quality faders and switches. Expandable to 48 faders.
- DSP Edit/Assign section.
- 8-character LED scribble strips (displays that identify what instrument each fader affects).
- Edit Pack option features two touch-sensitive motorized surround joystick panners, 8 high-resolution output meters to view up to 7.1 surround mixes, MachineControl, custom keyboard, and trackball.
- 16 Focusrite mic preamps.

Control|24 Control Surface has these features:

- Fixed at 24 faders.
- 4-character scribble strips that show channel or plug-in information.
- 6 regular resolution output meters.
- No audio inputs or outputs.

Optimizing Your Computer for Digital Audio

Once you've chosen some recording software and installed it, you'll want to make your computer run as fast and glitch-free as possible. You need to optimize its settings for best results. Appendix B suggests some ways to speed up your hard drive, reduce software interruptions, and reduce the CPU usage. If you follow those tips, you will have a faster system that handles more plug-ins and more tracks at once. Also, when you play tracks or burn CDs, clicks and drop-outs in the audio will be less likely.

Using a DAW

Let's say you purchased some DAW hardware and software, and tweaked your computer to make it fast and reliable. It's ready to rock. Here are some tips on setting up and using your DAW.

Connections

First, connect your audio equipment to the DAW as described below.

1. If you want to use a mixer or a standalone multitrack recorder in your studio, make the connections shown in Figure 13.15. If your mixer is analog, connect it to the audio-interface analog connectors. If your mixer is digital, connect it to the audio-interface digital connectors.

2. If you are not using a mixer, make the connections shown in Figure 13.16.

3. You may need some adapter cables between devices. The following adapter cable connects a sound card's unbalanced I/O to a power

Figure 13.15 Connecting audio equipment to a DAW if a mixer is used.

Figure 13.16 Connecting audio equipment to a DAW if a mixer is not used.

amp, powered speakers, or mixer: Stereo 1/8-inch phone plug, to two unbalanced phone plugs, or two RCA phono plugs (whatever matches your equipment).

Starting from your computer's desktop view, select Start > Settings > Control Panel. Double-click Sound and Audio Devices > Audio tab. Set the Playback and the Recording default devices to the soundcard or interface you are using for audio editing.

Before you record, set a bit depth and sampling rate for the project, such as 24-bit/44.1 kHz. If you want to record from a mic or instrument, you will record their amplified signal on an audio track. If you want to record from a MIDI controller, you will record its MIDI signal on a MIDI track. Details on recording MIDI sequences are in Chapter 16.

Specify the input and output devices for each audio track, such as "sound-card left channel input" and "sound-card stereo outputs." Also, specify the input and output devices for each MIDI track, such as "MIDI-card omni input" and "sound-card stereo outputs."

Maintaining Quality

Here are some tips on keeping the audio quality high when using a DAW.

Set gain staging as follows: Open the application that shows the volume controls for your sound card or audio interface. Use the record volume control (if any) to set the recording level. When your mic preamp or mixer is peaking at 0 dB, set the software record volume control so the on-screen level meter peaks at 0 dB or slightly less. If your mic preamp or mixer has a VU meter, play a 1-kHz tone through the mixer at 0 VU, and set the software record level to –15 dBFS (decibels Full Scale). This allows headroom for transient peaks that are too fast to show up on the VU meter.

When you set recording levels, each track's meter should peak around –5 dBFS maximum (with the meter set to peak meter mode). This allows some headroom for surprises. Also, musicians generally play louder during a performance than during a level check. If you exceed 0 dBFS you'll hear digital clipping distortion.

If you are recording many tracks simultaneously, or at a high sample rate, this may generate more data than your CPU and buffers can handle—causing drop-outs. A drop-out is a short silence or a noise burst (glitch) in the recorded audio. To prevent drop-outs while recording, turn off the time counter, track metering, effects, and waveform preview. Increase buffer size. Turn off or reduce video acceleration. Do not zoom or scroll while recording. Record one song at a time in a concert recording. Also, follow the tips in Appendix B on optimizing your computer for digital audio.

In your recording program, go to the Preferences menu and set an appropriate buffer size. Small buffers tend to create drop-outs, or reduce the number of tracks that play without drop-outs, but the application responds faster. Large buffers reduce drop-outs but increase latency (monitoring delay). Latency is buffer size divided by sample rate. You might set the buffer/latency small (under 5 msec) while overdubbing and set it large during mixdown.

To avoid latency while overdubbing an instrument or vocal, plug the instrument or vocal mic into a mixer. Connect the direct output of that mixer channel to your audio interface input. Connect the interface output (a mix of the recorded tracks) to another mixer input. Monitor the mixer's output. Set up a mix of the live instrument and the interface output.

Consider recording at 44.1 kHz, because that is the digital format for CDs. Recordings made at 88.2 or 96 kHz consume about twice as much disk space. 24-bit recordings sound better than 16-bit recordings but consume 50% more disk space. Recording at 24 bits is considered by many to be a bigger sonic improvement than recording at 96 kHz.

After recording all the tracks, delete silent sections in each track to reduce leakage and background noise. To do that, zoom into a single track, highlight a silent portion, and select Edit > Cut (or a similar command). Another method is to mute a track up to where the sound starts, then unmute it there. Or slip-edit the beginning and end of a track to where the sound starts and ends.

During mixdown, keep the output level of the stereo mix bus around −3 dBFS maximum (in peak meter mode, not rms). Try to keep the main output fader at or near 0 dB, and adjust the channel faders to get the correct output level (without changing their balance). To achieve that, you might start with all channel faders set to −12 dB, then bring up a few channels that need to be louder. Keep your playback levels high but not clipping on each track and in each plug-in. It's easy to overload an equalizer if you apply boost, so reduce the equalizer's overall level if necessary so that no clipping occurs in the EQ plug-in.

For best quality, record at 24 bits and stay there throughout the project. Then turn on dither and export to a 16-bit file that you will master for a CD. If you are mastering to a 24-bit device, set the output word length to 24-bit and turn on dither. That will retain some of the 32-bit resolution of the DAW's DSP calculations.

Normalization is a DAW process that raises the gain of the entire program so that the highest peak reaches 0 dBFS (in peak meter mode, not rms). Do not normalize the program until final mastering.

To prevent hum and jitter, keep audio and digital cables away from power cords and power amps. Use cables designed for digital audio, and use short cables. Use internal sync on your A/D converter to reduce jitter.

Avoid A/D and D/A conversions. Once your signal is digital, try to keep it that way. After converting the analog mic signal to digital in your audio interface, do all your processing in the computer if possible, then burn a CD of the final product. The only D/A conversion will be when the CD plays back.

Avoid unnecessary processing. DSP such as level changes result in slight distortion. If you raise the level of a section or of an entire song then change your mind, don't reduce its level back where it was, because that is twice the processing. Return to the original recording instead.

As we've seen, there is a whole world of computer music applications. Download the free trial versions to see what you like. Set up your DAW system, read the software manual, and have fun creating music.

JUDGING SOUND QUALITY

Seat an engineer behind a mixing console and ask him or her to do a mix. It sounds great. Then seat another engineer behind the same console and again ask for a mix. It sounds terrible. What happened?

The difference lies mainly in their ears—their critical listening ability. Some engineers have a clear idea of what they want to hear and how to get it. Others haven't acquired the essential ability to recognize good sound. By knowing what to listen for, you can improve your artistic judgments during recording and mixdown. You are able to hear errors in microphone placement, equalization, and so on, and correct them.

To train your hearing, try to analyze recorded sound into its components—such as frequency response, noise, reverberation—and concentrate on each one in turn. It's easier to hear sonic flaws if you focus on a single aspect of sound reproduction at a time. This chapter is a guide to help you do this.

Classical versus Popular Recording

Classical and popular music have different standards of "good sound." One goal in recording classical music (and often folk music or jazz) is to accurately reproduce the live performance. This is a worthy aim because the sound of an orchestra in a good hall can be quite beautiful. The music was composed and the instruments were designed to sound best when heard live in a concert hall. The recording engineer, out of respect for the music, should always try to translate that sound to disk with as little technical intrusion as possible.

By contrast, the accurate translation of sound to disc is not always the goal in recording popular music. Although the aim may be to reproduce the original sound, the producer or engineer may also want to play with that sound to create a new sonic experience, or to do some of both.

In fact, the artistic manipulation of sounds through studio techniques has become an end in itself. Creating an interesting new sound is as valid a goal as re-creating the original sound. There are two games to play, each with its own measures of success.

If the aim of a recording is realism or accurate reproduction, the recording is successful when it matches the live performance heard in the best seat in the concert hall. The sound of musical instruments is the standard by which such recordings are judged.

When the goal is to enhance the sound or produce special effects (as in most pop-music recordings), the desired sonic effect is less defined. The live sound of a pop group could be a reference, but pop-music recordings generally sound better than live performances—recorded vocals are clearer and less harsh, the bass is cleaner and tighter, and so on. The sound of pop music reproduced over speakers has developed its own standards of quality apart from accurate reproduction.

Good Sound in a Pop-Music Recording

Currently, a good-sounding pop recording might be described as follows (there are always exceptions):

- Well-mixed
- Wide-range
- Tonally balanced
- Clean
- Clear
- Smooth
- Spacious

It also has:

- Presence
- Sharp transients
- Tight bass and drums
- Wide and detailed stereo imaging
- Wide but controlled dynamic range
- Interesting sounds
- Suitable production

The next sections explore each one of these qualities in detail so that you know what to listen for. Assume that the monitor system is accurate, so that any colorations heard are in the recording and not in the monitors.

A Good Mix

In a good mix, the loudness of instruments and vocals is in a pleasing balance. Everything can be clearly heard, yet nothing is obtrusive. The most important instruments or voices are loudest; less important parts are in the background.

A successful mix goes unnoticed. When all the tracks are balanced correctly, nothing sticks out and nothing is hidden. Of course, there's a wide latitude for musical interpretation and personal taste in making a mix. Dance mixes, for example, can be very severe sonically.

Sometimes you don't want everything to be clearly heard. On rare occasions you may want to mix in certain tracks very subtly for a subconscious effect.

The mix must be appropriate for the style of music. For example, a mix that's right for loud rock music usually won't work for a pop ballad. A rock mix typically has the drums way up front and the vocals only slightly louder than the accompaniment. In contrast, a pop ballad has the vocals loudest, with the drums used just as "seasoning" in the background.

Level changes during the mix should be subtle, or should make sense. Otherwise, instruments jump out for a solo and fall back in afterwards. Move faders slowly, or set them to preset positions during pauses in the music. Nothing sounds more amateurish than a solo that starts too quietly and then comes up as it plays—you can hear the engineer working the fader.

Wide Range

Wide range means extended low- and high-frequency response. Cymbals should sound crisp and distinct, but not sizzly or harsh; kick drum and bass should sound deep, but not overwhelming or muddy. Wide-range sound results from using high-quality microphones and adequate EQ.

You might want to combine "hi-fi" and "low-fi" sounds in a single mix. The low-fi sounds generally cover a narrow frequency range and might be distorted.

Good Tonal Balance

The overall tonal balance of a recording should be neither bassy nor trebly. That is, the perceived spectrum should not emphasize low or high frequencies. Low bass, mid-bass, midrange, upper midrange, and highs should be heard in equal proportions (Figure 14.1). Emphasis of any one frequency band over the other eventually causes listening fatigue. Dance club mixes, however, are heavy on the bass end to get the crowd moving.

Recorded tonal balance is inversely related to the frequency response of the studio's monitor system. If the monitors have a high-frequency rolloff, the engineer will compensate by boosting highs in the recording to make the monitors sound correct. The result is a bright recording.

Figure 14.1 Loudness versus frequency of a pop recording with good sound.

320

Before doing a mix, play over the monitors some commercial recordings whose sound you admire, to become accustomed to a commercial spectral balance. After your mix is recorded, play it back and alternately switch between your mix and a commercial recording. This comparison indicates how well you matched a commercial spectral balance. Of course, you may not care to duplicate what others are doing. An effective tool for this purpose is Harmonic Balancer (www.har-bal.com).

In pop-music recordings, the tonal balance or timbre of instruments does not necessarily have to be natural. Still, many listeners want to hear a realistic timbre from acoustic instruments, such as the guitar, flute, sax, or piano. The reproduced timbre depends on microphone frequency response, microphone placement, the musical instruments themselves, and equalization.

Clean Sound

Clean means free of noise and distortion. Hiss, hum, and distortion are inaudible in a good recording. Distortion in this case means distortion added by the recording process, not distortion already present in the sound of electric-guitar amps or Leslie speakers. There are exceptions to this guideline; some popular recordings have noise or distortion added intentionally.

Clean also means "not muddy" or free of low-frequency ringing and leakage. A clean mix is one that is uncluttered or free of excess instrumentation. This is achieved by arranging the music so that similar parts don't overlap, and not too many instruments play at once in the same frequency range. Usually, the fewer the instruments, the cleaner the sound. Too many overdubs can muddy the mix.

Clarity

In a clear-sounding recording, instruments do not crowd or mask each other. They are separate and distinct. As with a clean sound, clarity arises when instrumentation is sparse, or when instruments occupy different areas of the frequency spectrum. For example, the bass provides low frequencies, keyboards might emphasize mid-bass, lead guitar provides upper midrange, and cymbals fill in the highs.

In addition, a clear recording has adequate reproduction of each instrument's harmonics. That is, the high-frequency response is not rolled off.

Smoothness

Smooth means easy on the ears, not harsh, uncolored. Sibilant sounds are clear but not piercing. A smooth, effortless sound allows relaxation; a strained or irritating sound causes muscle tension in the ears or body. Smoothness is a lack of sharp peaks or dips in the frequency response, as well as a lack of excessive boost in the midrange or upper midrange. It is also low distortion, such as provided by a 24-bit recording.

Presence

Presence is the apparent sense of closeness of the instruments—a feeling that they are present in the listening room. Synonyms are clarity, detail, and punch.

Presence is achieved by close miking, overdubbing, and using microphones with a presence peak or emphasis around 5 kHz. Using less reverb and effects can help. Upper-midrange boost helps, too. Most instruments have a frequency range that, if boosted, makes the instrument stand out more clearly or become better defined. Presence sometimes conflicts with smoothness because presence often involves an upper-midrange boost, while a smooth sound is free of such emphasis. You have to find a tasteful compromise between the two.

Spaciousness

When the sound is spacious or airy, there is a sense of air around the instruments. Without air or ambience, instruments sound as if they are isolated in stuffed closets. (Sometimes, though, this is the desired effect.) You achieve spaciousness by adding reverb, recording instruments in stereo, using room mics, or miking farther away.

Sharp Transients

The attack of cymbals and drums generally should be sharp and clear. A bass guitar and piano may or may not require sharp attacks, depending on the song.

Tight Bass and Drums

The kick drum and bass guitar should "lock" together so that they sound like a single instrument—a bass with a percussive attack. The drummer and bassist should work out their parts together so they hit accents simultaneously, if this is desired.

To further tighten the sound, damp the kick drum and record the bass direct. Rap music, however, has its own sound—the kick drum usually is undamped and boomy, sometimes with short reverb added. Equalize the kick and bass in complementary ways so that they don't mask each other; for example:

Kick: Boost 60 to 80 Hz, cut 150 to 400 Hz, boost 3 kHz.

Bass: Cut 60 to 80 Hz, boost 120 to 150 Hz, boost 900 Hz.

Wide and Detailed Stereo Imaging

Stereo means more than just left and right. Usually, tracks should be panned to many points across the stereo stage between the monitor speakers. Some instruments should be hard-left or hard-right, some should be in the center, others should be half-left or half-right. Try to achieve a stereo stage that is well balanced between left and right (Figure 14.2). Instruments that occupy the same frequency range can be made more distinct by panning them to opposite sides of center.

You may want some tracks to be unlocalized. Backup choruses and strings should be spread out rather than appearing as point sources. Stereo keyboard sounds can wander between speakers. A lead-guitar solo can have a fat, spacious sound.

There should also be some front-to-back depth. Some instruments should sound close or up front; others should sound farther away. Use different miking distances or different amounts of reverb on various tracks.

Figure 14.2 An example of image placement between speakers.

If you want the stereo imaging to be realistic (for a jazz combo, for example), the reproduced ensemble should simulate the spatial layout of the live ensemble. If you're sitting in an audience listening to a jazz quartet, you might hear drums on the left, piano on the right, bass in the middle, and sax slightly right. The drums and piano are not point sources, but are somewhat spread out. If spatial realism is the goal, you should hear the same ensemble layout between your speakers. On some commercial CDs, the piano and drums are spread all the way between speakers—an interesting effect, but unrealistic.

Pan-potted mono tracks often sound artificial in that each instrument sounds isolated in its own little space. It helps to add some stereo reverberation around the instruments to "glue" them together.

Often, TV mixes are heard in mono. Hard-panned signals sound weak in mono relative to center-panned signals. So pan sound sources to 3 and 9 o'clock, not hard-right and hard-left.

Wide but Controlled Dynamic Range

Dynamic range is the range of volume levels from softest to loudest. A recording with a wide dynamic range becomes noticeably louder and softer, adding excitement to the music. To achieve this, don't add too much compression (automatic volume control). An overly compressed recording sounds squashed—crescendos and quiet interludes lose their impact, and the sound becomes fatiguing.

Vocals often need some compression or gain-riding because they have more dynamic range than the instrumental backup. A vocalist may sing too loudly and blast the listener, or sing too softly and become buried in the mix. A compressor can even out these extreme level variations, keeping the vocals at a constant loudness. Bass guitar also can benefit from compression.

Interesting Sounds

The recorded sound may be too flat or neutral, lacking character or color. In contrast, a recording with creative production has unique musical instrument sounds, and typically uses effects. Some of these are equalization, echo, reverberation, doubling, chorus, flanging, compression, distortion, and stereo effects.

Making sounds interesting or colorful can conflict with accuracy or fidelity. That's okay, but you should know the trade-off.

Suitable Production

The way a recording sounds should imply the same message as the musical style or lyrics. In other words, the sound should be appropriate for the particular tune being recorded.

For example, some rock music is rough and raw. The sound should be, too. A clean, polished production doesn't always work for high-energy rock "n" roll. There might even be a lot of leakage or ambience to suggest a garage studio or nightclub environment. The role of the drums is important, so they should be loud in the mix. The toms should ring, if that is desired.

New Age, disco, rhythm and blues, contemporary Christian, or pop music is slickly produced. The sound is usually tight, smooth, and spacious.

Actually, each style of music is not locked into a particular style of production. You tailor the sound to complement the music of each individual tune. Doing this may break some of the guidelines of good sound, but that's usually okay as long as the song is enhanced by its sonic presentation.

Good Sound in a Classical-Music Recording

As with pop music, classical music should sound clean, wide-range, and tonally balanced. But because classical recordings are meant to sound realistic—like a live performance—they also require good acoustics, a natural balance, tonal accuracy, suitable perspective, and accurate stereo imaging (see Chapter 18).

Good Acoustics

The acoustics of the concert hall or recital hall should be appropriate for the style of music to be performed. Specifically, the reverberation time should be neither too short (dry) nor too long (cavernous). Too short a reverberation time results in a recording without spaciousness or grandeur. Too long a reverberation time blurs notes together, giving a muddy, washed-out effect. Ideal reverberation times are around 1.2 seconds for chamber music or soloists, 1.5 seconds for symphonic works, and 2 seconds for organ recitals. To get a rough idea of the reverb time of a room, clap your hands once, loudly, and count the seconds it takes for the reverb to fade to silence.

A Natural Balance

When a recording is well balanced, the relative loudness of instruments is similar to that heard in an ideal seat in the audience area. For example, violins are not too loud or soft compared to the rest of the orchestra; harmonizing or contrapuntal melody lines are in proportion.

Generally, the conductor, composer, and musicians balance the music acoustically, and you capture that balance with your stereo mic pair. But sometimes you need to mike certain instruments or sections to enhance definition or balance. Then you mix all the mics. In either case, consult the conductor for proper balances.

Tonal Accuracy

The reproduced timbre or tone quality should match that of live instruments. Fundamentals and harmonics should be reproduced in their original proportion.

Suitable Perspective

Perspective is the sense of distance of the performers from the listener—how far away the stage sounds. Do the performers sound like they're eight rows in front of you, in your lap, or in another room?

The style of music suggests a suitable perspective. Incisive, rhythmically motivated works (such as Stravinsky's "Rite of Spring") sound best with closer miking; lush, romantic pieces (a Bruckner symphony) are best served by more distant miking. The chosen perspective depends on the taste of the producer.

Closely related to perspective is the amount of recorded ambience or reverberation. A good miking distance yields a pleasing balance of direct sound from the orchestra and ambience from the concert hall.

Accurate Imaging

Reproduced instruments should appear in the same relative locations as they were in the live performance. Instruments in the center of the ensemble should be heard in the center between the speakers; instruments at the left or right side of the ensemble should be heard from the left or right speaker. Instruments halfway to one side should be heard halfway off center, and so on. A large ensemble should spread from speaker to speaker, while a quartet or soloist can have a narrower spread.

It's important to sit equidistant from the speakers when judging stereo imaging, otherwise the images shift toward the side on which you're sitting. Sit as far from the speakers as they are spaced apart. Then the speakers appear to be 60 degrees apart, which is about the same angle an orchestra fills when viewed from the typical ideal seat in the audience (tenth row center, for example).

The reproduced size of an instrument or instrumental section should match its size in real life. A guitar should be a point source; a piano or string section should have some stereo spread. Each instrument's location should be as clearly defined as it was heard from the ideal seat in the concert hall.

Reproduced reverberation (concert-hall ambience) should surround the listener, or at least it should spread evenly between the speakers. Surround-sound technology is needed to make the recorded ambience surround the listener, although spaced-microphone recordings have some of this effect. Accurate imaging is illustrated in Figure 14.3.

There should be a sense of stage depth, with front-row instruments sounding closer than back-row instruments.

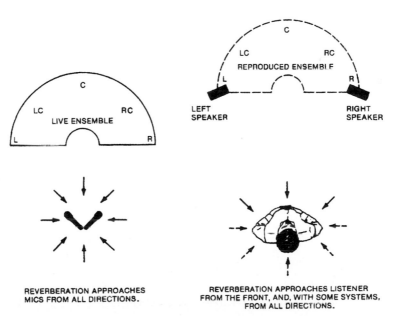

Figure 14.3 With accurate imaging, the sound-source location and size, and the reverberant field, are reproduced during playback.

Training Your Hearing

The critical process is easier if you focus on one aspect of sound reproduction at a time. You might concentrate first on the tonal balance—try to pinpoint what frequency ranges are being emphasized or slighted. Next listen to the mix, the clarity, and so on. Soon you have a lengthy description of the sound quality of your recording.

Developing an analytical ear is a continuing learning process. Train your hearing by listening carefully to recordings—both good and bad. Make a checklist of all the qualities mentioned in this chapter. Compare your own recordings to live instruments and to commercial recordings. Check out the Golden-Ear ear training CDs at www.moultonlabs.com/gold.htm.

A pop-music record that excels in all the attributes of good sound is The Sheffield Track Record (Sheffield Labs, Lab 20), engineered and produced by Bill Schnee. In effect, it's a course in state-of-the-art sound—required listening for any recording engineer or producer.

Another record with brilliant production is *The Nightfly* by Donald Fagen (Warner Brothers 23696-2), engineered by Roger Nichols, Daniel Lazerus, and Elliot Scheiner, produced by Gary Katz, and mastered by Bob Ludwig. This recording, and Steely Dan recordings by Roger Nichols, sound razor sharp, very tight and clear, elegant, and tasteful; and the music just pops out of the speakers.

The following listings are more examples of outstanding rock production, and set high standards:

Songs:

"I Need Somebody," by Bryan Adams; producer, Bob Clearmountain

"The Power of Love," by Huey Lewis & The News; producer, Huey Lewis & The News

Albums:

90125, by Yes; producer, Trevor Horn

Synchronicity, by The Police; producer, Hugh Padgham and The Police

Dark Side of the Moon, by Pink Floyd; producer, Alan Parsons

Thriller, by Michael Jackson; engineer, Bruce Swedien; producer, Quincy Jones

Avalon, by Roxy Music; engineer, Bob Clearmountain; producer, Roxy Music

Come Away With Me, by Norah Jones; engineer, Jay Newland

Genius Loves Company, by Ray Charles; several engineers

Give, by The Bad Plus; engineer, Tchad Blake

Live in Paris and *The Look of Love*, by Diana Krall; engineer, Al Schmitt

Smile, by Brian Wilson; engineer, Mark Linett

Then there are the incredibly clean recordings of Tom Jung (with DMP records) and George Massenberg. Some classical-music recordings with outstanding sound are on the Telarc, Delos and Chesky labels. You can learn a lot by emulating these superb recordings and many others.

Once you're making recordings that are competent technically—clean, natural, and well mixed—the next stage is to produce imaginative sounds. You're in command; you can tailor the mix to sound any way that pleases you or the band you're recording. The supreme achievement is to produce a recording that is a sonic knockout, beautiful and/or thrilling.

Troubleshooting Bad Sound

Now you know how to recognize good sound, but can you recognize bad sound? Suppose you're monitoring a recording in progress, or listening to a recording you've already made. Something doesn't sound right. How can you pinpoint what's wrong, and how can you fix it?

The rest of this chapter includes step-by-step procedures to solve audio-related problems. Read down the list of "bad sound" descriptions until you find one matching what you hear. Then try the solutions until your problem disappears. Only the most common symptoms and cures are mentioned; console maintenance is not covered.

This troubleshooting guide is divided into four main sections:

1. Bad sound on all recordings (including those from other studios)
2. Bad sound on playback only (the mixer output sounds all right)
3. Bad sound in a pop-music recording
4. Bad sound in a classical-music recording

Before you start, check for faulty cables and connectors. Also check all control positions; rotate knobs, and flip switches to clean the contacts, and clean connectors with De-Oxit from Caig Labs.

Bad Sound on All Recordings

If you have bad sound on all your recordings, including those from other studios, follow this checklist to find the problem:

- Upgrade your monitor system.
- Adjust tweeter and woofer controls on speakers.
- Adjust the relative gains of tweeter and woofer amplifiers in a bi-amped system.
- Relocate speakers.
- Improve room acoustics.
- Equalize the monitor system.
- Try different speakers.
- Upgrade the power amp and speaker cables.
- Monitor at a moderate listening volume, such as 85 dBSPL. We hear less bass and treble in a program if it is monitored at a low volume, and vice versa. If we hear too little bass due to monitoring at a low level, we might mix in too much bass.

Bad Sound on Playback Only

You might have bad sound on your playback only, but your mixer output sounds okay. If DAT tape playback has glitches or drop-outs, try these steps:

- Clean the recorder with a dry cleaning tape.
- Before recording, fast-forward the tape to the end and rewind it to the top.
- Use better tape.
- Format videocassettes nonstop from start to finish.

If a hard-drive recording has glitches or drop-outs on playback, try this:

- Increase the latency setting in your recording software.

- Follow the tips in the section Optimizing Your Computer for Digital Audio in Appendix B.

  If a digital recording sounds distorted, these suggestions might help:

- Keep the recording level as high as possible, but don't exceed 0 dBFS (decibels Full Scale).
- Avoid clipping in effects plug-ins.
- Record at a higher sampling rate or higher bit depth.
- Avoid sampling-rate conversion.
- Apply dither when going from a high bit depth to a lower bit depth.

Bad Sound in a Pop-Music Recording Session

Sometimes you have bad sound in a pop-music recording session.

Muddiness (Leakage)

If the sound is muddy from excessive leakage, try the following:

- Place the microphones closer to their sound sources.
- Spread the instruments farther apart to reduce the level of the leakage.
- Place the instruments closer together to reduce the delay of the leakage.
- Use directional microphones (such as cardioids).
- Overdub the instruments.
- Record the electric instruments direct.
- Use baffles (goboes) between instruments.
- Deaden the room acoustics (add absorptive material or flexible panels).
- Filter out frequencies above and below the spectral range of each instrument. Be careful or you'll change the sound of the instrument.
- Turn down the bass amp in the studio, or monitor the bass with headphones instead.

Muddiness (Excessive Reverberation)

If the sound is muddy due to excessive reverberation, try these steps:

- Reduce the effects-send levels or effects-return levels. Or don't use effects until you figure out what the real problem is.

- Place the microphones closer to their sound sources.
- Use directional microphones (such as cardioids).
- Deaden the room acoustics.
- Filter out frequencies below the fundamental frequency of each instrument.

Muddiness (Lacks Highs)

If your sound is muddy and lacks highs, or has a dull or muffled sound, try the following:

- Use microphones with better high-frequency response, or use condenser mics instead of dynamics.
- Change the mic placement. Put the mic in a spot where there are sufficient high frequencies. Keep the high-frequency sources (such as cymbals) on-axis to the microphones.
- Use small-diameter microphones, which generally have a flatter response off-axis.
- Boost the high-frequency equalization or cut slightly around 300 Hz.
- Change musical instruments; replace guitar strings; replace drum heads. (Ask the musicians first!)
- Use an enhancer signal processor, but watch out for noise.
- Use a direct box on the electric bass. Have the bassist play percussively or use a pick if the music requires it. When compressing the bass, use a long attack time to allow the note's attack to come through. (Some songs don't require sharp bass attacks—do whatever's right for the song.)
- Damp the kick drum with a pillow, folded towel or blanket, and mike it next to the center of the head near the beater. Use a wooden or plastic beater if the song and the drummer allow it.
- Don't plug an electric guitar directly into a mic input. Use a direct box or a high-impedance input.

Muddiness (Lacks Clarity)

If your sound is muddy because it lacks clarity, try these steps:

- Consider using fewer instruments in the musical arrangement.
- Equalize instruments differently so that their spectra don't overlap.
- Try less reverberation.

- Using equalizers, boost the presence range of instruments that lack clarity. Or cut 1 to 2 dB around 300 Hz.
- In a reverb unit, add about 30 to 100 msec of pre-delay.
- Pan similar-sounding instruments to opposite sides.

Distortion

If you hear distortion when monitoring the mics in a pop-music recording, try the following:

- Increase input attenuation (reduce input gain), or plug in a pad between the microphone and mic input.
- Readjust gain-staging: Set faders and pots to their design centers (shaded areas).
- If you still hear distortion, switch in the pad built into the microphone (if any).
- Check connectors for stray wires and bad solder joints.
- Unplug and plug-in connectors. Clean them with Caig Labs De-Oxit or Pro Gold.

Tonal Imbalance

If you have bad tonal balance—the sound is boomy, dull, or shrill, for example—follow these steps:

- Change musical instruments; change guitar strings; change reeds, etc.
- Change mic placement. If the sound is too bassy with a directional microphone, you may be getting proximity effect. Mike farther away or roll off the excess bass.
- Use the 3:1 rule of mic placement to avoid phase cancellations. When you mix two or more mics to the same channel, the distance between mics should be at least three times the mic-to-source distance.
- Try another microphone. If the proximity effect of a cardioid mic is causing a bass boost, try an omnidirectional mic instead.
- If you must place a microphone near a hard, reflective surface, try a boundary microphone on the surface to prevent phase cancellations.
- If you're recording a singer/guitarist, delay the vocal mic signal by about 1 msec.

333

- Change the equalization. Avoid excessive boost. Maybe cut slightly around 300 Hz if the sound is muddy, or cut around 3 kHz if the sound is harsh.
- Use equalizers with a broad bandwidth, rather than a narrow, peaked response.

Lifelessness

If your pop-music recording has a lifeless sound and is unexciting, these steps might help you solve it:

- Work on the live sound of the instruments in the studio to come up with unique effects.
- Add effects: reverberation, echo, exciter, doubling, equalization, etc.
- Use and combine recording equipment in unusual ways.
- Try overdubbing little vocal licks or synthesized sound effects.

If your sound seems lifeless due to dry or dead acoustics, try these:

- If leakage is not a problem, put microphones far enough from instruments to pick up wall reflections. If you don't like the sound this produces, try the next suggestion.
- Add reverb or echo to dry tracks. (Not all tracks require reverberation. Also, some songs may need very little reverberation so that they sound intimate.)
- Use omnidirectional microphones.
- Add hard, reflective surfaces in the studio, or record in a hard-walled room.
- Allow a little leakage between microphones. Put mics far enough from instruments to pick up off-mic sounds from other instruments. Don't overdo it, though, or the sound becomes muddy and track separation becomes poor.

Noise (Hiss)

Sometimes your pop-music recording has extra noise on it. If your sound has hiss, try these:

- Check for noisy guitar amps or keyboards.
- Switch out the pad built into the microphone (if any).
- Reduce mixer input attenuation (increase input gain).

- Use a more sensitive microphone.
- Use an impedance-matching adapter (a low- to high-Z step-up transformer) between microphones and phone-jack mic inputs.
- Use a quieter microphone (one with low self-noise).
- Increase the sound pressure level at the microphone by miking closer. If you're using PZMs, mount them on a large surface or in a corner.
- Apply any high-frequency boost during recording, rather than during mixdown.
- If possible, feed recorder tracks from mixer direct outs or insert sends instead of group or bus outputs.
- Use a lowpass filter (high-cut filter).
- As a last resort, use a noise gate.

Noise (Rumble)

If the noise is a low-frequency rumble, follow these steps:

- Reduce air-conditioning noise or shut off the air conditioning temporarily.
- Use a highpass filter (low-cut filter) that is set around 40 to 80 Hz.
- Use microphones with limited low-frequency response.
- See the section Noise (Thumps) below.

Noise (Thumps)

- Change the microphone position.
- Change the musical instrument.
- Use a highpass filter set around 40 to 80 Hz.
- If the cause is mechanical vibration traveling up the mic stand, put the mic in a shock-mount stand adapter, or place the mic stand on some carpet padding. Try to use a microphone that is less susceptible to mechanical vibration, such as an omnidirectional mic or a unidirectional mic with a good internal shock mount.
- Use a microphone with a limited low-frequency response.
- If the cause is piano pedal thumps, also try working on the pedal mechanism.

Hum

Hum is a subject in itself. See the section Hum Prevention at the end of Chapter 4.

Pop

Pops are explosive breath sounds in a vocalist's microphone. If your pop-music recording has pops, try these solutions:

- Place the microphone above or to the side of the mouth.
- Place a foam windscreen (pop filter) on the microphone.
- Stretch a silk stocking over a crochet hoop, and mount it on a mic stand a few inches from the microphone (or use an equivalent commercial product).
- Place the microphone farther from the vocalist.
- Use a microphone with a built-in pop filter (ball grille).
- Use an omnidirectional microphone, because it is likely to pop less than a directional (cardioid) microphone.
- Switch in a highpass filter (low-cut filter) set around 80 Hz.

Sibilance

Sibilance is an overemphasis of "s" and "sh" sounds. If you are getting sibilance on your pop-music recording, try these steps:

- Use a de-esser signal processor or plug-in. Or use a multiband compressor, and compress the range from 5 to 10 kHz.
- Place the microphone farther from the vocalist.
- Place the microphone toward one side of the vocalist, rather than directly in front.
- Cut equalization in the range from 5 to 10 kHz.
- Change to a duller sounding microphone.

Bad Mix

Some instruments or voices are too loud or too quiet. To improve a bad mix, try the following:

- Change the mix. (Maybe change the mix engineer!)
- Compress vocals or instruments that occasionally get buried.

- Change the equalization on certain instruments to help them stand out.
- During mixdown, continuously change the mix to highlight certain instruments according to the demands of the music.
- Change the musical arrangement so that different musical parts don't play at the same time. That is, consider having a call-and-response arrangement (fill-in-the-holes) instead of everything playing at once, all the time.

Unnatural Dynamics

When your pop-music recording has unnatural dynamics, loud sounds don't get loud enough. If this happens, try these steps:

- Use less compression or limiting.
- Avoid overall compression.
- Use multiband compression on the stereo mix instead of wideband (full-range) compression.

Isolated Sound

If some of the instruments on your recording sound too isolated, as if they are not in the same room as the others, follow these steps:

- In general, allow a little crosstalk between the left and right channels. If tracks are totally isolated, it's hard to achieve the illusion that all the instruments are playing in the same room at the same time. You need some crosstalk or correlation between channels. Some right-channel information should leak into the left channel, and vice versa.
- Place microphones farther from their sound sources to increase leakage.
- Use omnidirectional mics to increase leakage.
- Use stereo reverberation or echo.
- Pan effects returns to the channel opposite the channel of the dry sound source.
- Pan extreme left-and-right tracks slightly toward center.
- Make the effects-send levels more similar for various tracks.
- To give a lead-guitar solo a fat, spacious sound, use a stereo chorus. Or send its signal through a delay unit, pan the direct sound hard left, and pan the delayed sound hard right.

Lack of Depth

If the mix lacks depth, try these steps:

- Achieve depth by miking instruments at different distances.
- Use varied amounts of reverberation on each instrument. The higher the ratio of reverberant sound to direct sound, the more distant the track sounds.

Bad Sound in a Classical-Music Recording

Check the following procedures if you have problems recording classical music:

Too Dead

If the sound in your classical recording is too dead—there is not enough ambience or reverberation—try these measures to solve the problem:

- Place the microphones farther from the performers.
- Use omnidirectional microphones.
- Record in a concert hall with better acoustics (longer reverberation time).
- Turn up the hall mics (if used).
- Add artificial reverberation.

Too Close

If the sound is too detailed, too close, or too edgy, follow these steps:

- Place the microphones farther from the performers.
- Place the microphones lower or on the floor (as with a boundary microphone).
- Roll off the high frequencies.
- Use mellow-sounding microphones (many ribbon mics have this quality).
- Turn up the hall mics (if used).
- Increase the reverb-send level.

Too Distant

If the sound is distant and there is too much reverberation, these steps might help:

- Place the microphones closer to the performers.
- Use directional microphones (such as cardioids).
- Record in a concert hall that is less live (reverberant).
- Turn down the hall mics (if used).
- Decrease the reverb-send level.

Stereo Spread Too Narrow or Too Wide

If your classical-music recording has a narrow stereo spread, try these steps:

- Angle or space the main microphone pair farther apart.
- If you're doing mid-side stereo recording, turn up the side output of the stereo microphone.
- Place the main microphone pair closer to the ensemble.

If the sound has excessive stereo spread (or "hole-in-the-middle"), try the following:

- Angle or space the main microphone pair closer together.
- If you're doing mid-side stereo recording, turn down the side output of the stereo microphone.
- In spaced-pair recording, add a microphone midway between the outer pair, and pan its signal to the center.
- Place the microphones farther from the performers.

Lack of Depth

Try the following to bring more depth into your classical-music recording:

- Use only a single pair of microphones out front. Avoid multimiking.
- If you must use spot mics, keep their level low in the mix.
- Add more artificial reverberation to the distant instruments than to the close instruments.

Bad Balance

If your classical-music recording has bad balance, try the following:

- Place the microphones higher or farther from the performers.
- Ask the conductor or performers to change the instruments' written dynamics. Be tactful!

- Add spot microphones close to instruments or sections needing reinforcement. Mix them in subtly with the main microphones' signals.

Muddy Bass

If your recording has a muddy bass sound, follow these steps:

- Aim the bass-drum head at the microphones.
- Put the microphone stands and bass-drum stand on resilient isolation mounts (such as a carpet pad), or place the microphones in shock-mount stand adapters.
- Roll off the low frequencies or use a highpass filter set around 40 to 80 Hz.
- Use artificial reverb with a shorter decay time at low frequencies.
- Record in a concert hall with less low-frequency reverberation.

Rumble

Sometimes your classical-music recording picks up rumble from air conditioning, trucks, and other sources. Try the following to clear this up:

- Check the hall for background rumble problems.
- Temporarily shut off the air conditioning.
- Record in a quieter location.
- Use a highpass filter set around 40 to 80 Hz.
- Use microphones with limited low-frequency response.
- Mike closer and add artificial reverb.

Distortion

If your classical-music recording has distortion, try the following:

- Switch in the pads built into the microphones (if any).
- Increase the mixer input attenuation (turn down the input trim).
- Check connectors for stray wires or bad solder joints.
- Avoid sample-rate conversion.
- Apply dithering when going from 24 to 16 bits.

Bad Tonal Balance

Bad tonal balance expresses itself in a sound that is too dull, too bright, or colored. If your recording has this problem, follow these steps:

- Change the microphones. Generally, use flat-response microphones with minimal off-axis coloration.
- Follow the 3:1 rule mentioned in Chapter 7.
- If a microphone must be placed near a hard, reflective surface, use a boundary microphone on the surface to prevent phase cancellations between direct and reflected sounds.
- Adjust equalization.
- Place the mics at a reasonable distance from the ensemble (too-close miking sounds shrill).
- Avoid mic positions that pick up standing waves or room modes. Experiment with small changes in mic position.

This chapter describes a set of standards for good sound quality in both popular- and classical-music recordings. These standards are somewhat arbitrary, but the engineer and producer need guidelines to judge the effectiveness of the recording. The next time you hear something you don't like in a recording, the lists in this chapter will help you define the problem and find a solution.

15

SESSION PROCEDURES, EDITING, MASTERING, AND CD BURNING

"We're rolling. Take One." These words begin the recording session. It can be an exhilarating or an exasperating experience, depending on how smoothly you run it.

The musicians need an engineer who works quickly yet carefully. Otherwise, they may lose their creative inspiration while waiting for the engineer to get it together. And the client, paying by the hour, wastes money unless the engineer has prepared for the session in advance.

This chapter describes how to conduct a multitrack recording session. These procedures should help you keep track of things and run the session efficiently.

There are some spontaneous sessions, especially in home studios, that just "grow organically" without advance planning. The instrumentation is not known until the song is done! You just try out different musical ideas and instruments until you find a pleasing combination.

In this way, a band that has its own recording gear can afford to take the time to find out what works musically before going into a professional studio. In addition, if the band is recording itself where it practices, the microphone setup and some of the console settings can be more or less permanent. This chapter, however, describes procedures usually followed at professional studios, where time is money.

Preproduction

Long before the session starts, you're involved in preproduction—planning what you're going to do at the session, in terms of overdubbing, track assignments, instrument layout, and mic selection.

Instrumentation

The first step is to find out from the producer or the band what the instrumentation will be and how many tracks will be needed. Make a list of the instruments and vocals that will be used in each song. Include such details as the number of tom toms, whether acoustic or electric guitars will be used, and so on.

Recording Order

Next, decide which of these instruments will be recorded at the same time and which will be overdubbed one at a time. It's common to record the instruments in the following order, but there are always exceptions:

1. Loud rhythm instruments—bass, drums, electric guitar, electric keyboards
2. Quiet rhythm instruments—acoustic guitar, piano
3. Lead vocal and doubled lead vocal (if desired)
4. Backup vocals (in stereo)
5. Overdubs—solos, percussion, synthesizer, sound effects
6. Sweetening—horns, strings

The lead vocalist usually sings a reference vocal or scratch vocal along with the rhythm section so that the musicians can get a feel for the tune and keep track of where they are in the song. The vocalist's performance in this case is recorded but probably is redone later.

In a MIDI studio, a typical order might be:

1. Drum machine (playing programmed patterns)
2. Synthesizer bass sound
3. Synthesizer chords

4. Synth melody
5. Synth solos, extra parts
6. Vocals and miked solos

Track Assignments

Now you can plan your track assignments. Decide what instruments will go on which tracks of the multitrack recorder. The producer may have a fixed plan already.

What if you have more instruments than tracks? Decide what groups of instruments to put on each track. In a 4-track recording, for example, you might record a stereo mix of the rhythm section on tracks 1 and 2, then overdub vocals and solos on tracks 3 and 4. Or you might put guitars on track 1, bass and drums on track 2, vocals on track 3, and keyboards on track 4.

Remember that when several instruments are assigned to the same track, you can't separate their images in the stereo stage. That is, you can't pan them to different positions—all the instruments on one track sound as if they're occupying the same point in space. For this reason, you may want to do a stereo mix of the rhythm section on tracks 1 and 2, for instance, and then overdub vocals and solos on tracks 3 and 4.

It's possible to overdub more than four parts on a 4-track recorder. To do this, bounce or ping-pong several tracks onto one. Your recorder manual describes this procedure.

If you have many tracks available, leave several tracks open for experimentation. For example, you can record several takes of a vocal part using a separate track for each take, so that no take is lost. Then combine the best parts of each take into a single final performance on one track. Most DAWs let you do these extra takes on virtual tracks. It's also a good idea to record the monitor mix on one or two unused tracks. The recorded monitor mix can be used as a cue mix for overdubs, or to make a recording for the client to take home and evaluate.

Session Sheet

Once you know what you're going to record and when, you can fill out a session sheet (Figure 15.1). This simple document is adequate for home studios. "OD" indicates an overdub. Note the recorder-counter time for each take, and circle the best take.

345

```
┌─────────────────────────────────────────────────┐
│  SONG: Escape to Air Island                       │
│                                                   │
│  TRACK      INSTRUMENT        MICROPHONE          │
│  1          BASS              DIRECT              │
│  2          KICK              AKG D-112           │
│  3          DRUMS             CROWN GLM-100        │
│  4          LEAD VOC OD       STUDIO PROJECTS B1   │
│  5          HARM. VOC OD      STUDIO PROJECTS B1   │
│  6          LEAD GUIT OD      SHURE SM57          │
│  7          KEYS L            DIRECT              │
│  8          KEYS R            DIRECT              │
│                                                   │
│  TAKE 1   03:21 - 06:18   FS                      │
│       2   06:25 - 09:24   INC                     │
│      (3)  10:01 - 13:02                            │
└─────────────────────────────────────────────────┘
```

Figure 15.1   A session sheet for a home studio.

Production Schedule

In a professional recording studio, the planned sequence of recording basic tracks and overdubs is listed on a production schedule (Figure 15.2).

Track Sheet

Another document used in a pro studio is the track sheet or multitrack log (Figure 15.3). Write down which instrument or vocal goes on which track. The track sheet also has blanks for other information. If you are using a DAW, you can enter this information by typing it on-screen.

Microphone Input List

Make up a microphone input list similar to that seen in Table 15.1.

Later you will place this list by the mic snake box and by the mixing console.

Be flexible in your microphone choices—you may need to experiment with various mics during the session to find one giving the best sound with the least console equalization. During lead-guitar overdubs, for example, you can set up a direct box, three close-up microphones, and one distant microphone—then find a combination that sounds best.

Find out what sound the producer wants—a "tight" sound; a "loose, live" sound; an accurate, realistic sound. Ask to hear recordings having

```
Tape Speed:  15 ips               Artist:  Muffin
8 Track                           Producer:  B. Brauning
Noise Reduction:  Dolby c


1.  Song:  "Mr. Potato Head."
    Instrumentation:  Bass, drums, electric rhythm guitar,
    electric lead guitar, acoustic piano, sax, lead vocal.
    Comments:  Record rhythm section together with reference
    vocal. Overdub sax, acoustic, piano, and lead vocal later.

2.  Song:  "Sambatina."
    Instrumentation:  Bass, drums, acoustic guitar, percussion,
    synthesizer.
    Comments:  Record rhythm section with scratch acoustic
    guitar. Overdub acoustic guitar, percussion, and
    synthesizer.

3.  Song:  "Mr. Potato Head."
    Overdubs:  (1) acoustic piano, (2) lead vocal, (3) sax.

4.  Song:  "Sambatina."
    Overdubs:  (1) acoustic guitar, (2) synthesizer,
    (3) percussion.

5.  Mix:  "Mr. Potato Head."
    Comments:  Add 80-msec delay to toms.
               Double lead guitar in stereo.
               Increase reverb on sax during solo.

6.  Mix:  "Sambatina."
    Comments:  Add flanger to bass on intro only.
               Manually flange percussion.
```

Figure 15.2 A production schedule.

Table 15.1 A Microphone Input List

Input	Instrument	Microphone
1	Bass	Direct
2	Kick	EV N/D868
3	Snare	AKG C451
4	Overhead L	Shure SM81
5	Overhead R	Shure SM81
6	High toms	Sennheiser MD421-II
7	Floor tom	Sennheiser MD421
8	Electric lead guitar	Shure SM57
9	Electric lead guitar	Shure SM57
10	Piano treble	Crown PZM-6D
11	Piano bass	Crown PZM-6D
12	Scratch vocal	Beyer M88

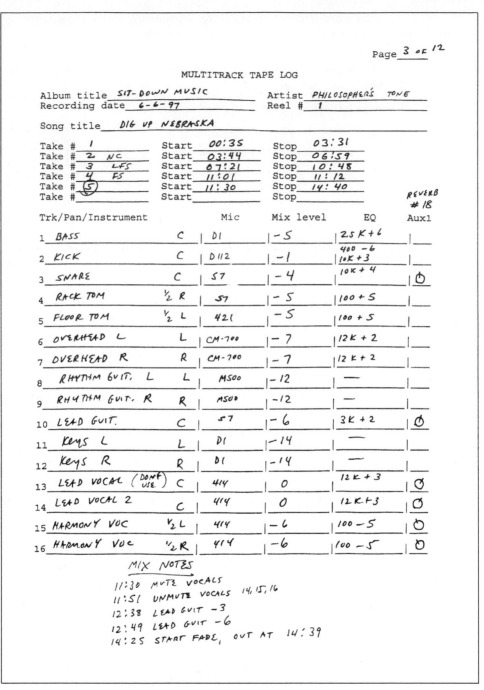

MULTITRACK TAPE LOG

Album title SIT-DOWN MUSIC   Artist PHILOSOPHER'S TONE
Recording date 6-6-97   Reel # 1

Song title DIG UP NEBRASKA

Take # 1   Start 00:35   Stop 03:31
Take # 2 NC   Start 03:44   Stop 06:59
Take # 3 LFS   Start 07:21   Stop 10:48
Take # 4 FS   Start 11:01   Stop 11:12
Take # (5)   Start 11:30   Stop 14:40   REVERB
Take #   Start   Stop   # 18

Trk/Pan/Instrument		Mic	Mix level	EQ	Aux1
1 BASS	C	DI	– 5	2.5 K + 6	
2 KICK	C	D 112	– 1	400 – 6 / 10K + 3	
3 SNARE	C	57	– 4	10K + 4	◔
4 RACK TOM	½ R	57	– 5	100 + 5	
5 FLOOR TOM	½ L	421	– 5	100 + 5	
6 OVERHEAD L	L	CM-700	– 7	12K + 2	
7 OVERHEAD R	R	CM-700	– 7	12 K + 2	
8 RHYTHM GUIT. L	L	M500	– 12	—	
9 RHYTHM GUIT. R	R	M500	– 12	—	
10 LEAD GUIT.	C	57	– 6	3K + 2	◔
11 KEYS L	L	DI	– 14	—	
12 KEYS R	R	DI	– 14	—	
13 LEAD VOCAL (DON'T USE)	C	414	0	12K + 3	◔
14 LEAD VOCAL 2	C	414	0	12 K + 3	◔
15 HARMONY VOC	½ L	414	– 6	100 – 5	◔
16 HARMONY VOC	½ R	414	– 6	100 – 5	◔

MIX NOTES

11:30 MUTE VOCALS
11:51 UNMUTE VOCALS 14, 15, 16
12:38 LEAD GUIT – 3
12:49 LEAD GUIT – 6
14:25 START FADE, OUT AT 14:39

Figure 15.3 A track sheet (multitrack recorder log).

the kind of sound the producer desires. Try to figure out what techniques were used to create those sounds, and plan your mic techniques and effects accordingly. Tips on choosing a microphone are given in Chapter 6.

Instrument Layout Chart

Work out an instrument layout chart, indicating where each instrument will be located in the studio, and where baffles and isolation booths will be used (if any). In planning the layout, make sure that all the musicians can see each other and are close enough together to play as an ensemble.

Setting Up the Studio

About an hour before the session starts, clean up the studio to promote a professional atmosphere. Lay down rugs and place AC power boxes according to your layout chart.

Now position the baffles (if any) on top of what has gone before. Put out chairs and stools according to the layout. Add music stands and music-stand lights. Run headphone cables from each artist's location to the headphone junction box in the studio.

Place mic stands approximately where they will be used. Wrap one end of a microphone cable around each microphone-stand boom, leaving a few extra coils of cable near the mic-stand base to allow slack for moving. Run the rest of the cable back to the mic input panel or snake box. Plug each cable into the appropriate wall-panel or snake-box input according to your mic input list.

Some engineers prefer to run cables in reverse order, connecting to the input panel first and running the cable out to the microphone stand. That procedure leaves less of a confusing tangle at the input panel where connections might be changed.

Now set up the microphones. Check each mic to make sure its switches are in the desired positions. Put the mics in their stand adapters, connect the cables, and balance the weight of each boom against the microphone.

Finally, connect the musicians' headphones for cueing. Set up a spare cue line and microphone for last-minute changes.

Setting Up the Control Room

Having prepared the studio, run through the checklist below to make sure the control room is ready for the session:

1. Pull all the patch cords from the patch panel.
2. If necessary, patch console bus 1 to recorder track 1, bus 2 to track 2, and so on.
3. Check out all the equipment to make sure it's working.
4. Label the blank recording medium with the artist, date, and reel number. If you're recording on a DAW, type in this information on screen along with the file name.
5. If you're using an MDM, insert the blank tape.
6. Normalize or zero the console by setting all switches and knobs to "off," "zero," or "flat" so they have no effect.
7. Switch on phantom powering for condenser microphones.
8. Set the console input-selector switches (if any) to "mic" or "line" as needed.
9. Attach a designation strip of masking tape across the front of the console. Referring to your mic input list, write on the strip the instrument each fader affects (bass, kick, guitar, etc.). Also label the sub-masters and monitor-mixer knobs according to what is assigned to them.
10. Turn up the monitor system. Carefully bring up each fader one at a time and listen to each microphone. You should hear normal studio noise. If you hear any problems such as dead or noisy microphones, hum, bad cables, or faulty power supplies, correct them before the session.
11. Verify the mic input list. Have an assistant scratch each mic grille with a fingernail and identify the instrument the microphone is intended to pick up. If you have no assistant, listen on headphones as you turn up one mic at a time and listen for background noise.
12. Check all the cue headphones by playing a tone or music through them and listening while wiggling each cable.

Session Overview

This is the typical sequence of events:

1. For efficiency, record the basic rhythm tracks for several songs on the first session.
2. Do the overdubs for all the songs in a dubbing session.
3. Mix all the tunes in a mixdown session.
4. Edit the tunes and master the album.

After the musicians arrive, allow them 1/2 hour to 1 hour free setup time for seating, tuning, and mic placement. Show them where to sit, and work out new seating arrangements if necessary to make them more comfortable.

Once the instruments are set up, you may want to listen to their live sound in the studio and do what you can to improve it. A dull-sounding guitar may need new strings, a noisy guitar amp may need new tubes, and so on. Adjust the studio lighting for the desired mood.

Recording

Follow the mixer recording procedures described in Chapter 12. Before you start recording, you might want to make connections to record the monitor mix. This recording is for the producer to take home to evaluate the performance.

When you're ready to record the tune, briefly play a metronome to the group at the desired tempo, or play a click track (an electronic metronome) through the cue system. Or just let the drummer set the tempo with stick clicks.

Start recording. Note the recorder counter time. Hit the slate button (if any) and announce the name of the tune and the take number.

Have someone play the keynote of the song (for tuning other instruments later). Then the group leader or the drummer counts off the beat, and the group starts playing.

The producer listens to the musical performance while the engineer watches levels and listens for audio problems. As the song progresses, you may need to make small level adjustments. As stated before, the recording levels are set as high as possible without causing distortion. Balancing the instruments at this time is done with the monitor mixer. The monitor mix affects only what is being heard, not what is being recorded.

The assistant engineer (if any) runs the multitrack recorder and keeps track of the takes on the track sheet, noting the name of the tune, the take number, and whether the take was complete (Figure 15.3). Use a code to indicate whether the take was a false start, nearly completed, a "keeper," and so on.

While the song is in progress, don't use the solo function, because the abrupt monitoring change may disturb the producer. The producer should stop the performance if a major flub (mistake) occurs but should let the minor ones pass.

At the end of the song, the musicians should be silent for several seconds after the last note. Or, if the song ends in a fade-out, the musicians should continue playing for about 30 seconds so there is enough material for a fade-out during mixdown.

After the tune is done, you can either play it back or go on to a second take. Set a rough mix with the aux knobs. If you connected your multitrack to the insert jacks, use the faders to set a rough mix with EQ and effects. The musicians will catch their flubbed notes during playback; you just listen for audio quality.

Now record other takes or tunes. Pick the best takes. To protect your hearing, try to limit tracking sessions to four hours or less with five hours maximum.

Overdubbing

After recording the basic or rhythm tracks for all the tunes, add overdubs. A musician listens to previously recorded tracks over headphones and records a new part on an open track. Follow the overdubbing and punching in/out procedures in Chapter 12.

Composite Tracks

If several open tracks are available, you can record a solo overdub in several takes, each on a separate track or virtual track. This is referred to as "comping" or "recording composite tracks." After recording all the takes, play back the solo and assign all the overdubbed tracks to a remaining open track set in record mode. You will bounce all the solo tracks to a composite track. Match the levels of the different takes. Then switch the overdubbed tracks on and off (using muting), recording just the best parts of each take. Finally, erase or archive the original tracks to free them up for other instruments.

If you are recording on hard disk or MiniDisc, usually you can record several takes on virtual tracks, then comp those virtual tracks during mixdown. If you are recording on a computer DAW, you can simply cut and paste the best parts of each take onto a single track, then archive the source tracks. Cakewalk Sonar Producer lets you record several takes on one track, view all the take waveforms as "lanes" in that track, select the best parts of each take, and create a composite track of those parts.

Drum Overdubs

Drum overdubs are usually done right after the rhythm session because the microphones are already set up, and the overdubbed sound will match the sound of the original drum track.

Overdubbing in the Control Room

To aid communications among the engineer, producer, and musician, have the musician play in the control room while overdubbing. You can patch a synth or electric guitar into the console through a direct box, and feed the direct signal to a guitar amp in the studio via a cue line. Pick up the amp with a microphone, and record and monitor the mic signal.

Breaking Down

When the session is over, tear down the microphones, mic stands, and cables. Put the microphones back in their protective boxes and bags. Wind the mic cables onto a power-cable spool, connecting one cable to the next. Wipe off the cables with a damp rag if necessary. Some engineers hang each cable in big loops on each mic stand. Others wrap the cable "lasso style" with every other loop reversed. You learn this on the job.

Put the labeled recording in its box. Also in the box, or in a file folder, put the designation strips, track sheet, and take sheet. Label the box and folder. Normally the studio keeps the multitrack master for possible future work unless the group wants to buy or rent it.

Log the console settings by writing them in the track sheet or reading them slowly into a portable cassette recorder. At a future session you can play back the tape and reset the console the way it was for the original session. Consoles or DAWs with automated mixing can store and recall the control settings.

Mixdown

After all the parts are recorded, you're ready for mixdown. Prepare the console and recorders, erase noises, and play the multitrack recording while adjusting balances, panning, equalization, reverberation, and effects. Automate changes in the mix if possible. Once you've rehearsed the mix to perfection, record it onto a 2-track recorder or your computer hard drive. Follow the mixdown procedures in Chapter 12. Repeat for all the song mixes.

Mastering

After recording the mixes, you can burn a CD of the mixes as they are, or you can master an album or demo of those mixes. Mastering is the last creative step before burning the final CD used for duplication. In mastering you edit out noises or false starts before the beginning of each song, put the songs in the desired order, insert a few seconds of silence between songs, and make each song the same loudness and the same tonal balance. The object is a consistent sound from track to track, so that everything flows better and the album sounds unified. You also might try to make the CD as loud or "hot" as possible (without destroying its sound quality).

At this point, you have three choices:

1. If the CD will be just a reference, not a demo or album, you can burn a CD of the unedited mixes and stop there.
2. If the CD will be a demo or album, burn a CD of the unedited mixes, then send it to a mastering engineer for mastering.
3. Or you can edit and master the mixes yourself, then burn a CD. It will be the CD master for duplication.

Let's look at each option above.

Burning a Reference CD

You can use the CD-burning software that came with your computer CD burner, or use other software. Some examples of CD-burning programs are Roxio's Toast, Jam, and Easy CD Creator; Ahead Software's Nero; Stomp's Click 'N Burn; Steinberg's Clean!; My MP3 and Get it on CD; Golden Hawk Technologies' CDRWIN; and Cakewalk's Pyro. Some DAW software includes a CD-burning application.

Here's the procedure:

1. The CD burning software puts 2 seconds of silence between songs. If you want to have extra seconds of silence at the end of a song, record the silence as part of the song's wave file.

2. Start the CD-R recording software.

3. Select which wave files you want to put on the CD-R, drag them to the playlist, and arrange them in the desired order. The total playing time must be less than the CD-R length (74 or 80 minutes).

4. Set the recording settings. You can either burn a disc on-the-fly or create a disk image. When recording on-the-fly, the computer grabs the sound files from random locations on hard disk and puts them in order as the CD-R does a burn. When recording a disk image, the computer rewrites the sound files to a single, contiguous space on your hard disk and puts them in order. Then you copy the disk image from hard disk to CD-R. Disk image is less likely to produce CD errors than On-the-fly.

You have a choice of transfer speed. Some CD recorders and blank CDs can record up to 52 times normal speed. Normal speed is 172 KBps (Bytes is capital B), double speed is 344 KBps, etc. A 52X recorder can burn a 74-minute CD in less than 2 minutes. Some CD burners automatically select the optimum speed based on what the blank CD-R can handle. High speeds do not degrade sound quality seriously, but they tend to increase errors—which the CD player may or may not correct accurately. The recommended speed to prevent errors is 2X to 4X when making a final master CD to be sent out for replication. But some CD recorders and blank CD-Rs have fewer errors at higher speeds.

You might want to normalize each track. This raises the highest peak in the track to maximum level: 0 dBFS (decibels Full Scale). Normalizing does not make the tracks the same loudness, because loudness depends on the average signal level, not the peak level.

5. Set the software to Disc-at-Once Mode. Start recording the CD-R. The wave files will transfer in order to the CD-R disc. To prevent glitches, do not multitask while the CD is recording.

CD-R discs cannot be erased and used over again, so try to make everything right before you burn a disc. You could do a practice burn on a CD-RW.

355

6. As soon as the recording is done, the display will indicate that the table of contents is being written. Eventually, the system will beep and eject the disc. To prevent error-causing fingerprints, be sure to handle the disc only by the edges. Pop the disc in an audio CD player, press Play, and check that all the tracks play correctly.

Sending Out Your CD for Mastering

You might prefer to send your CD of mixes to a good mastering engineer. This person can listen to your program with fresh ears, then suggest processing for your album that will make it sound more commercial. He or she is likely to have a better monitor system and better equipment than yours. They have heard hundreds of recording projects done by others, and know how to make your CD sonically competitive.

If you plan to have your program mastered outside, do not apply any signal processing to your finished mixes such as editing, level changes, compression, normalization, fades, or EQ. Let the mastering house do it with their better equipment and software. Also, leave some headroom by recording the finished mixes at about −3 dBFS maximum. Deliver your mixes in the highest possible resolution, such as 96 to 192 kHz and 24 to 32 bits. If possible, create a data CD rather than an audio CD. That is, copy the mixes' wave files to CD, rather than creating CD audio tracks (.cda files). The resulting data CD can be read by a CD-ROM but cannot be played on a CD player.

To copy the mixes as data, use the data copying feature of your CD burning program. Drag the wave files of the mixes to the playlist, then record the CD as described in the previous section. If your CD burning program does not support data CDs, just create an audio or music CD.

Mastering Your Own Album

Mastering can be done in your multitrack recording software, or in an audio production program such as Steinberg Nuendo, Sony Sound Forge, BIAS Peak, Digidesign MasterList CD, TC Works Spark, 1K Multimedia T-Racks 24, or Magix Samplitude.

First, discuss the order of the songs on the recording. For the first song, use a strong, accessible, up-tempo tune. Alternate keys or tempos from song to song. The last tune should be as good as or better than the first to leave a good final impression.

Mastering engineer Bob Katz offers these suggestions for song sequencing: Create the album in sets of one to three songs of the same tempo. Make sure the songs flow from one to the next. Here's a suggested song order:

1. An up-tempo, exciting song that hooks the listener.
2. After a short space, an up-tempo or mid-tempo song.
3. After three or four songs, slow down the tempo.
4. Reach a climax near the end of the album.
5. The last song should sound relaxed and intimate, perhaps using fewer members of the band.

Once you have decided on the order of songs, you're ready to master the demo or album as described below.

1. If you already recorded your mixes directly onto hard disk as stereo wave files, skip to Step 4.
2. If your song mixes are wave files on CD-R, put the CD-R in your CD-ROM drive and copy each song to your hard drive. Then go to Step 4. If your song mixes are audio tracks (.cda files) on a CD-R, convert each track to a wave file on your hard drive. This process requires CD-audio-to-wave conversion software. Skip to Step 4.
3. If your mixes are on a standalone 2-track recorder without a USB or FireWire port, plug the recorder's digital output into the digital input of your computer audio interface. If your interface lacks a digital input, use analog out to analog in. Launch the recording software. Set levels if necessary, start recording on your hard drive, and play the 2-track recording containing your mixes. All the mixes will copy in real time to your hard drive to create a single long wave file.

If your 2-track recorder has a USB or FireWire port, plug it into your computer's matching port. The recorder will appear as a hard drive to your computer. Copy the wave file(s) of the mixes to your computer hard drive.

4. Now that all the song mixes are on your hard drive, you will convert them to clips in an audio editing program. Start the software. If your song mixes are in one long wave file, import the file to a single stereo track. The waveform of the mixes appears. Zoom into the beginning of the first song and play it. Using a mouse, mark the start and end

Figure 15.4 Highlighting a song to create a region or clip.

of the song to highlight it (Figure 15.4). Be sure not to include spaces, noises, or outtakes on either side of the song. Save the highlighted song as a clip or region. Repeat for the other songs.

If each song mix is a separate wave file, import the first song's wave file to a track. The waveform of the song appears as a clip or segment of audio. Using a mouse, slip-edit or trim the beginning and end of the song clip to remove extra space and noises. Repeat for the other songs.

It's convenient to put each song clip on a different track, one after another (Figure 15.5). That way you can easily adjust the spacing between songs, and apply different processing to each song as needed.

5. Now that all the songs are in place and trimmed, add a fade-out at the end of certain songs if desired. If you want to crossfade between

Figure 15.5   Placing song clips on successive tracks makes mastering easier.

two songs, overlap their clips near the transition point, and use the crossfade function in your DAW.

6. Next you can adjust the spacing or gap between songs. Two to three seconds of silence between songs is typical, but go by ear. Use a longer space if you want to change moods between songs. Use a shorter space to make similar songs flow together. A short space also works well after a long fade-out because the fade-out itself acts like a long pause between songs.

7. Click on and play part of each song's waveform to check for loudness. Make all the songs equally loud by adjusting the track's fader before each song. If necessary, do a snapshot of each fader setting.

Here's one method to match song levels. If most songs are equal in level, set the fader to zero before each of those songs in order to avoid processing. Then, insert a fader snapshot or fader-level envelope before other songs: turn down any louder songs and turn up quieter songs to match the rest of the songs. Do this by ear.

CAUTION: If you increase a song's level, make sure the peaks in the song don't clip.

Another way to match song levels is to find the song with the highest peaks, and leave its level alone Adjust the levels of other songs so they match the loudness of the song with the highest peaks. Again, do this by ear. You may need to turn up the intro of a song so it works in context with the previous song. Also see www.har-bal.com.

8. Apply EQ to songs that need it. Put songs with different EQs on different tracks. If necessary, touch up levels after adding EQ.

9. If desired, apply multiband compression, limiting and normalization to the entire mix in order to get a "hot" or loud CD. Don't overdo it, or your project will be distorted and fatiguing to listen to. Squashed dynamics can suck the life out of music. If you apply only peak limiting and normalization, you'll get a hot CD without altering the musical dynamics. *Play CD track 40.*

The idea is to knock down the peaks in the waveform because they do not contribute to perceived loudness—the average level does. Once you limit the peaks about 6 dB, you can normalize (raise the overall level) and thus create a louder program. Normalizing to 100% of full scale can create errors with some D/A converters, so you might normalize to 99%.

According to mastering engineer Bob Katz, hot CDs and quiet CDs end up at the same level when processed with radio-station dynamic processors. Overcompressed material does not sound louder on the air; it sounds more distorted. So ignore the "loudness wars" and avoid excessive compression.

10. DAW software includes a feature called "Build mix to new soundfile," "Export Audio," or something similar. After your program is edited, export (save) the mastered song mixes to a 16-bit/44.1 kHz stereo wave file. If your recording was 24-bit, first turn on dithering. That retains much of the 24-bit sound quality when converting to 16-bit format.

Transferring the Mastered Program to CD-R

To get your mastered songs onto a CD, you need professional CD burning software that can insert song-start IDs within the single wave file of mastered songs. The software can adjust the pause or gap length down to zero, and can set the Start ID of each song anywhere in the program; that is, it allows PQ subcode editing. Some programs that do this are Gear Software Gear CD-RW and Gear Pro, Digidesign MasterList CD, Sony Media Software CD Architect, Adaptec Jam, and Goldenhawk Technology CDRWIN.

I'll describe CDRWIN as an example (www.goldenhawk.com). It creates CD tracks from a single wave file by following a cue sheet: a text file that lists the start time of each song. When the CD is recorded, the software creates a Start ID for each song based on the cue sheet. With this method you can create CD song-start IDs that occur even during a continuous program, such as a live concert recording. Here's the procedure:

1. Note the start time of each song in the DAW editing software's Edit Decision List or playlist. Or check the start time of each song clip by right-clicking on it or by looking at its beginning on the project time line. The start time of each song is specified in your DAW in minutes:seconds:frames, and there are 30 frames per second. Write down the start time of each song.

2. In the cue sheet, make each song's start time about one-third of a second (10 frames) before the actual start time of each song. That way the CD player's laser has time to find the track before playing audio. For example, if a song actually starts at 12:47:28, make its start time in your cue sheet 12:47:18. The cue sheet should also include the name and filepath of the mastered program's wave file.

3. Place a blank CD-R in your CD burner. Do not put a paper label on the CD before recording because the label can cause jitter.

4. Open your CD burning program and load in the cue sheet text file. Set the recording speed to 2X to 4X speed to reduce errors. Some CD-recorders and blank CD-Rs have fewer errors at higher speeds.

5. Start recording on CD. The CD-burning program writes the cue-sheet start times in the CD-R's table of contents, and copies the mastered program's wave file to the CD-R as separate.cda (CD-audio) tracks. Each track starts at the time you specified in your cue sheet. To prevent glitches, do not multitask while the CD is burning.

6. When the CD is finished, handle it by its edges to prevent error-causing fingerprints. Play the CD from start to finish to check for glitches. Press the track-advance button to make sure each song starts at the right time.

Another way to create master CDs is to use the Alesis MasterLink: a CD recorder with a built-in hard drive and mastering tools. With it you can record your stereo mixes, edit and master them, and burn a finished CD. Sample rates can be up to 96 kHz and word length can be 16-, 20-, or 24-bit. The MasterLink includes sample-rate conversion and noise shaping to alter sample rates and bit resolution as needed.

Several editing/mastering tools are offered with MasterLink, such as 16 different playlists, gain control, editing start and end points, joining or splitting song sections, EQ, compression, normalization, and peak limiting.

If you're intending the CD for commercial release, I recommend using a professional CD checker such as the Clover QA1010 or EC-2. It will catch unacceptably high error rates. Their Web site has a lot of useful information on CD-Rs (www.cloversystems.com).

You can identify the CD either by writing on the label side with a felt-tip marker or by sticking on a paper label. Special pens are available to write on CDs without damaging the plastic. Use a label applicator device and label design software such as Neato's MediaFACE, Roxio EZ CD Creator, or Stomp's Click 'N Design 3D. Some software can even print liner notes, including the track numbers, titles, and timing. If you are duplicating CDs yourself, you might want to use a CD printer rather than labels.

Master Log

Type or print a log describing the CD-R master (such as shown in Table 15.2) and include it with your master.

Table 15.2 A CD Master Log

CD Master Log
Album title: Don't Press That Button (demo)
Artist: Puff Daddy Bartlett
Mastering date: 2-26-06

Track	Start MM:SS:FF	Duration	Title
1	00:00:00	03:23	60 Gigabytes to Go
2	03:24:20	03:15	Unplugged Plug-In
3	06:41:00	02:49	Digital Dropout Blues
4	09:32:05	02:55	Late and See
5	12:29:11	03:07	Buffer the Vampire

Total running time: 15:47

Also include the CD liner notes information such as song lyrics, instrumentation, composers, credits, arrangers, publishers, and artwork. Include contact information for the artist, engineer, and mastering engineer. Stay in contact with the duplication house, especially about artwork.

Before you send out the master, be sure to make another copy that you keep in your studio. This copy can be used if the master is lost or damaged.

Note that the master CD-R doesn't leave the studio until all studio time is paid for! When this is done, send the CD-R to the duplication house.

The document "Recommendations for Delivery of Recorded Music Projects" provides examples of session sheets, tracks sheets, and so on. It also recommends media for delivering and backing up master recordings. It can be found at http://www.aes.org/technical/documents/ AESTD1002.1.03–10_1.pdf.

It's amazing how the long hours of work with lots of complex equipment have been concentrated into that little CD-R—but it's been fun. You crafted a product you can be proud of. When played, it recreates a musical experience in the ears and mind of the listener—no small achievement.

16

THE MIDI STUDIO: EQUIPMENT AND RECORDING PROCEDURES

Welcome to the computerized world of MIDI equipment—samplers, synthesizers, sampling keyboards, drum machines, and sequencers. Because other texts explain this equipment in detail, brief definitions serve the purpose here. See Appendix D for books on MIDI.

Musical Instrument Digital Interface (MIDI) is a standard connection between electronic musical instruments and computers that allows them to communicate with each other. Some of the things you can do with MIDI are

- Make a keyboard, MIDI guitar, or breath controller produce a sound like any instrument.
- Create the effect of a band playing. To do this, you record keyboard performances into a computer memory, edit the recording note by note if you wish, and have the recording play through synthesizers and a drum machine in sync.
- Combine or layer the sounds of two electronic musical instruments by playing them both with one keyboard.

- Automate a mixdown, or automate effects changes.
- Automate the playback of sound effects and music for video productions.

The MIDI signal is a stream of digital data—not an audio signal—running at 31,250 bits per second. It sends information about the notes you play on a MIDI controller, such as a piano-style keyboard or drum pads. Up to 16 channels of information can be sent on a single MIDI cable.

There are three types of MIDI ports on MIDI devices:

1. MIDI IN receives data going into the device.
2. MIDI OUT sends out data generated by the device.
3. MIDI THRU is like MIDI OUT, but duplicates the data that is at the MIDI IN port.

Connect MIDI OUT from the sending device to MIDI IN of the receiving device. For example, connect a keyboard controller's MIDI OUT to the MIDI IN connector of a sequencer or a sound module. Use MIDI THRU to connect two or more receiving devices in a row. For example, connect a keyboard controller's MIDI OUT to sequencer MIDI IN, and connect sequencer MIDI THRU to sound module MIDI IN.

MIDI-Studio Components

The following equipment typically is used in a MIDI studio:

- MIDI controller
- Sequencer
- Synthesizer
- Sampler and sample CDs
- Drum machine
- Power amplifier and speakers
- Personal computer
- MIDI computer interface
- Mixer
- Recorder-mixer (optional)
- 2-track recorder
- Effects

- Audio cables
- MIDI cables
- Equipment stand

You have learned about most of these in previous chapters, but a review might help at this point.

A **MIDI controller** is an instrument that generates MIDI data when you play on it. Examples are a piano-style keyboard, drum pads, MIDI guitar, or a MIDI breath controller. A synthesizer or drum machine can act as a controller.

Some newer keyboard controllers are compact and lightweight, and connect to your computer via a USB cable. Examples are Korg's 37-key microControl, M-Audio's 25-key Oxygen 8, Edirol's 49-key PC-300, and Novation's RemMote25 mini-synth. Because they have fewer keys than a standard keyboard, they use octave up/down buttons.

A **sequencer** is a device or program that lets you record, edit, and play back MIDI data. A recording done on a sequencer is called a sequence, which is a MIDI song. Unlike an audio recorder, a sequencer does not record audio. Instead, it records the key number of each note you play, note-on signals, note-off signals, and other parameters such as velocity, pitch-bend, and so on. A sequencer captures a performance, not the sound. Then any sound you want can be played by that performance. The recorded performance (sequence) can be modified to fix wrong notes, etc.

A sequencer plays MIDI files (.mid files). They are sequencer recordings of your own performances, or are MIDI files that can be downloaded from the Web.

The sequencer can be a standalone unit (Figure 16.1), a circuit built into a keyboard instrument, or a computer running a sequencer program. Like a multitrack audio recorder, a sequencer can record 8 or more tracks, with each track containing a performance of a different instrument.

A **synthesizer** is a musical instrument that creates sounds electronically. It can play MIDI data, either from your MIDI controller, a MIDI sequence, or a MIDI file downloaded from the Web. Synthesizers come in four forms: piano-style, sound module, software, or a synth chip on a sound card.

- A piano-style synth has a piano-style keyboard and built-in sound generators (Figure 16.2). You might want to use more than one synth to expand your palette of sounds.

Figure 16.1 A sequencer/drum machine.

- A **sound module** or **tone module** is a synthesizer without a keyboard. This standalone device is triggered by a sequencer or a controller.
- A **soft synth** is a synthesizer that is simulated in software. It runs in your CPU. The GUI of the software looks like a hardware synthesizer (Figure 16.2). A wavetable soft synth plays samples of real instruments, which sound more realistic than an FM soft synth.
- Another option is a **synth chip**, which is built into many sound cards.

A **patch** in a synthesizer is a sound preset (an instrumental timbre), such as a synthesized piano, bass, or snare drum. A **multitimbral** synthesizer can play two or more patches at once. A **polyphonic synthesizer** can play several notes at once (chords) with a single patch.

A **sampler** is a device that records sound events, or samples, into computer memory and plays them back when activated by a sequencer MIDI file or MIDI controller. A **sample** is a digital recording of one note of a real sound source: a flute note, a bass pluck, a drum hit, etc. A sample also can be a digital recording of a short segment of another recording. The sampling process is described in Chapter 9 under "Digital Recording."

A **soft sampler** is software that plays samples and lets you map them along your keyboard. Unlike a hardware sampler, a soft sampler has no memory limit on the number of samples that are accessible. It lets you import any sound, such as a wave file of a single note that you recorded,

Figure 16.2 A hardware-type synthesizer.

or a file from a sample CD. In contrast, a soft synth is limited to its supplied patches.

A **soundfont** is an audio sample (an instrument patch) in a special.sf2 format. It's like a wave file, but also includes a key range so that when you play a MIDI note number (keyboard key), it plays the sample pitch assigned to that note number. Soundfonts also include velocity switching, note envelope, looping, release sample, filter, and low-frequency oscillator (LFO) settings. A single soundfont can contain many wave files of different pitches. You can import a soundfont into a sample player. To use soundfonts, you need an EMU or SoundBlaster sound card, a MIDI controller, MIDI interface, and MIDI sequencing software.

Often a sampler is built into a sample-playing keyboard, which resembles an electronic piano. It contains samples of several different musical instruments. When you play on the keyboard, the sample notes are heard. The higher the key you press, the higher the pitch of the reproduced sample.

You can buy CDs with samples for use in your own projects. You copy the samples to your hard drive, load them into a software sample player, then trigger them with a sequencer or MIDI controller. For example, TASCAM's GigaStudio is sampling software with a huge sample library. Other libraries are available from www.primesounds.com and www.sonomic.com. Several DAW recording programs have samples included.

You can also download samples from the Web. Two sources of piano samples are Steinberg Grand VST 2.0 ($199 at www.steinberg.net) and Maxim Digital Audio Piano (freeware at www.mda-vst.com). Some samples on the Web come with their own sample-player software.

A **drum machine** is a device that plays built-in samples of all the sounds of a drum set and percussion (Figure 16.1). It also is a sequencer that records and plays back drum patterns played or programmed with built-in keys or drum pads. Some units can sample sounds. Most recorder-mixers have a drum machine built in.

A **power amplifier and speakers** (or **powered speakers**) let you hear what you're performing and recording. Usually these are small monitor speakers set up in a Nearfield arrangement (about 3 feet apart and 3 feet from you).

A **computer** is used mainly to run a sequencer program, which replaces the standalone sequencer and its tiny LCD screen. The computer monitor screen displays much more information at a glance, making editing easier and more intuitive (Figure 16.3).

A step up from a sequencer program is a **MIDI/digital audio recording program**. It records both MIDI sequences and digital audio tracks on your hard drive and keeps them in sync. In other words, this program lets you add audio signals such as a vocal, sax, etc., to a MIDI sequence. This software and your computer form a Digital Audio Workstation (DAW). Details are given in Chapter 13 on Computer Recording.

Figure 16.3 Screen shot of a sequencer program.

A **film-sound program** is a MIDI/digital audio program that includes a window for viewing videos. It lets you enter a list of sound effects or music with the time each occurs, and runs through the list automatically, triggering the effects and music at the right times, in sync with the video.

A **librarian program** manipulates patches or samples and stores them on computer disks. A **voice editor** program lets you create your own patches. A **notation program** converts your performance to standard musical notation. You can edit the notes, add lyrics and chords, and print out a copy.

A **MIDI computer interface** plugs into a user port in your computer and converts MIDI signals into computer data and vice versa. You need this only if you're using a computer in your system. Many sound cards and audio interfaces include MIDI ports; others have a joystick port that accepts MIDI when you add a joystick-to-MIDI cable.

If you have two or more synthesizers, or a synth and a drum machine, you need a **mixer** to blend their audio outputs into a single stereo signal.

A **recorder-mixer** (standalone DAW) combines a multitrack audio recorder, MIDI sequencer, and mixer into a single chassis.

A **2-track recorder** records the stereo mix of all your sound sources. The recorder can DAT, MiniDisc, or hard drive.

Audio effects include compression, reverb, echo, gating, chorus, and so on. MIDI effects are effects that process MIDI signals rather than audio signals. Some MIDI effects are arpeggiate, transpose, and delay.

Audio cables carry audio signals and typically have a 1/4-inch phone plug on each end. They connect synths, sound modules, and drum machines to your mixer line inputs.

MIDI cables carry MIDI signals and are used to connect synths, drum machines, and computers together so that they can communicate with each other. A MIDI cable is a 2-conductor shielded cable with a 5-pin DIN plug on each end. Pins 4 and 5 are the MIDI signal, pin 2 is shield, and pins 1 and 3 are not connected.

An **equipment stand** is a system of tubes, rods, and platforms that supports all your equipment in a convenient arrangement. It allows comfort, shorter cable lengths, and more floor area for other activities.

A **keyboard workstation** includes several MIDI components in one chassis: a keyboard, a sample player, a sequencer, and perhaps a synthesizer, and disk drive. That's everything you need to compose, perform, and record instrumental music. Some workstations include drum sounds so that you can get by without a separate drum machine.

MIDI Recording Procedures

The rest of this chapter describes recording procedures for several different MIDI studio setups, from simple to complex. The more complex procedures are based on the simpler ones, so it helps to read all the procedures here. Also, read your instruction manuals thoroughly and simplify them into step-by-step procedures for various operations. Note that each piece of MIDI gear has its own idiosyncrasies, and the instructions may have errors or omissions. If you have questions, call or e-mail tech support for your equipment.

2-Track Recording of a Synthesizer Performance

This is the simplest method of recording. You plug a MIDI interface into your computer, plug your synth into the interface, and run a sequencer program on the computer (Figure 16.4). An alternative to the computer and sequencer program is a standalone sequencer. You play chords and melody, record this MIDI data with your sequencer, and play back the sequence through your synthesizer. The basic steps include:

1. Select tempo, metronome, and MIDI track.
2. Assign a patch to the MIDI track.
3. Start recording.
4. Play a tune on your keyboard.
5. Play back the sequencer recording to hear it. Your performance will be duplicated by the synthesizer.

Figure 16.4 A synthesizer connected via a MIDI interface to a computer running a sequencer program.

6. Quantize the track if desired.
7. Punch in/out to correct mistakes.
8. Edit the sequence recording.
9. Arrange the song by combining various sequences.
10. Enter any program changes (changes in timbre).
11. Play the composition and set recording levels.
12. Start the sequencer playing, and record the synth output on a 2-track recorder. That recording is the final product.
    Below are the details for each step.

Select Tempo, Metronome, and Sequencer Track

Choose a tempo in your sequencer. Set the metronome to count off two measures. Select or click on the MIDI sequencer track you want to record on.

Assign a Patch to the MIDI Track

Sounds are stored in banks, and each bank contains several different patches (sounds), such as a fretless bass, grand piano, sax, drum set, and so on. The patch is in your synthesizer, which is a standalone synth, sound module, sound-card chip, soft synth, or soft sampler. Here's how to set up a MIDI track to play the desired patch:

1. If you are using a soft synth, insert it into an audio track.
2. Enable a MIDI track.
3. Set the input of the MIDI track to your MIDI interface device, omni mode. This mode makes the track respond to all MIDI channels.
4. Set a unique MIDI channel for that track. For example, set MIDI track 1 to channel 1.
5. Click on the output of the MIDI track. If you are using a soft synth, select its name from the pop-up list. Then the track's MIDI signal will play the soft synth of your choice. If you are using a standalone synth or sound module, select your MIDI interface device as the output of the MIDI track. That way, the track's MIDI signal will go through the MIDI interface to your standalone synth, which will play the patch for that track.
6. Choose a bank and patch. That is, choose the synth sound that you want to play, such as a drum set or fretless bass. If necessary, set the

patch to channel 1 so that MIDI track 1 (which is also set to channel 1) will play the correct patch.

Start Recording

Using your mouse, click on the RECORD key on-screen or on your computer keyboard. You'll hear a metronome ticking at the tempo you set.

Play Music on Your Keyboard

Listen to the sequencer's metronome and play along with its beat. The sequencer keeps track of the measures, beats, and pulses. As you play, MIDI data from your keyboard goes from keyboard MIDI OUT to interface MIDI IN, and is recorded as a MIDI file or sequence. When the song is done, click on STOP. The sequencer stops recording and should go to the beginning of the sequence (the top of the tune).

Another way to record your performance is in step-time, one note at a time. If the part is difficult to play rapidly, you also can set the sequencer tempo very slow, record while playing the synth at that tempo, and then play back the sequence at a faster tempo.

Play Back the Sequencer Recording

Click on PLAY. You'll hear the sequence playing through your synthesizer. If you change the patch and click on PLAY, you'll hear the same performance played by a different instrument.

Quantize the Track

Quantizing is the process of automatically correcting the timing of each note to the nearest note value (quarter note, eighth note, and so on). If you wish, quantize the performance by the desired amount. **Caution**: Quantizing can de-humanize the performance, making it too rhythmically perfect. It's better to adjust the timing of certain notes only, and to the smallest note value that works. Quantizing is essential if you want to use a notation program.

Punch-In/Out to Correct Mistakes

To correct mistakes, you can punch-in to record mode before the mistake, record a new performance, and then punch-out of record mode. Here's one way to do it:

1. Go to a point in the song a few bars before the mistake.
2. Just before you get to the mistake, punch-in to record mode and play a new, correct performance.
3. As soon as you finish the correction, punch-out of record mode.

Alternatively, you can use autopunch. With this feature, the computer punches in and out automatically at preset measures; all you have to do is play the corrected musical part. Perform an autopunch as follows:

1. Using the computer keyboard or mouse, set the punch-out point (the measure, beat, and pulse where you want to go out of record mode).
2. Set the punch-in point (just before the part you want to correct).
3. Set the cue point (where you want the track to start playing before the punch).
4. Click on PLAY.
5. When the screen indicates punch-in mode, or when the appropriate measure comes up, play the corrected part.
6. The sequencer punches out automatically at the specified point in the song.

These punch-in routines were done in real time. You can also punch-in/out in step-time:

1. Go to a point in the song just before the mistake.
2. Set the sequencer to step-time mode.
3. Step through the sequence note by note, and punch-in to record mode at the proper point.
4. Record the proper note in step-time.
5. Punch-out of record mode.

Edit the Sequence Recording

You might find it easier to edit the MIDI performance. Go to the MIDI edit screen, which resembles a piano-roll. It's a grid showing pitch verses time. The pitch of each note is represented by its height on the grid, and the duration of each note is represented by it length. You can grab incorrectly played notes and put them at the correct pitch and timing, delete unwanted notes, copy and paste phrases of notes, and so on.

Arrange the Song by Combining Sequences

Now your sequenced performance is perfect, so you can put together your composition. Many songs have repeated sections: The verse and chorus are each repeated several times. If you wish, you can record the verse and chorus once. Then copy the verse section and paste it every place it occurs in the song. Do the same for the chorus.

You can rearrange song sections and append one section to another by pressing a few keys on the computer. You can also have any section repeated. In this way, you might build a song by having the computer play sections A, A, B, A, C, A, B, B.

Enter Program Changes

To add variety to the song, you might want to have the synth play different programs (patches) at different parts of the song. For example, play a piano on the first verse, organ on the second, and marimba on the chorus. One way is to press different presets (program numbers) as you record the sequence.

Another way is to record these program changes on another MIDI track, which is called the controller track. Be sure to set the controller track to the same channel as the performance track, and turn off any patch on the controller track. Enter the appropriate program numbers at the right time on your synthesizer. Putting the program changes on a separate track makes it easy to edit them. You can punch-in new program changes just as you can punch-in new performances. When all the program changes are correct, you can bounce or copy them to the performance track if you wish.

Some sequencers do not record the program settings on your synth. They record only program changes. Consequently, when you play the sequence into a synth you just turned on, you might hear the wrong sounds. To prevent this problem, insert a few blank measures at the beginning of the tune and record your initial program changes there, according to the sounds you want to hear at the beginning of the tune. To do this, follow the procedure below:

1. Insert two or four blank measures at the beginning of your composition.

2. Set each track's patch to the wrong program number. If you want patch #17, for example, set it to #16. This way, you can key in a program change later.

3. Set your sequencer to punch-in mode so that you record only on the blank measures at the beginning of the tune.

4. When the punch-in starts, key in the correct program numbers. You can perform these program changes in several passes, one track at a time.

5. When the sequence plays back, it sets the synth automatically to the correct patches at the beginning of the tune.

An alternative to this procedure is to record a system exclusive or sysex dump—data about patch settings and so forth—into the sequencer. This works only if your synth and sequencer implement the sysex dump.

Play the Composition and Set Recording Levels

If you're using a standalone synth or sound module, plug its audio output (mono or stereo) into the line inputs of your 2-track recorder. Hit the PLAY key on your computer keyboard, and set the recording level for your recorder to –3 dB maximum. Later, you can normalize the recording (bring its level up to 0 dB maximum). If you're working with a DAW, set the track levels so that the level at the stereo mix bus reaches about –3 dB maximum (in peak-reading meter mode).

Record the Synth Output

Once your levels are set, put your 2-track recorder in record mode, and start the sequencer. In a DAW simply export the stereo mix to a wave file. This produces the finished product: a stereo recording of your song.

Multitrack Recording of a Synthesizer Performance

With this method, you play the parts for several different instruments (patches) on the same keyboard, and record each performance on a separate track of your sequencer software. During playback, the sequencer plays the desired patches (instruments) in your synth. It sounds like a band playing. You record the synthesizer's output, or export the soft-synth mix to a wave file, and that recording is the final product.

Each track and patch is set to a corresponding MIDI channel. For example, suppose both track 1 and the bass patch are set to channel 1. Then track 1's performance in the sequencer plays the bass patch in the synthesizer. Track 2 will play another patch (piano, flute, or whatever).

Some sequencers are designed so that, on power-up, track 1 goes to channel 1, track 2 goes to channel 2, and so on. You simply select a track to record on and select a patch for that track. The channel assignments are already taken care of.

Here's a short summary of the procedure. Record a drum part on MIDI track 1. Then go back to the start of the sequence, play the drum part, and add a bass line on track 2 in sync with the drums. Then go back to the top and add a piano on track 3, and so on.

In the sequencer program, you set up a different patch (instrument sound) on each MIDI track. Then you record a performance on each MIDI track. Play back the multitrack recording through the synthesizer, which plays all the patches simultaneously. Or you can use several synths, one for each part, if necessary. Set each track to a different MIDI channel, and if necessary set each instrument or patch to the same channel that its track is set to.

Refer back to Figure 16.4 to see the connections. Here is an outline of the steps for recording:

1. Start recording on the first sequencer track.
2. Play music on your keyboard.
3. Play back the recording.
4. Punch in/out or edit to correct mistakes.
5. Record overdubs on other tracks.
6. Edit the composition.
7. Mix the tracks.
8. Record the mix.

Again, these steps require a closer look. The following sections present details for each one.

Record the First Sequencer Track

Adjust the metronome tempo and count-off on your sequencer as desired. Select MIDI track 1. If necessary, set MIDI track 1 to MIDI channel 1.

Select the first patch you want to hear on your synthesizer (for example, a drum set). If necessary, set the synthesizer patch to MIDI channel 1. Click on RECORD on your computer screen.

Play Music on Your Keyboard

Listen to the sequencer's metronome and play along with its beat, or record in step-time. The sequencer keeps track of the measures, beats, and pulses. When you click on STOP, the sequencer stops recording and goes to the beginning of the sequence (the top of the tune).

Play Back the Recording

Click on PLAY on the computer screen. You'll hear the sequence playing through your synthesizer.

Punch-In/Out or Edit the Sequence to Correct Mistakes

As described in the previous section, you can correct mistakes by punching into record mode before the mistake, recording a new performance, and then punching out of record mode. You might prefer to edit the sequence instead. Also, you can quantize the track to make it rhythmically correct.

Record Overdubs on Other Tracks

With your first track recorded and corrected, you're ready to record other tracks. Select MIDI track 2. If necessary, set MIDI track 2 to MIDI channel 2.

On your synthesizer, select the next patch or instrument timbre you want to use (for example, bass). If necessary, set it to MIDI channel 2. You might want to adjust the timbre of the patch with the parameter controls on the synthesizer.

Then click on RECORD on your computer screen. Play the piano keys on your keyboard while listening to the prerecorded drums on track 1.

Repeat this procedure for other tracks and patches.

Edit the Composition

Now your sequenced performance is okay, so you can put together your composition. As described in the previous section, you can rearrange song sections and append one section to another by making selections on your computer screen. You also can have any section replayed. Key in program changes at the beginning of the song and anywhere else you want the patch on a track to change.

Mix the Tracks

Now that your song is recorded and arranged, you'll adjust the relative volumes of the tracks to achieve a pleasing balance.

If your multitimbral synthesizer doesn't have separate outputs for each patch, you have to adjust the mix at the sequencer. To do this, adjust the volume (key-velocity scaling) of each track with your computer. This only works if your keyboard is velocity-sensitive.

After you adjust the volume of each track in this way, click on PLAY on your computer screen to play the sequence. The desired mix of patches plays on your synth. Some synths let you add internal effects to the overall mix.

If your synth has several individual outputs—one for each patch—connect them to a mixer and set up a stereo mix with panning and effects.

Record the Mix

If your synth has a single output (mono or stereo), use your 2-track recorder to record the mix off that output. Plug your synthesizer's audio output into the line inputs of your 2-track recorder. Or, if your synth has several individual outputs connected to a mixer, record off the mixer stereo outputs.

Click PLAY on your computer screen and set the recording level for your recorder. Then put your 2-track recorder in record mode and start the sequencer. With a DAW, export the mix to a stereo wave file. This produces a stereo recording of your song.

Recording with a Keyboard Workstation

Each workstation operates in a different way, but here is a typical recording procedure:

1. Set up for recording a song.
2. Record the first musical part.
3. Do step recording (optional).
4. Overdub more parts.
5. Punch-in.
6. Set effects.
7. Store the song.

The following sections present more detailed instructions for each step.

Set Up for Recording a Song

1. Press SEQ (for SEQUENCER) on the front panel.
2. A sequencer menu appears on the LCD screen. You can move a cursor to select various parameters, and press the up or down buttons to set the value of each parameter.
3. Set the time signature (in the Initialize menu).
4. Select the song number.
5. Set the tempo.
6. Select the track number (track 1 to start).
7. Select the program number for the desired sound (for example, a drum set).

Record the First Musical Part

1. Press RECORD and START. Listen to two measures of metronome clicks and then start playing.
2. When you finish, press STOP.
3. To hear what you just played, press START.
4. If you want to re-record the part, press REC and START, and play the part again. You also could edit the performance, do punch-ins, and so on.

Step Recording

Instead of performing a musical part in real time, you might prefer to enter the notes one at a time, in step-time. Here's the basic procedure:

1. Select the track number and the measure number where you want to start.
2. Press RECORD and START.
3. Set the length of the first note (1/32 to 1/1).
4. If necessary, specify triplets, dotted notes, key dynamics, style of playing, and rests.
5. Press the desired note or chords on the keyboard.
6. Release all the keys; the recording proceeds to the next step.
7. After entering all the notes, press STOP.

Overdub More Parts

1. To record the next track, set the track number to the desired track (in this case, track 2).
2. Select the program number for the desired sound (for example, a bass).
3. Press RECORD and START. As you listen to track 1 playing the drum part, play a bass part on track 2.
4. Continue this procedure (steps 1 through 3), adding a new instrument each time.

Punch-In

You can correct mistakes easily in each track by punching in, either manually or automatically. The procedure follows:

1. Play the song to find the measures needing correction.
2. Select punch-in mode.
3. Page up one page, set the punch-in measure and the punch-out measure, page down one page.
4. Set the measure number to a point a few bars before the punch-in.
5. Press RECORD and START. You'll hear the song playing.
6. When the punch-in measure comes up, play the corrected part.
7. Press STOP when done.

Set Effects

You can go to the effects menus to set overall effects: hall reverb, chorus, flanging, echo, distortion, and so on. Press the correct number on the numeric keypad to get to the effects menus. (Note that these are built-in keyboard effects, not outboard studio effects.)

Save the Song File

Save the completed song in multitrack form to a plug-in RAM card or to an external sequencer and disk drive. To prevent data overload, you might have to do the external sequencer recording one track at a time with other tracks muted. If you're satisfied with the final results, record the stereo output signals of the workstation to a 2-track recorder.

In addition to these basic operations, you can:

- Bounce tracks (copy one track's performance to another track so it will play another patch)
- Edit each note event
- Create and copy patterns-for a drum or bass part, for example
- Modify track and song parameters
- Insert/delete/erase measures
- Modify sounds and effects (in great detail)
- Change the instrument (patch) that each track plays

Recording with a Drum Machine and Synth

This system combines a standalone synthesizer with a standalone drum machine. Figure 16.5 shows how to connect the cables. The sequencer shown in Figure 16.5 could be a computer with a MIDI interface running a sequencer program.

Understanding Synchronization

The drum machine has a built-in sequencer that records what you tap on its pads. Suppose you record a drum pattern with its built-in sequencer, and you record a synthesizer melody with an external sequencer. How do you synchronize the drum patterns in the drum machine with the synthesizer melody in the sequencer? In other words, how do you get the two devices to play in sync when both have different patterns recorded in different memories?

To synchronize the machines, you use a single MIDI clock (timing reference) that sets a common tempo for all the equipment. The MIDI

Figure 16.5 Connections to make a sequencer drive a synth and drum machine.

383

clock is a series of bytes in the MIDI data stream that conveys timing information. The clock is like a conductor's baton movements, keeping all the performers in sync at the same tempo. The clock bytes are added to the MIDI performance information in the MIDI signal. The clock signal is 24, 48, or 96 pulses per quarter note (ppq). That is, for every quarter note of the performance, 24 or more clock pulses (bytes) are sent in the MIDI data stream.

Decide which device you want to be your master timing reference—the sequencer or the drum machine. Set the master device to internal clock and set the slave device to external clock or MIDI clock. Then the slave will follow the tempo of the master.

To make this happen, the master sends clock pulses from its MIDI OUT connector. The slave receives those clock pulses at its MIDI IN connector. The slave also passes clock pulses through its MIDI THRU connector to other slave devices down the chain.

If a slave device lacks a MIDI THRU, enable "Echo MIDI in" in the slave device. Then the incoming pulses are duplicated at the MIDI OUT connector.

In the setup shown in Figure 16.5, the sequencer's clock drives both the drum machine and the synthesizer. In other words, the sequencer is the master tempo setter, and the drum machine and synth follow along. The drum machine's internally recorded patterns play in sync with the synthesizer's sequencer-recorded melody.

Basic Recording Procedure

Once you work out the synchronization problem, you are ready to begin recording with this system. There are two basic methods. The first uses the following steps:

1. Record drum patterns into the drum machine.
2. While listening to the drum patterns, record a synth part with your sequencer.
3. Sync the drums and synth by setting MIDI clocks and channels.
4. Press the PLAY key on the sequencer.

   Use these steps for the second method:

1. Record drum patterns into the drum machine.
2. Copy the MIDI sequence of those patterns onto one track of your sequencer.

3. While listening to the drum track, record a synth part on another track.
4. Play both sequencer tracks.

Here's how to record drum patterns, then sync the drums with the synth.

Recording Drum Patterns

Often the first step in composing a song is to record a drum pattern. There are many ways to do this; the following is one suggested procedure.

1. On the drum machine, set the tempo, time signature, and pattern length in measures. For this example, the pattern is 2 bars long.
2. Start recording, and play the hi-hat key in time with the metronome beat.
3. At the end of 2 bars, the hi-hat pattern you tapped repeats over and over (loops).
4. While this is happening, you can add a kick drum beat.
5. While the hi-hat and kick drum are looping, add a snare drum back beat, and so on.
6. Mix the recording by adjusting the faders or keys on the drum machine for each instrument.

Next, you repeat the process for a different rhythmic pattern—say, a drum fill—and store this as Pattern 2. Then develop other patterns. Finally, you make a song by repeating patterns and chaining them together as described in the drum machine's instruction manual. A song is just a list of patterns in order.

It's a good idea to add a count-off (a few measures of clicks) at the beginning so that later overdubs can start at the correct time.

Some musicians like to program a simple repeating drum groove first. While listening to this, they improvise a synth part. After recording the synth part, they redo the drum part in detail, adding hand claps, tom-tom fills, accents, and so on.

Synching Drums and Synth

Now you're ready to add a synth part and synchronize it with the drum track. The following procedure refers to a sequencer; it could also be a computer running a sequencer program.

1. Record a synth part with the sequencer.
2. Set the drum machine to external clock or MIDI clock.
3. Set the sequencer to internal clock.
4. Set MIDI channels: set the drum machine to channel 1; set the sequencer synth track and the synthesizer to channel 2. In this way, the sequencer's recorded performance will play only the synthesizer. The MIDI clock still controls both devices, even though they are set to different channels.
5. Press the PLAY key on the sequencer. As the sequencer plays its recorded synth melody, the sequencer's clock pulses drive the drum machine and synthesizer at the same tempo. The drum machine plays its internally recorded patterns while the synth plays the sequencer track.

Another way to synchronize a drum machine and a synthesizer is to record the drum patterns on one track of your sequencer. The advantage is whenever you rearrange parts of the music in the sequencer, you also rearrange the drum part. So you don't have to change drum patterns each time you repeat or delete a verse or a chorus. Follow this procedure to record the drum patterns into your sequencer:

1. Record a drum pattern with the drum machine's internal sequencer.
2. Enable the drum machine's clock out and MIDI data out.
3. On your sequencer, turn off the MIDI-THRU feature (if it has one).
4. Set the sequencer to external clock or MIDI clock mode, and set an open track in record mode.
5. Hit the PLAY key on the drum machine. The sequencer records the drum pattern on the open track.

To play back the drum patterns you just recorded, follow this procedure:

1. Set the drum machine to external clock mode.
2. Set the sequencer to internal clock.
3. Set the drum machine's track and the drum machine to the same MIDI channel.
4. Load an empty pattern into the drum machine so that the machine plays only the sequencer track.

5. Put the sequencer in play mode. The drum machine plays its sequencer track at the sequencer's tempo, and other synths connected to the sequencer play their tracks on their channels.

Recording with MIDI/Audio Recording Software

You can record MIDI sequences and digital audio tracks, edit them, and mix them, all with your PC or Mac computer. You need MIDI/audio recording software, an audio interface, and a MIDI interface. Many audio interfaces (such as sound cards) have MIDI connectors built in.

The recording software has both MIDI tracks and audio tracks. The MIDI tracks are recorded by a sequencer in the software, and the audio tracks are recorded as wave files.

This DAW lets you add several tracks of vocals, sax, guitar, or any audio signal to your MIDI sequences. While the recorded MIDI tracks play, you overdub audio tracks, which are recorded to your hard drive. Chapter 13 covers DAWs in detail.

Figures 16.6 and 16.7 show the system connections.

Several things happen when you play back what you recorded:

Figure 16.6 Connections for a MIDI/audio recording system with a standalone synthesizer (mixer not shown).

Figure 16.7 Connections for a MIDI/audio recording system with a soft synth.

1. The sequencer's MIDI signal comes from the MIDI OUT connector in the audio/MIDI interface.

2. If you're using an external synthesizer, sound module or drum machine, that MIDI signal plays notes from those devices. If you're using a soft synth (virtual synth) in your computer software, the MIDI sequence plays that soft synth.

3. If you're using a soft synth, the digital audio tracks blend with the audio from the soft synths, and this mix plays from the audio interface line output.

4. If you're using an external standalone synthesizer, you feed its audio output (and the audio-interface line output) into a mixer. You listen to the mix over powered monitors, and record the mix on a 2-track recorder.

5. If you're using a soft synth, you export the digital audio/MIDI mix to a stereo file on your hard drive.

Here is a typical procedure for using a MIDI/audio recording system:

Record MIDI Tracks

1. Set up a MIDI track to record. Select the input device for the track. Set the input channel to Omni so it will record any channel from your MIDI controller. Repeat for all the tracks you want to record.

2. Select an output device for the MIDI track from a drop-down menu. If you want to play the track on an external sound module or synth, set the output device to your MIDI interface. If you want to play the track with a soft synth in the computer, such as a GM General MIDI synth, specify which sound bank and patch you want that track to play. Repeat for all the tracks you want to record.

3. Set the output channel to Ch. 1 for track 1, Ch. 2 for track 2, and so on. Set the tempo and time signature.

4. If you want an external sound module or synthesizer to play the sounds, set it to the desired patch number so that you hear the instrument you want to hear.

5. Start recording, and play the first MIDI part on your MIDI controller or synth. If the part is difficult, you can record it at a slow tempo and play it back at normal speed. Or record it in step-time. You can edit wrong notes or chords. You could also enter the notes on a musical scale, one at a time.

6. Repeat steps 1 through 3 for the other MIDI instruments.

Overdub Audio Tracks

1. Plug a mic into a mic preamp or mic input of a mixer. Connect the preamp or mixer line output to your sound-card line input. Or if you are using an outboard audio interface, connect its USB or FireWire port to your computer. Select an audio track, and set its input source to the audio interface channel that the mic signal is plugged into.

2. Set the microphone input trim on your preamp or mixer, and set the recording level in your recording software or its volume-controls application.

3. Go to the beginning of the tune and hit PLAY in your DAW. The MIDI sequences that you recorded earlier should start playing. (You may need to press the PLAY key on a drum machine first if it is an outboard device, rather than a part of software.)

4. While listening to the MIDI tracks playing through headphones, record the vocal on an audio track. Then overdub more vocals and non-MIDI instruments on other open tracks.

Edit and Mix

1. You might record a few takes of the vocal part and then cut and paste selected portions to create a perfect take. For example, record one good chorus and copy it in each chorus section in your song. You also can edit individual MIDI notes or audio notes to correct their pitch or timing.

2. After all your tracks are recorded, use the on-screen mixer to set up a mix of the audio tracks and MIDI instruments. Adjust levels, panning, and effects (plug-ins).

3. Play the song several times to perfect the mix and to set up automation.

4. When you're satisfied with the mix, export or save it as a stereo wave file.

Refer to these chapters for more detail: Chapter 12, the sections on mixing procedures and automation; Chapter 13, the sections on editing and mastering.

Using Effects

No matter how you record with MIDI, effects are an important part of the mix. To keep the sound lively, try to vary the effects throughout the song, or use several types of effects at once.

For example, suppose you have a multitimbral synth, and you want to add a different effect to each patch. Whether or not you can do this depends on your synth. If it has a separate output for each patch, you can use a different effect on each patch. But if your synth has only a single output (mono or stereo) and you run it through an effects device, the same effect is on all the patches.

If your song includes program changes (patch changes), you can have the effects change when the patch changes. Set up a MIDI multieffects processor so that each synth program change corresponds to the desired effect. When the synth program changes, the effect changes also.

What if you want the effect, but not the synth patch, to change during a mix? Reserve a track and channel just for effects program changes. You don't hear these program changes in your synth, but you do hear the effects change. During a mixdown, it's usually easier to change effects automatically with your sequencer, rather than manually.

If your synth is a sampling keyboard, each sample could have reverberation or some other effect already on it; in that case each sample can have a different effect. The effect is not recorded in the sequencer; rather, the effect is part of the sampled sound. Note that the sampled reverberation cuts off every time you play a new note. Although this sounds unnatural, you can use it for special effect.

Because effects are audio signals, audio recorders can record effects but sequencers can't. If an effect is an integral part of the sound of an instrument, it's probably best to record it with the instrument on the multitrack audio recorder. If the effect is overall ambience or reverb (to put the band in a concert hall), however, then it's best to add it to almost everything during mixdown.

Some recording software lets you convert a MIDI track to an audio track, then apply audio effects to that track. Other software uses a separate MIDI track and audio track for the same instrument patch. You apply effects to the audio track.

MIDI effects (MFX) are nonaudio processes applied to MIDI signals, such as an arpeggiator, echo/delay, chord analyzer, quantize, transpose MIDI event filter, or velocity change. They can be used as real-time, nondestructive plug-ins in MIDI tracks.

Loop-based Recording

Let's turn to a different aspect of computer recording. It's possible to compose, record, and perform music entirely in software. You might start by creating a variety of loops or grooves, which are constantly repeated rhythmic or musical patterns.

To make a loop in 4/4 time, use an editing program to trim the start of the loop waveform just before beat 1, and trim the end of the loop after beat 4, and just before the next beat 1. To avoid a click in the audio, the trim points should be at zero crossings where the waveform crosses the 0-volt line.

There are four types of loop audio files based on their ability to have their tempo changed:

- A loop made from a standard digital audio file. It has a fixed tempo, so you must build your song around that tempo.

- A processed audio file. The tempo or pitch of the loop can be changed by a time-stretch or pitch-shift algorithm in your audio editing program. But if you need to change the tempo of a song, you need to adjust all the loops you stretched.

- A file with REX-based time-stretching. The transients in the audio file are cut into slices, whose spacing depends on the tempo. REX files follow tempo changes in your composition.

- An acidized (RIFF) WAV file. Pitch and tempo information are in the file header, and the audio is sliced at transients as with REX files. Acidized files follow tempo and key changes. Slowing down the loop can add artifacts, so it helps to start with a slow loop.

You drag and drop (or import) loops into audio editing programs. There, you can copy and paste loops and play them along with audio and MIDI tracks, such as soft-synth parts, vocals, and acoustic instruments. Then add effects and do a mix.

You can also create loops externally, then import them into a DAW recording program. For example, compose a repetitive beat of drums, synth, and samples in a sampler/sequencer box. Record the beat onto a CD, then use ripper software to convert the CD's beat track to a wave file. Import the wave file into a stereo track in your DAW. On other DAW tracks, you might overdub vocals, doubled vocals, rap sections, and harmony.

Some loop programs offer groove quantizing, which transfers the timing and dynamics from one groove to another. It allows human-like variations in timing and key velocity.

Loop libraries are collections of sampled drum and bass beats that let you loop (repeat), change tempo, etc. The beats come in MIDI files and wave files. Some examples are BitHeadz Phrazer, Club Tracks, Beatboy, Fruity Loops, Keyfax Twiddly Bits, Vamtech Drumtrax, FXpansion Session Drummer, Discrete Drums, APO Multimedia Mix It, Multiloops Naked Drums, Pocket Fuel RADS series, Smart Loops Percussion Kit, Ilio/Spectrasonics Groove Control and BackBeat, Wizoo VST Drum Sessions, www.bigfishaudio.com, and Cakewalk loop libraries. Complete looped rhythm tracks are available online for purchase. Just add vocals.

Many DAW recording programs include loop-creation software, which is also available separately. For example, Propellerhead Software has a variety of programs to create and modify loops. They also offer soft synths, drum machines, and sequencers. Here are some of their products:

ReCycle starts with a loop and lets you change its tempo and pitch, and replace and process sounds within the loop. By detecting peaks in the waveform, **ReCycle** automatically breaks a loop into parts or slices. A slice might be a snare-drum hit, a kick-drum hit, or a kick-and-snare hit. When a loop's tempo is changed, the start point of each slice moves in time so that the beats occur at the right time. (You might need to touch this up manually.) If the tempo is slowed down, ReCycle creates a decay after each drum hit to fill in the gaps. You can delete slices, change their length, attack, decay, and pitch, and add compression or EQ. Then you can import the improved loop as an REX2 file into an audio track in your sampler or sequencer program. There you can control all aspects of the loop.

Reason is a group of synthesizers and samplers, a drum machine, ReCycle-based loop player, mixer, effects, pattern sequencer, and more (Figure 16.8). It's all-in-one and easy to learn.

Figure 16.8 One screen in Reason software.

Reason Adapted is a lite version that is distributed as part of some software bundles.

ReBirth is a software emulation of two analog bass-line synths and two classic drum machines. Also included are delay, distortion, compression, and filtering. It integrates into sequencer software and allows real-time audio streaming.

Reload is a utility that converts AKAI S1000 and S3000 formatted media into formats (such as wave files) that can be used with Reason, ReCycle, and other audio applications

ReWire is a useful feature found in some loop programs. It transfers or streams audio data between two computer applications in real time, almost like a cable. This allows programs to communicate with each other and synchronize together.

Other loop-based tools to compose and record music are listed below.

Ableton Live lets you compose with soft synths, record hi-resolution multitrack audio, and play loops in a live performance. With Live you can create arrangements and modify grooves, timing, pitch, volume, and effects.

Sony Media Software's ACID products let you select audio loops from Windows Explorer, drag them into a recording program's track view, and arrange them into multitrack projects. The tempo and key of each loop are automatically matched to your project's music in real time, using the slicing technique described before. ACID uses time-stretching algorithms to lengthen sounds when the tempo is slowed down. Prerecorded ACID loops are available.

Other loop-intensive DAWs include Image-Line Software FL Studio (PC), Cakewalk Plasma (PC), and Apple GarageBand. Two great articles on loops by Craig Anderton were in the July 2004 and August 2002 issues of *EQ Magazine*.

"No Sound" MIDI Troubleshooting

Suppose you load a MIDI file and hit PLAY, but hear nothing. Or you play notes on your MIDI controller, but there is no sound. Here are some possible causes:

- Your computer is not communicating with your MIDI interface. Try replugging the interface.

- If you're using a MIDI controller and MIDI sequencing software, the MIDI sequencer track is not selected and Record-enabled.
- The MIDI sequencer track is on a different channel than the synth or sound module.
- The MIDI sequencer track has no MIDI channel assigned.
- The MIDI sequencer track is assigned a channel already used by another MIDI sequencer track.
- MIDI OUT or MIDI THRU is not connected to MIDI IN somewhere in your system.
- The wrong sound bank was selected in a synth.
- The power amp, mixer, or synth output is off or turned down.
- The synth audio output is not connected to power amp.
- In the sequencer's MIDI setup menu, your MIDI interface is not selected as the input and/or output device.
- The sequencer track volume is not turned up.
- You started playing the file in the middle of a long note, rather than at its beginning.
- The MIDI driver is buggy. Download the latest update from the interface manufacturer.

Summary

MIDI is a computer code that transmits and stores information about the position and motion of controls, such as keys on a piano-style keyboard. In this way, MIDI lets you record and play back a musical performance that is independent of the sound of that performance. Similarly, MIDI lets you record and play back mixer-control settings, which allows for an automated mixdown.

17

ON-LOCATION RECORDING OF POPULAR MUSIC

Sooner or later you'll want to record a band—maybe your own—playing in a club or concert hall. Many bands want to be recorded in concert because they feel that's when they play best. Your job is to capture that performance on a recorder and bring it back alive.

There are several ways to record live:

- Record with two mics out front into a 2-track recorder.
- Using the PA mixer, record off a spare MAIN output.
- Using a recorder-mixer, record a stereo mic at the front of house (FOH) position on tracks 1 and 2, while recording a feed from the house mixer on tracks 3 and 4. Mix the tracks later.
- Feed the PA mixer INSERT jacks to a multitrack recorder.
- Feed the PA mixer INSERT jacks to a recording mixer, and from there to a 2-track or multitrack recorder.
- Use a mic splitter on stage to feed the PA snake and recording snake. Record to multitrack or 2-track.

We'll start by explaining simple two-microphone techniques and work our way up to elaborate multitrack setups.

Two Mics Out Front

Let's start with the simplest, cheapest technique: two mics and a 2-track recorder. The sound will be distant and muddy compared to using a mic on each instrument and vocal. Not exactly CD quality! But you'll hear how your band sounds to an audience.

Recording this way is much simpler, faster, and cheaper than multimic, multitrack recording. Still, if time and budget permit, you'll get better sound with a more elaborate setup.

Equipment

Here's what you need for 2-mic recording:

- A stereo mic, or two mics of the same model number. Your first choice might be cardioid condenser mics. The cardioid pickup pattern cuts down on room reverb and noise. The condenser type generally sounds more natural than the dynamic type. Another option is a pair of boundary mics such as PZMs. Simply tape them to the ceiling several feet in front of your group.

- A 2-track recorder of your choice. Either use a unit with mic inputs, or use a separate mic preamp.

- Blank recording medium. Bring enough to cover the duration of the recording.

- Two long mic cables.

- Two mic stands.

- Headphones. You could monitor with speakers in a separate room, but headphones are more portable and they sound the same in any environment. Closed-cup headphones partly block out the live sound of the band so you can better hear what's going on tape. Ideally you'd set up in a different room than the band is in, so you can clearly hear what you're recording.

Mic Placement

Use a pair of mic stands or hang the mics out of the reach of the audience. Aim the two mics at the group about 12 feet away, and space them about 5 to 15 feet apart (Figure 17.1). Place the mics far apart (close to the PA speakers) to make the vocals louder in the recording. Do the opposite

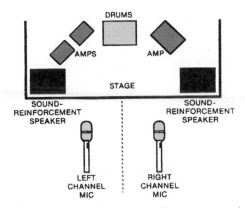

Figure 17.1 Recording a musical group with two spaced microphones.

to make them quieter. The stereo imaging will be vague, but at least you can control the balance between instruments and vocals.

To record a small folk group or acoustic jazz group, set up two mics of the same model number in a stereo arrangement of your choice. Place the mics about 3 to 10 feet from the group. The balance may not be the best, but the method is simple.

Recording

After setting the recording level, leave it alone as much as possible. If you must change the level, do so slowly and try to follow the dynamics of the music.

If the playback sounds distorted—even though you did not exceed a normal recording level—the mics probably overloaded the mic preamps in the recorder. A mic preamp is a circuit that amplifies a weak mic signal up to a useable level. With loud sound sources such as rock groups, a mic can put out a signal strong enough to cause distortion in the mic preamp.

Some recorders have a **pad** or **input attenuator**. It reduces the mic signal level before it reaches the preamp, and so prevents distortion. You can build a pad (Figure 17.2), or buy some plug-in pads from your mic dealer. Some condenser mics have a switchable internal pad that reduces distortion in the mic itself. If you have to set your record-level knobs very low (less than one-third of the way up) to get a 0 dB or 0 VU recording level, that shows you probably need to use a pad.

Figure 17.2 Balanced and unbalanced microphone pads.

Recording from the Sound-Reinforcement Mixer

You can get a fairly good recording by plugging into the main output of the band's sound-reinforcement mixer, also called the FOH or PA mixer. Connect the main output(s) of the mixer to the line input(s) of a 2-track recorder. Use the mixer output that is ahead of any graphic equalizer that is used to correct the speakers' frequency response (Figure 17.3).

Mixers with balanced +4 dBu outputs can produce a signal that is too high in level for the recorder's line input, causing distortion. This is probably occurring if your record-level controls have to be set very low. To reduce the output level of the mixer, turn it down so that its signal peaks around −12 on the mixer meters, and turn up the PA power amplifier to compensate. That practice, however, degrades the mixer's signal-to-noise ratio.

A better solution is to make a 12-dB pad (Figure 17.4). The output level of a balanced-output mixer is 12 dB higher than the normal input level of a recorder with an unbalanced input.

Drawbacks

The recorded mix off the sound-reinforcement mixer might be poor. At the FOH position, the sound mixer hears a combination of the band's

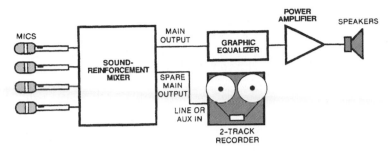

Figuro 17.3 Recording from the sound-reinforcement mixer.

NOTE: If mixer output is balanced by a grounded center-tapped transformer or is electronically balanced, leave pin 3 unconnected to prevent distortion. If mixer output is floating-transformer coupled, connect pin 3 to shield.

Figure 17.4 A 12-dB pad for matching a balanced output to an unbalanced input.

amps and drums, the stage monitors, and the house speakers. The sound mixer tries to get a good mix of all these elements. That means the signal is mixed to augment the band's on-stage instruments and vocals—not to sound good by itself. A recording made from the FOH mixer is likely to sound too strong in the vocals and too weak in the bass.

Recording with a 4-Tracker

A 4-track portable studio or recorder-mixer can do a good job of capturing a band's live sound. With this method, you place a stereo mic (or a pair of mics) by the FOH mixer. Record the mic on tracks 1 and 2 while recording a spare FOH mixer output on tracks 3 and 4 (Figure 17.5). After the concert, mix the four tracks together.

The FOH mics pick up the band as the audience hears it: lots of room acoustics, lots of bass, but rather muddy or distant. The FOH mixer

401

Figure 17.5 Recording two mics and an FOH mix on a 4-track recorder-mixer.

output sounds close and clear, but typically is thin in the bass. When you mix the FOH mics with the FOH-mixer signal, the combination has both warmth and clarity.

Consider using a stereo mic at the FOH position. Good stereo mics cost more than $500, but they provide great stereo in a portable, convenient package. You can also set up two mics of the same model number in a stereo arrangement. For example, angle two cardioid mics 90 degrees apart and space them 1 foot apart horizontally. Or place two omni mics 2 feet apart.

Plug the mics into your 4-track's mic inputs 1 and 2. Adjust the trim to prevent distortion, and set the recording level. Find a spare main output on the FOH mixer, and plug it into your 4-track's line inputs 3 and 4. You may need to use the 12-dB pad described earlier. Set recording levels and record the gig.

Back in your studio, mix the four tracks to stereo. Tracks 1 and 2 provide ambience and bass; tracks 3 and 4 provide definition and clarity.

You might hear an echo because the FOH mics pick up the band with a delay (sound takes time to travel to the mics). A typical delay is 20 to 100 msec. To remove the echo, delay the FOH mix by the same amount. Patch a delay unit into the insert jacks for tracks 3 and 4. As you adjust the delay time up from zero, the echo will get shorter until the signals are aligned in time. You may be surprised at the quality you get with this simple method.

Recording Off the FOH Mixer Aux Output

On the FOH mixer, find an unused aux send output. Plug in a Y-cord: One end goes into a PA mixer's aux-send connector; the other end has

Figure 17.6 Recording off the FOH mixer aux outputs.

two connectors that mate with your 2-track recorder's line inputs (Figure 17.6). If you have two spare aux busses, you could plug a cable into both of them, and set up a stereo mix.

Put on some good closed-cup headphones and plug them into your 2-track recorder to monitor the recording. Adjust the aux-send knob for each instrument and vocal to create a good recording mix. Record the gig.

The advantages of this method are

- It's simple. All you need is a recorder, a cable, and headphones.
- The recorded sound is close-up and clear.
- If the mix is done well, the sound quality can be very good.

The disadvantages are

- It's hard to hear what you're mixing. You may need to do several trial recordings. Set up a mix, record, play back, and evaluate. Redo the mix and try another recording.
- As you adjust the aux knobs, you might get in the way of the FOH mixer operator.
- The recording will be dry (without effects or room ambience). However, you could plug two room mics into the FOH mixer and add them to the recording mix. Do not assign these mics to the FOH output channels.
- If the aux send is pre-EQ, there will be no EQ on the mics. If the aux send is post-EQ, there will be EQ on the mics, but it may not be appropriate for recording.

Recording an aux mix works best where the setup is permanent and you have time to experiment. Some examples are recording a church service and recording a regularly scheduled show in a fixed venue.

Feeding the FOH Mixer Insert Sends to a Multitrack Recorder

This is an easy way to record, and it offers excellent sound quality. Plug one or more multitrack recorders into the insert-send jacks on the back of the FOH mixer. Set recording levels with the FOH mixer input trims. After the concert, mix the tracks back in your studio. The multitrack recorder can be an MDM, hard-disk recorder, recorder-mixer, or multi-channel audio interface plugged into a laptop computer.

Connections

Suppose you want to record one instrument or vocal on each track. In each FOH mixer input channel locate the INSERT jacks. Connect that jack to a recorder track input (Figure 17.7). INSERT jacks are usually pre-fader, pre-EQ. So any fader or EQ changes that the FOH sound mixer does will not show up on your recording.

If the FOH mixer has separate Send and Return INSERT jacks, connect the Send to the recorder track input, and connect the recorder track output to the Return. Some boards use a single stereo INSERT jack with TRS (Tip/Ring/Sleeve) connections. Usually the tip is send and the ring is return. In the stereo phone plug that you plug into the INSERT jack, wire tip and ring together, and also to the cable hot conductor.

On some FOH mixers with a TRS INSERT jack, you can use a mono (tip/sleeve; TS) phone plug. Plug it in halfway to the first click so you don't break the signal path—the signal still goes through the FOH mixer. If you plug in all the way to the second click, the signal does not go through the FOH mixer—just to your multitrack recorder.

Figure 17.7   Feeding a multitrack recorder from the FOH mixer INSERT jacks.

What if you want to record several instruments on one track, such as a drum mix? Assign all the drum mics to one or two output busses in the FOH mixer. Plug the BUS OUT insert jack to the recorder track input. Use two busses for stereo.

Monitor Mix

Although you can hear what you're recording through the PA speakers, you may want to set up a monitor mix over headphones. Here are two ways to do this:

1. Connect all the multitrack recorder outputs to unused line inputs on your mixer. Use those faders to set up a monitor mix. Assign them to an unused bus, and monitor that bus with headphones.
2. Set up a monitor mix with some unused aux knobs. Monitor the aux send bus over headphones.

Setting Levels

Set recording levels with the FOH mixer's TRIM or INPUT ATTEN knobs. This affects the levels in the house mix, so be sure to discuss your trim adjustment in advance with the FOH mixer operator. If you turn down an input trim, the FOH mixer operator must compensate by turning up that channel's fader and monitor send.

Set recording levels before the concert during the sound check (if any!). It's better to set the levels a little too low than too high, because during mixdown you can reduce noise but not distortion. A suggested starting level is –10 dBFS (decibel Full Scale), which allows for surprises. Signals exceeding 0 dBFS will be distorted by digital clipping.

Keep a log as you record, noting the counter times of tunes, sonic problems, and so on. Refer to this log when you mix.

Mixdown

After the recording is finished, mix down the tracks back in the studio, spending as much time as you need to perfect the mix. You can even overdub parts that were flubbed during the live performance, taking care to match the overdubbed sound to the original recording.

405

Feeding the FOH Mixer's Insert Sends to a Recording Mixer

The previous method has a drawback: You have to adjust the FOH mixer's trim controls, and this changes the FOH mix slightly. A way around this is to connect the FOH mixer's direct outs or insert sends to the line inputs of a separate recording mixer. Connect the recording mixer insert sends to the multitrack recorder(s) (Figure 17.8). Set recording levels with the recording mixer.

This method has some compromises. You need more cables and another mixer. Also, the signal goes through more electronics, so it is not quite as clean as connecting straight to the multitrack recorder.

Using the recording mixer's faders, you can set up a monitor mix. If you can hear the monitor mix well enough over headphones, you can even omit the multitrack recorder, and attempt a live mix to 2-track.

In a live mix, never turn off a mic completely unless you know for sure that it's not going to be used. Otherwise, you'll invariably miss cues. Turn down unused mics about 10 dB.

Splitting the Microphones

With this method, you plug each mic into a microphone splitter on stage. The mic splitter sends each mic's signal to two paths: the FOH mixer snake and the recording mixer snake. Some splitters have a third output

Figure 17.8 Connecting the FOH mixer to recording mixer to multitrack recorder.

to feed a stage monitor mixer. Back at the recording mixer where the snake is plugged in, assign each mic to a different track.

In most mic splitters, the signals are transformer isolated to prevent ground loops and interaction between the mixers. There is a ground-lift switch on each channel (Figure 17.9). Set it to the position where you monitor the least hum. Usually the mic-cable shields are grounded only to one console, which provides phantom power. The cable shields going to the other mixer are floated (disconnected) at the splitter with ground-lift switches.

Multitrack Recording in a Van

Here's the ultimate setup. Each mic is split three ways to feed the snake boxes for the recording, reinforcement, and monitor consoles. A long multiconductor snake is run to a recording truck or van parked outside the concert hall or club.

In the van, the snake connects to a mixing console, which is used to submix groups of mics and route their signals to a multitrack recorder.

Figure 17.9 Transformer-isolated microphone splitter.

Sometimes two multitrack recorders are run in parallel to provide a backup in case one fails.

Preparing for the Session

So far this chapter has given an overview of several on-location recording methods. The rest of this chapter explores the details of on-location pre-session procedures.

Ready to record a live gig? The recording will go a lot smoother if you plan what you're going to do. So sit down, grab a pen, and make some lists and diagrams as described here. We'll go over the steps to plan a recording.

Preproduction Meeting

Call or meet with the sound-reinforcement company and the production company putting on the event. Find out the date of the event, location, phone numbers, and e-mails of everyone involved; when the job starts; when you can get into the hall; when the second set starts; and other pertinent information. Decide who will provide the split, which system will be plugged in first, second, and so on. Draw block diagrams for the audio system and communications (comm) system. Determine who will provide the comm headphones.

If you're using a mic splitter, work out the splitter feeds. The mixer getting the direct side of the split provides phantom power for condenser mics that are not powered on stage. If the house system has been in use for a long time, give them the direct side of the split.

Overloud stage monitors can ruin a recording, so work with the sound-reinforcement people toward a workable compromise. Ask them to start with the monitors quiet because the musicians always want them turned up louder.

Make copies of the meeting notes for all participants. Don't leave things unresolved. Know who is responsible for supplying what equipment.

Figure 17.10 shows a typical equipment layout worked out at a preproduction meeting. There are three systems in use: sound reinforcement, recording, and monitor mixing. The mic signals are split three ways to feed these systems.

Figure 17.10 Typical layout for recording a concert.

Site Survey

If possible, visit the recording site in advance and go through the following checklist:

- Check the AC power to make sure the voltage is adequate, the third pin is grounded, and the waveform is clean.

- Listen for ambient noises: ice machines, coolers, 400-Hz generators, heating pipes, air conditioning, nearby discos, etc. Try to have these noise sources under control by the day of the concert.

- Sketch dimensions of all rooms related to the job. Estimate distances for cable runs.

- Turn on the sound-reinforcement system to see if it functions okay by itself (no hum, and so on). Turn the lighting on at various levels with the sound system on. Listen for buzzes. Try to correct any

409

problem so that you don't document bad sound-reinforcement sound on your recording.

- Determine locations for any audience/ambience mics. Keep them away from air-conditioning ducts and noisy machinery.
- Plan your cable runs from stage to recording mixer.
- If you plan to hang mic cables, feel the supports for vibration. You may need microphone shock mounts. If there's a breeze in the room, plan on taking windscreens.
- Make a file on each recording venue including the dimensions and the location of the circuit breakers.
- Determine where the control room will be. Find out what surrounds it—any noisy machinery?
- Visit the site when a crowd is there to see where there may be traffic problems.

Mic List

Now write down all the instruments and vocals in the band. If you want to put several mics on the drum kit, list each drum that you want to mike. As for keyboards, decide whether you want to record off each keyboard's output, or off the keyboard mixer (if any).

Next, write down the mic or direct box you want to use on each instrument (see Table 17.1).

Make copies of this mic list. At the gig, you'll place one list by the stage box, and the other by each mixing console. The list will act as a guide to keep things organized.

Track Sheet

Next, decide what will go on each track of your multitrack recorder. If you have enough tracks, your job is easy: Just assign each instrument or vocal to its own track: bass to track 1, kick to track 2, and so on.

What if you have more instruments than tracks? Suppose you have an 8-track recorder, but you have 15 instruments and vocals (including each part of the drum set). You'll need to assign several instruments or vocals to the same track. That is, you will set up a submix.

Let's say the drum kit includes a snare, kick drum, two rack toms, two floor toms, a hi-hat, and cymbals. If you want to mike everything

Table 17.1   Mic List (Example)

Input	Instrument	Mic
1	Bass	DI
2	Kick	AKG D112
3	Snare	Shure Beta 57
4	Hi-hat	Crown CM-700
5	Small rack tom	Shure SM57
6	Large rack tom	Shure SM57
7	Small floor tom	Senn. MD421
8	Large floor tom	Senn. MD421
9	Overhead L	AT 4051A
10	Overhead R	AT 4051A
11	Lead guitar	Shure SM57
12	Rhythm guitar	Shure SM57
13	Keyboard mixer	DI
14	Lead vocal	Beyer M88
15	Harmony vocal	Crown CM-311A

Table 17.2   Track Sheet (Example)

Track	Instrument
1	Drum mix L
2	Drum mix R
3	Bass
4	Lead guitar
5	Rhythm guitar
6	Keys mix
7	Lead vocal
8	Harmony vocal

individually, that's nine mics. But you don't need to use up nine tracks. At the sound check, assign or group those mics to busses 1 and 2 to create a stereo drum mix. Connect busses 1 and 2 to recorder tracks 1 and 2. Control the overall level of the drum mix with submaster faders 1 and 2 (also called bus faders or group faders).

Use tracks 3 through 8 for amps and vocals (as shown in Table 17.2 below). Feed tracks 3 through 8 from insert sends.

Block Diagram

Now that your track assignments are planned, you can figure out what equipment you'll need. Draw a block diagram of your recording setup from input to output (such as Figure 17.11). Include mics, mic cables, mic stands and booms, DI boxes, insert cables, multitrack recorder(s), outlet strip and extension cord, and recording media. You might bring your own mixer and snake, or use those from the house system. On your diagram, label the cable connectors on each end so you'll know what kinds of cables to bring. It's a good idea to keep a file of system block diagrams for various recording venues.

In Figure 17.11, the block diagram shows a typical recording method: feeding FOH console insert jacks to a multitrack recorder. We'll use this example in the rest of this chapter.

Equipment List

Generate a list of recording equipment from your block diagram. Based on Figure 17.11, you'd need the following recording gear:

Figure 17.11 Example block diagram of recording setup.

- Direct boxes, mics, mic cables, mic stands, and booms (unless these are part of the house system)
- Insert cables (for example, 8 stereo phone plugs to 8 RCA connectors)
- Multitrack recorder
- Outlet strip and extension cord
- Recording medium (bring enough for the duration of the gig)

Don't forget the incidentals: a cleaning tape, pen, notebook, flash-light, guitar picks, heavy-duty guitar cords, drum keys, spare tape, mic pop filters, gaffer's tape, guitar tuner, ear plugs, audio-connector adapters, audio cable ground-lift adapters, in-line pads, in-line polarity reversers, spare cables, gooseneck lights for the console, spare batteries, water, food—and aspirin!

Bring a tool kit with screwdrivers, pliers, soldering iron and solder, AC-outlet checkers, fuses, a pocket radio to listen for interference, ferrite beads of various sizes for RFI suppression, canned air to shoot out dirt, cotton swabs and pipe cleaners, and De-Oxit from Caig Labs to remove oxide from connectors.

Check off each item on the list as you pack it. After the gig, you can check the list to see whether you reclaimed all your gear.

Preparing for Easier Setup

You want to make your setup as fast and easy as possible. Here are some tips to help this process.

Protective Cases

Mount your console and recorders in protective carrying cases. Install casters or swivel wheels under racks and carrying cases so you can roll them in. Rolling is so much easier than lifting and carrying. You might permanently install the multitrack recorders in SKB carrying cases that act as racks. When a remote job comes up, just grab them and go.

A very helpful item is a dolly or wheeled cart to transport heavy equipment into the venue. Consider getting some lightweight tubular carts. Being collapsible, they store easily in your car or truck. One maker of equipment carts is Rock 'n' Roller, who advertises in the *Musician's Friend* catalog at www.musiciansfriend.com. Another cart is the Remin Kart-a-Bag at www.kart-a-bag.com.

Pack mics, headphones, and other small pieces in cloth bags, trunks, or milk crates.

> **CAUTION**: Keep tapes and hard drives separated from magnets, such as in headphones, monitor speakers, and dynamic mics. (However, tape erasure is much less of a problem with digital media such as DAT than it is with analog tapes.)

You might want to build a mic container: a big box full of foam rubber with cutouts for all the mics. Or construct a wheeled cabinet with drawers for mics, DIs, and speaker cables.

In an article in the September-October 1985 issue of *db magazine*, remote recording engineer Ron Streicher offers these suggestions:

> Especially for international travel, make sure your documentation is up to date and matches the equipment you're carrying. Make a list of everything you take: all the details, such as each pencil, razor blade, connector, etc. Also make sure your insurance is up to date. You need insurance for en route as well as at the destination.

> I organize my cases so I know where every item is. They're ready to go anytime and make setup much faster. The cables are packed with their associated equipment, not in a cables case. I check everything coming and going, and try to have 100% redundancy, such as a small mixer to substitute for the large console.

Mic Mounts

If you'll be recording a singer/guitarist, take a short mic mount that clamps onto the singer's mic stand. Put the guitar mic in the mount. Also bring some short mounts to clamp onto drum rims and guitar amps. By using these mounts, you eliminate the weight and clutter of several mic stands.

Some examples of short mounts are the Mic-Eze units by Ac-cetera (www.ac-cetera.com). They have standard 5/8-27 threads and mic clamps that either spring shut or are screw-tightened. Flex-Eze is two clamps joined by a short gooseneck, Max-Eze is two clamps joined by a rod, and Min-Eze is two clamps joined by a swivel.

Snakes and Cables

You can store mic cables on a cable spool, available in the electrical department of a hardware store. Wrap one mic cable around the spool, plug it into the next cable and wrap it, etc. No more tangled cables!

A snake can be wrapped around a large reel or can be coiled in a trunk. Commercial snake reels are made by such companies as Whirlwind, ProCo, and Hannay (www.hannay.com).

Use wire ties to join cables that you normally run together, such as PA sends and returns.

Snake hookup is quicker if the snake has a multipin twist-lock connector (such as Whirlwind W1 or W2). This connector plugs into a mating connector that divides into several male XLRs. Those XLRs plug into the mixing console. Leave the XLRs in the console carrying case. You'll find that the snake is easier to handle without the XLR pigtails on it.

For a clean, rapid hookup of drum mics, put a small snake near the drum kit, and run it to the main stage box. Or, put the main stage box by the drum kit. Snakes are made by such companies as Whirlwind, ProCo, and Horizon.

Check that all your mic cables are wired in the same polarity: pin 2 hot on both ends.

You might want to use 3-conductor shielded mic cables. Connect the shield to ground only at the male XLR end. Also use cables with 100% shielding. Those measures enhance the shielding capability of the shield and reduce pickup of lighting buzzes.

In XLR-type cable connectors, do not wire pin 1 to the shell, or you may get ground loops when the shell contacts a metallic surface.

Label all your cables on both ends according to what they plug into; for example, DSP-9 effects in, track 12 out, power amp in, snake aux 2 out. Or you might prefer to number the cables near their connectors. Cover these labels with clear heat-shrink tubing.

Label both ends of each mic cable with the cable length. Put a drop of glue on each connector screw to temporarily lock it in place.

Rack Wiring

You can speed the console wiring by using a small snake between the rack and the console, and between the multitrack recorders and the console. When packing, plug the snakes into the rack gear and multitracks, and coil the snake inside the rack and the multitrack carrying case. In other words, have all your equipment pre-wired. At the gig, pull out a bundled harness and plug it into the console jacks.

You might be feeding your multitrack from the insert jacks of the house console. If so, use a snake with TRS stereo phone plugs at the console end. Carry some TRS-to-dual-mono phone adapters to handle consoles that have separate jacks for insert send and return.

Some engineers prefer to make a clearly marked interface panel on the rear of the racks and plug into the panels. This is easier than trying to find the right connectors on each piece of equipment.

Small bands might get by with all their equipment in a single, tall rack. Mount a small mixer on top, wired to effects in the middle, with a coiled snake and power cord on the bottom.

Small snakes for rack and multitrack connections are made by Hosa, Horizon, and ProCo, among others.

Other Tips

Here are some more helpful hints for successful on-location recordings.

- Plan to use a talk-back mic from the board to the stage monitor speakers during sound checks. You might bring a small instrument amp for talkback so that you can always be heard.
- Hook up and use unfamiliar equipment before going on the road. Don't experiment on the job!
- Consider recording with redundant (double) systems so you have a backup if one fails.
- Walkie-talkies are okay for pre-show use, but don't use them during the performance because they cause RF interference. Use hard-wired communications headsets. Assistants can relay messages to and from the stage crew while you're mixing.
- During short set changes, use a laptop computer Local Area Network to show what set changes and mic-layout changes are coming up next; transmit this information to the monitor mixer and sound-reinforcement mixer.
- Don't put tapes or hard drives through airport X-ray machines because the transformer in these machines is not always well shielded. Have the tapes or hard drives inspected by hand.

- Hand carry your mics on airplanes. Arrange to load and unload your own freight containers, rather than trusting them to airline freight loaders. Expect delays here and at security checkpoints.
- Get a public-liability insurance policy to protect yourself against lawsuits.
- Call the venue and ask directions to the load-in door. Make sure that someone will be there at setup time to let you in. Ask the custodian not to lock the circuit-breaker box the day of the recording.
- A few days before the session, check out the parking situation.
- Just before you go, check out all your equipment to make sure it's working.
- Arrive several hours ahead of time for parking and setup. Expect failures—there's always something going wrong, something unexpected. Allow 50% more time for troubleshooting than you think you'll need. Have backup plans if equipment fails.

In general, plan everything in advance so you can relax at the gig and have fun!

At the Session: Setup

Okay! You've arrived at the venue. After parking, offload your gear to a holding area, rather than onstage, because gear on stage will likely need to be moved.

Learn the names of the house sound-crew members, and be friendly. These people can be your assets or your enemies. Think before you comment to them! Try to remain in the background and do not interfere with their normal way of doing things (for example, take the secondary side of the split).

Power Distribution System

At the job, you need to take special precautions with power distribution, interconnecting multiple sound systems, and electric guitar grounding.

Consider buying, renting, or making your own single-phase power distribution system (distro). It will greatly reduce ground loops and increase reliability. One source of AC power distribution equipment is www.furmansound.com. Figure 17.12 shows a suggested AC power distribution system.

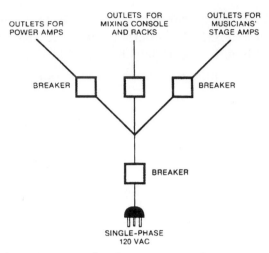

Figure 17.12 An AC power distribution system for a touring sound system.

The amp rating of the distro's main breaker box should exceed the current drain of all the equipment that will be plugged into the distro system.

Power Source

If you're using a remote truck, find a source of power that can handle the truck's power requirements, usually at a breaker panel. Some newer clubs have separate breaker boxes for sound, lights, and a remote truck. Find out whether you'll need a union electrician to make those connections. Label your breakers.

Check that your AC power source is not shared with lighting dimmers or heavy machinery; these devices can cause noises or buzzes in the audio.

The industry-standard power connector for high-current applications is the Cam-lok, a large cylindrical connector. Male and female Cam-loks join together and lock when you twist the connector ring. Distro systems and power cables with Cam-lok connectors can be rented from rental houses for film, lighting, electrical equipment, or entertainment equipment. One such rental house is Mole-Richardson at www.mole.com.

Use an adapter from Cam-lok to bare wires. Pull the panel off the breaker box, insert the bare wires, and connect the Cam-lok to your truck's power.

> **CAUTION**: Have an electrician do the wiring if you don't know what you're doing. A union electrician might be required anyway. Some breaker boxes have Cam-loks already built in.

To reduce ground-loop problems, get on the same power that the house sound system is using. From that point, run your distribution system, or at least run one or two thick (14- or 16-gauge) extension cords, to your recording system. These cords may need to be 100 to 200 feet long. Plug AC outlet strips into the extension cord; then plug all your equipment into the outlet strips.

If your recording system is one or two multitrack recorders that will connect to the FOH console, simply plug into the same outlet strip that the FOH console is using.

Measure the AC line voltage. If the AC voltage varies widely, use a line voltage regulator (power conditioner) for your recording equipment. If the AC power is noisy, you might need a power isolation transformer.

Check AC power on stage with a circuit checker. Are grounded outlets actually grounded? Is there low resistance to ground? Are the outlets correct polarity? There should be a substantial voltage between hot and ground, and no voltage between neutral and ground.

Some recording companies have a gasoline-powered generator ready to use if the house power fails. If there are a lot of lighting and dimmer racks at the gig, you might want to put the truck on a generator to keep it isolated from the lighting power.

Interconnecting Multiple Sound Systems

If you encounter an unknown system where balanced audio cables are grounded at both ends, you might want to use some cable ground-lift adapters (Figure 17.13) to float (remove) the extra pin-1 ground connection at equipment inputs.

If you hear hum or buzz when the systems are connected, first make sure that the signal source is clean. You might be hearing a broken snake shield or an unused bass-guitar input.

Figure 17.13 A ground-lift adapter for balanced line-level cables.

If hum persists, experiment with flipping the ground-lift switches on the splitter and on the direct boxes. On some jobs you need to lift almost every ground. On others you need to tie all the grounds. The correct ground-lift setting can change from day to day because of a change in the lighting. Expect to do some trial-and-error adjustments.

If the house system has serious hum and buzz problems, offer help. You can hear buzzes in your quiet truck that they can't hear over the main system with noise in the background.

Often, a radio station or video crew will take an audio feed from your mixing console. In this case, you can prevent a hum problem by using a console with transformer-isolated inputs and outputs. Or you can use a 1:1 audio isolation transformer between the console and the feeds. For best isolation, use a distribution amp with several transformer-isolated feeds. Lift the cable shield at the input of the system you're feeding. Some excellent isolation transformers are made by Jensen (www.jensen-transformers.com).

From the September/October 1989 db *magazine*, Guy Charbonneau, owner of Le Mobile, Hollywood, has this to say:

> The truck uses 3-phase, filtered 240V power with a 25kVA transformer having six different taps. I don't use the neutral; it carries a lot of current from lighting systems. The truck chassis is grounded. I use no ground lifts 99 percent of the time; I carry through the shield with the sound company. In some clubs, I bring 50 amps to the stage on a 220V line with a distribution box. All the musicians' instruments and the club console plug in there. This prevents ground loops and AC line noises from coffee machines and dishwashers. When working with a big P.A. company, I just ask for their split. Often a Y-adapter works. Source: AES

Connections

After unpacking, place one mic list by the stage box so you know what to plug in where. Place a duplicate list by each console. Attach a strip of white tape just below the mixer faders. Use this strip to write down the instrument that each fader affects.

Based on the mic list you wrote, you might plug the bass DI into snake input 1, plug the kick mic into snake input 2, and so on. Label fader 1 "BASS," label fader 2 "KICK," etc. Also plug in equipment cables according to your block diagram.

Have an extra microphone and cable offstage ready to use if a mic fails.

Don't unplug mics plugged into phantom power because this will make a popping noise in the sound-reinforcement system.

Running Cables

To reduce hum pickup and ground-loop problems associated with cable connectors, try to use a single mic cable between each mic and its snake-box connector.

Avoid bundling mic cables, line-level cables, and power cables together. If you must cross mic cables and power cables, do so at right angles and space them vertically.

Plug each mic cable into the stage box; then run the cable out to the each mic and plug it in. This leaves less of a mess at the stage box. Leave the excess cable at each mic stand so you can move the mics. Don't tape the mic cables down until the musicians are settled.

It's important that audience members do not trip over your cables. In high-traffic areas, cover cables with rubber floor mats or cable crossovers (metal ramps). At least tape them down with gaffer's tape pressed lengthwise onto the cables.

It helps to set up a closed-circuit TV camera and TV monitor to see what's happening on stage. You need to know when mics get moved accidentally, or when singers use the wrong mic, etc.

Recording-Console Setup

Here's a suggested procedure for setting up the recording system efficiently:

1. If the console is set up in a dressing room or locker room, add some acoustic absorption to deaden the room reflections. You might bring a carpet for the floor, plus acoustic foam or moving blankets for the walls.

2. Turn up the recording monitor system and verify that it is clean.

3. Plug in one mic at a time and monitor it to check for hums and buzzes. Troubleshooting is easier if you listen to each mic as you connect it, rather than plugging them all in, and trying to find a hum or buzz.

4. Check and clean up one system at a time: first the sound-reinforcement system, then the stage-monitor system, then the recording system. Again, this makes troubleshooting easier because you have only one system to troubleshoot.

5. Use as many designation strips as you need for complex consoles. Label the input faders bottom and top. Also label the monitor-mix knobs and the meters.

6. Monitor the reverb returns (if any) and check for a clean signal.

7. Make a short test recording and listen to the playback.

8. Verify that left and right channels are correct, and that the pan-pot action is not reversed audibly.

9. If you are setting up a separate recording monitor mix, do a preliminary pan-pot setup. Panning similar instruments to different locations helps you identify them.

Mic Techniques

Usually the miking is left up to the sound-reinforcement company. But there are some mic-related problems you should know about, such as feedback, leakage, room acoustics, and noise. Here are some ways to control these problems:

- Use directional microphones, such as cardioid, supercardioid, or hypercardioid. These mics pick up less feedback, leakage, and noise than omnidirectional mics at the same miking distance.

- With vocal mics, aim the null of the polar pattern at the floor monitors. The null (area of least pickup) of a cardioid is at the rear of the mic—180 degrees off-axis. The null of a supercardioid is 125 degrees off-axis; hypercardioid is 110 degrees.

- Mike close. Place each mic within a few inches of its instrument. Ask vocalists to sing with lips touching the mic's foam pop filter.
- Use direct boxes. Bass guitar and electric guitar can be recorded direct to eliminate leakage and noise in their signals. However, you might prefer the sound of a miked guitar amp. You could record the guitar direct from its effects boxes. Then use a guitar-amp emulator during mixdown. Note that sequencers and some keyboards have high-level outputs, so their DI boxes need transformers that can handle line level.
- Use contact pickups. On acoustic guitar, acoustic bass, and violin, you can avoid leakage by using a contact pickup. Such a pickup is sensitive only to the instrument's vibration, not so much to sound waves. The sound of a pickup is not as natural as a microphone, but a pickup may be your only choice. Consider using both a pickup and a microphone on the instrument. Feed the pickup to the house and monitor speakers, and feed the mic to the recording mixer.
- To reduce breath pops with vocal mics, be sure to use foam pop filters. Allow a little spacing between the pop filter and the mic grille. It also helps to switch in a low-cut filter (100-Hz highpass filter).

When you're recording a band that has been on tour, should you use their PA mics or your own mics? In general, go with their mics. The artists and PA company have been using their mics for a while and may not want to change anything. Most mics currently used in PA are good quality anyway, unless they are dirty or defective.

If you're not happy with their choice, you could add your own instrument mics. Let the PA people listen to the sound in the recording truck, or in headphones. If it sounds bad because of their mic choice, ask "Would it be okay if we tried a different mic (or mic placement)?" Usually it's all right with them—it's a team effort.

Electric Guitar Grounding

While setting up mics, you need to be aware of a safety issue with the electric guitar. Electric-guitar players can receive a shock when they touch their guitar and a mic simultaneously. This occurs when the guitar amp is plugged into an electrical outlet on stage, and the mixing console (to which the mics are grounded) is plugged into a separate outlet across the room. If you're not using a power distro, these two power points may be

423

at widely different ground voltages. So a current can flow between the grounded mic housing and the grounded guitar strings.

> **CAUTION**: Electric guitar shock is especially dangerous when the guitar amp and the console are on different phases of the AC mains.

It helps to power all instrument amps and audio gear from the same AC distribution outlets. If you lack a power distro, run a heavy extension cord from a stage outlet back to the mixing console (or vice versa). Plug all the power-cord ground pins into grounded outlets. That way, you prevent shocks and hum at the same time.

If you're picking up the electric guitar direct, use a transformer-isolated direct box and set the ground-lift switch to the minimum-hum position.

Using a neon tester or voltmeter, measure the voltage between the electric-guitar strings and the metal grille of the microphones. If there is a voltage, flip the polarity switch on the amp. Use foam windscreens for additional protection against shocks.

Audience Microphones

If you have enough mic inputs, you can use two audience mics to pick up the room acoustics and audience sounds. This helps the recording to sound "live." Without audience mics, the recording may sound too dry, as if it were done in a studio.

One easy method is to aim two cardioid mics at the audience. Put them on tall mic stands, on the stage floor, on either side of the stage. If those mic stands must not be seen, try hanging mics (Figure 17.14).

Some engineers pick up the audience and hall with two mics placed at the FOH mix position. That's an easy, effective way to capture the crowd. But the signal from these mics is delayed relative to the on-stage mics. If the FOH mix position is far from the stage (say, 50 feet or more), this delay will cause an echo when the audience mics are mixed with the stage mics. Prevent this by placing the audience mics fairly near the stage. Or, during mixdown, delay the stage-mic signals so that they coincide with the audience-mic signals. If you're using the audience mics only to capture applause, the delay is not an issue.

Here is another way to prevent this echo: Record the stage-mic mix on two tracks of a DAW. Record the audience mics on two other tracks.

SIDE VIEW OF MIC PLACEMENT

Figure 17.14 Some audience miking techniques.

Slide the stage-mic tracks to the right in time (that is, delay them) so that they coincide with the audience-mic tracks.

What if you don't have enough tracks for the audience mics? Record them on a 2-track recorder. Load this recording into your DAW along with the stage-mic mix. Align the two recordings in time as described above.

If the audience mics are run through the FOH mixer, leave the audience mics unassigned in that mixer to prevent feedback.

To get more isolation from the house speakers in the audience mics, use several mics hung close to the audience. Some engineers put up four audience mics maximum; some use eight to ten. Use directional mics and aim the rear null at the house speakers.

Another option is not miking the audience, or not using the audience tracks. Instead, during mixdown, you could simulate an audience with audience-reaction CDs. Simulate room reverb with an effects unit.

Setting Levels and Submixes

Now that the mics are set up, you might have time for a sound check. That's when you set recording levels. Have the band play a loud song. Locate a mixer input module that is directly feeding a recorder track. Set the input trim (mic preamp gain) to get the desired recording level on each track. On a multitrack recorder, you might set each track's level to peak around −10 dB, which allows for surprises if someone plays louder during the live gig.

Most of the mixer channels feed recorder tracks directly from the insert sends. On those channels, any fader moves during the gig will not affect the levels going onto the multitrack recorder. Why? In most mixers,

425

the insert send is pre-fader. That is, the signal at the insert send jack is not affected by the fader. However, you may encounter FOH consoles where some insert sends are tied up with signal processors. You must use those channels' direct-out jacks instead, which are usually post-fader (unless they can be switched to pre-fader).

Now let's set up the drum submix. (Ideally, you would do this with the PA turned down, and monitor over headphones or Nearfield monitors in a separate room.) Assign each drum mic to busses 1 and 2, and pan each mic as desired. Put the faders for busses 1 and 2 at design center—the shaded area about one-half to three-quarters of the way up. Set each drum-mic fader to about −10 dB. Set a rough drum-kit mix with the input trims while keeping the mixer meters around 0 dB or 0 VU. Fine-tune the drum mix with the faders, and set the recording level with the drum-mix bus faders.

Here's another way to create the drum submix. Have the drummer hit each drum repeatedly, one at a time, as you adjust the input trims to prevent clipping. For example, ask the drummer to bang on the kick drum. Turn down the kick drum's input trim all the way. Slowly bring it up until the clip LED (overload light) flashes. Then turn down the input trim about 10 dB to allow some headroom.

When all the drum trims are set, set a drum mix with the faders, and set the recording level with the bus 1 and 2 faders.

> **CAUTION**: Any changes to the drum mix or drum-mix level will show up on your recording.

Recording

If your recording will be synched later with a video tape, record a SMPTE time-code feed on a spare track on each recorder.

A few minutes before the band starts playing, start recording. Keep a close eye on recording levels. If a track is going into the red, slowly turn down its input trim and note the recorder-counter time where this change occurred.

> **CAUTION**: If you are recording off the FOH mixer, turning down its input trim will affect the house levels. The FOH mixer operator will need to turn up the corresponding monitor send and channel fader.

This is a touchy situation that demands cooperation. Ideally, you set enough headroom during the sound check so you won't have to change levels. But be sure the house mixer operator knows in advance that you might need to make changes. Ask the operator whether she wants to adjust the gain trims for you, so she can adjust corresponding levels at the same time. Thank the operator for helping you get a good recording.

If you are recording with a splitter and mic preamps on stage, assign someone to watch the levels and adjust them during the concert. Preamps with meters allow more precise level setting than preamps with clip LEDs.

Keep a track sheet and log as you record. For each song in the set list, note the recorder-counter time when the song starts, or press the Set Locate button on your multitrack recorder. Later, during mixdown, you can go to those counter times or locate points to find songs you want to mix. Also note where any level changes occurred so you can compensate during mixdown. It helps to note a counter time when the signal level was very high. When you mix the recording you can start at that point in setting your overall mix levels.

Teardown

After the gig, pack your mics away first because they may be stolen or damaged. Refer to your equipment list as you repack everything. Note equipment failures and fix broken equipment as soon as possible.

After you haul your gear back to the studio, it's time for mixing and editing.

Some of the information this chapter was derived from two workshops presented at the 79th convention of the Audio Engineering Society in October 1985. These workshops were titled "On the Repeal of Murphy's Law-Interfacing Problem Solving, Planning, and General Efficiency On-Location," given by Paul Blakemore, Neil Muncy, and Skip Pizzi and "Popular Music Recording Techniques," given by Paul Blakemore, Dave Moulton, Neil Muncy, Skip Pizzi, and Curt Wittig.

18

ON-LOCATION RECORDING OF CLASSICAL MUSIC

Perhaps your civic orchestra or high school band is giving a concert, and you would like to make a professional recording. Or maybe there's an organist or string quartet playing at the local college, and they want you to record them.

This chapter explains how to make professional-quality recordings of these ensembles. It describes the necessary equipment, microphone techniques, and session procedures.

Incidentally, recording classical-music ensembles is a great way for the beginning recording engineer to gain experience. With just two microphones and a 2-track recorder, much can be learned about acoustics, microphone placement, level setting, and editing—all essential skills in the studio.

Equipment

As a minimum, you need the following equipment to record classical music on-location:

- 2-track recorder
- Microphones

- Cables
- Mic stands and stand adapters
- Headphones (or powered speakers)
- Mic preamps (optional)
- Mixer (optional)

If you plan to record overseas, you may need a power converter that converts 50 Hz, 220V AC power to 60 Hz, 110V. Or power your equipment from batteries, and recharge the batteries overnight with a 220V/110V converter. You also need some AC power outlet adapters.

The 2-Track Recorder

Two-track recordings can be made on MiniDisc, CD-R, portable hard drive, laptop computer with a USB or FireWire port, DAT, or Flash memory recorder. Make sure that your recording medium can handle the length of the concert.

Microphones

Next on your list of equipment are some quality microphones. You need two or three of the same model number, or a stereo mic. Good mics are essential, for the microphones—and their placement—determine the sound of your recording. You should spend at least $100 per microphone, or rent some good ones, for professional-quality sound.

For classical-music recording, the preferred microphones are condenser types with a wide, flat frequency response and very low self-noise (less than 21 dB equivalent SPL, A-weighted). (Self-noise is explained in Chapter 6.)

These mics are available with an omni- or unidirectional pickup pattern. An omnidirectional mic is equally sensitive to sounds arriving from any direction, so it helps to add liveness (reverberation) to a recording made in an acoustically dead hall. Omni condenser mics have excellent low-frequency response, so they are a good choice for recording pipe organ or bass drum.

A unidirectional microphone (such as a cardioid) is most sensitive to sounds approaching the front of the microphone, and partly rejects sounds approaching the sides and rear. It helps reduce excessive reverberation in the recording. You need a pair of unidirectional mics if you want to do coincident or near-coincident stereo miking (see Chapter 7).

Stands versus Hanging

You can mount the microphones on stands or hang them from the ceiling with nylon fishing line. Stands are much easier to set up, but are more visually distracting at live concerts. Stands are more suitable for recording rehearsals or sessions with no audience present.

The mic stands should have a tripod folding base and should extend at least 14 feet high. You can purchase "baby booms" to extend the height of regular mic stands. Many camera stores have telescoping photographic stands that are lightweight and compact.

A useful accessory is a stereo bar or stereo microphone adapter. This device mounts two microphones on a single stand for stereo recording.

Hiding Microphones

In some live concerts—especially those that are videotaped—the microphones must not be seen. You might be able to hang some miniature condenser mics, or place boundary mics on the stage floor. If the musical ensemble is large (e.g., an orchestra), and you lay the mics on the stage floor, this placement usually overemphasizes the front row of the ensemble and results in a muffled sound. But if the ensemble is small (e.g., a string quartet or small choir), floor placement can work very well. You can also mount boundary mics on the ceiling or on the front edge of a balcony. These placements tend to sound too distant, but they may be your only option.

Monitors

For monitoring you can use either high-quality loudspeakers or headphones. The headphones should be closed-cup, circumaural (around the ear) types to block out the sound of the musicians. You want to hear only what's being recorded. Of course, the headphones should have a wide-range, smooth response for accurate monitoring.

Loudspeakers give more accurate stereo imaging than headphones. So you might want to set up monitor speakers in a control room separate from the concert hall. Place a pair of Nearfield monitors about 3 feet apart and 3 feet from you, on stands behind the mixer. An alternative is to use high-end consumer or professional loudspeakers placed several feet from the walls to weaken early reflections. You could add absorptive material such as acoustic foam to the walls behind and to the side of the speakers. For the best stereo imaging, sit exactly between the speakers, and as far from them as they are spaced apart.

Mic Cables

You have to sit far from the musicians to clearly monitor what you're recording. To do that, you need a pair of 50-foot mic cables. Longer extensions are needed if the mics are hung from the ceiling, or if you want to monitor in a separate room.

Mic Preamp or Mixer

You need a mixer when you want to record more than one source—an orchestra and a choir, for instance, or a band and a soloist. You might put a pair of microphones on the orchestra and another pair on the choir. The mixer blends the signals of all four mics into a composite stereo signal. It also lets you control the balance (relative loudness) among microphones. You also need a mixer if you want to use spot microphones or house microphones. Spot (accent) microphones are placed close to each orchestra section or soloist. House microphones are placed about 25 feet from the ensemble, back in the hall, to pick up hall reverberation.

For the cleanest sound, consider using some high-quality, stand-alone mic preamps instead of the preamps built into recorders and mixers. Place each preamp on stage; then run its line-level output signal back to your recording gear.

Other miscellaneous equipment you may need includes a power extension cord, an outlet strip, DAT dry cleaning tape, spare mic cables, pen and notebook, and gaffer tape or vinyl mats to keep cables in place.

Stereo Microphone Techniques

As a starting point, you place two or three mics several feet in front of the group, raised up high (Figure 18.1). The mic placement controls the acoustic perspective or sense of distance to the ensemble, the balance among instruments, and the stereo imaging.

Recall from Chapter 7 that there are four mic techniques commonly used for stereo recording: coincident pair, near-coincident pair, spaced-pair, and baffled omni techniques. Below is a review:

Figure 18.1   Typical microphone placement for on-location recording of a classical-music ensemble.

- Coincident-pair: Two directional mics angled apart with their grilles nearly touching and their diaphragms aligned vertically.
- Near-coincident pair: Two directional mics angled apart and spaced a few inches apart horizontally.
- Spaced pair: Two or three matched microphones of any pattern aiming straight ahead toward the ensemble and spaced several feet apart horizontally.
- Baffled omni: Two omnidirectional mics that are ear spaced and separated by a padded baffle or a hard-surface sphere.

Recall from Chapter 6 that boundary microphones can be mounted on clear plastic panels about 2 feet square. You can space these panels apart for spaced-pair stereo, or place them with one edge touching to form a "V." Aim the point of the V at the ensemble. This near-coincident arrangement provides excellent stereo imaging. Also available is a stereo boundary microphone that is smaller than the two panels and provides excellent stereo imaging, extended low-frequency response, and mono compatibility.

Preparing for the Session

Once you have the equipment, you are ready to go on location. First, ask the musical director what groups and soloists will be playing, where they will be located, and how long the program is.

If possible, plan to record in a venue with good acoustics. It should have low background noise and adequate reverberation time for the music being performed (typically 1.5 to 2 seconds). This is very important, because it can make the difference between an amateur-sounding recording and a commercial-sounding one. Try to record in an auditorium or spacious church rather than in a band room or gymnasium. If you're forced to record in a hall that is relatively dead, you might want to add artificial reverberation to the recording back in the studio.

Next, get all your equipment ready. If you're recording to DAT, fast-forward and rewind the blank DAT tape and clean the heads with a dry cleaning tape. Label the recording medium with the name of the artist, location, and date of the session. Check all cables and equipment for proper operation.

Keep your equipment inside your home or studio until you're ready to leave. A recorder left outside in a cold car may become sluggish if the lubricant stiffens, and batteries may lose some voltage.

Session Setup

Allow an extra hour or so for setup and for fixing broken cables, etc. There's always something unexpected in any new recording situation.

When you first arrive at the recording venue, locate some AC power outlets where you want to set up. Check that these outlets are "live." If not, ask the custodian to turn on the appropriate circuit breaker. Always check in with union technicians if the session is at a union venue.

Find a table or folding chairs on which to set your equipment. You might even sit in an audience seat with a portable recorder on your lap. Plug into the AC outlets and let your equipment warm up. Leave a few turns of AC cord near the outlet, and tape down the cord so that it isn't pulled out accidentally.

Then take out your microphones and place them in the desired stereo miking arrangement. As an example, suppose you are recording an orchestra rehearsal with two crossed cardioids on a stereo bar (the near-coincident method). Screw the stereo bar onto a mic stand, and mount two cardioid microphones on the stereo bar. For starters, angle them 110 degrees apart and space them 7 inches apart horizontally. Aim them down so that they point at the orchestra when raised.

You may want to mount the microphones in shock mounts or put the stands on sponges to isolate the mics from floor vibration.

As a starting position, place the mic stand behind the conductor's podium, about 12 feet in front of the front-row musicians. Connect mic cables. Raise the microphones about 14 feet off the floor. This prevents overly loud pickup of the front row relative to the back row of the orchestra.

Mic techniques for piano recitals and other solo instruments are covered in Chapter 8.

Leave some extra turns of mic cable at the base of each stand so you can reposition the stands. This slack also allows for people pulling on the cables accidentally. Try to route the mic cables where they won't be stepped on, or cover them with mats.

Make connections in one of the following ways:

- If you are using two mics, and your recorder has high-quality mic preamps, plug the mics directly into the recorder mic inputs.
- If you prefer to use an outboard mic preamp, plug the mics into the preamp, and plug the preamp output into the recorder line inputs.
- If you're using multiple mics and a mixer, plug the mics into the mixer mic inputs, and plug the mixer stereo outputs into the recorder line inputs.

Now put on your headphones, turn up the recording-level controls, and monitor the signal. When the orchestra starts to play, set the recording levels to peak around – 12 dBFS (decibels Full Scale).

Microphone Placement

Nothing has more effect on the production style of a classical-music recording than microphone placement. Miking distance, stereo positioning, and spot miking all influence the recorded sound character.

Distance

The microphones must be placed closer to the musicians than a good live listening position would be. If you place the mics out in the audience where the live sound is good, the recording probably will sound muddy and distant when played over speakers. That's because the recorded reverberation is condensed into the space between the playback speakers, along with the direct sound of the orchestra. Close miking (5 to 20 feet from the front row of the ensemble) compensates for this effect by increasing the ratio of direct sound to reverberant sound.

The closer the mics are to the orchestra, the closer it sounds in the recording. If the instruments sound too close, too edgy, too detailed—if the recording lacks hall ambience—the mics are too close to the ensemble. Move the mic stand 1 or 2 feet farther from the orchestra and listen again.

If the orchestra sounds too distant, muddy, or reverberant, the mics are too far from the ensemble. Move the mic stand a little closer to the musicians and listen again.

Eventually you'll find a "sweet spot" where the direct sound of the orchestra is in a pleasing balance with the ambience of the concert hall. Then the reproduced orchestra will sound neither too close nor too far.

Another way to vary the direct/reverb ratio is to mix the main pair of mics with distant house mics. Place the main pair just behind the conductor. Mix in a second pair of house mics placed about 23 to 52 feet behind the main pair. The house mics are cardioids aiming at the upper rear corners of the hall to pick up reverb, spaced about 12 feet apart and about 30 feet above the floor. Even a small amount of house mics in the mix will increase the sense of distance of the orchestra. Adjusting the direct/reverb ratio is best done back in the control room during mixdown.

Stereo-Spread Control

Now concentrate on the stereo spread. If the spread heard over headphones is too narrow, that means the mics are angled or spaced too close together. Increase the angle or spacing between mics until localization is accurate. Angling the mics farther apart makes the instruments sound farther away; spacing the mics farther apart does not, but may make the images less focused.

If the instruments that are slightly off-center are heard far-left or far-right in your headphones, your mics are angled or spaced too far apart. Move them closer together until localization is accurate.

You localize sounds differently with headphones than with speakers. For this reason, coincident-pair recordings have less stereo spread over headphones than over loudspeakers. Take this into account when monitoring.

You can test the stereo localization accuracy of your chosen stereo miking method. If you have time, record yourself speaking from various positions on stage while announcing your position: far-left, half-left, center, half-right, and far-right. Listen to the monitor system to check whether the image of your voice is reproduced in corresponding positions. Generally, the far-left and far-right positions should be reproduced at the left and right loudspeakers, respectively.

Soloist Pickup and Spot Microphones

Sometimes a soloist plays in front of the orchestra. By raising or lowering the stereo mic pair, you can control the balance between soloist and ensemble. If the soloist is too loud relative to the orchestra (as monitored), raise the mics. If the soloist is too quiet, lower the mics. You may want to add a spot mic about 3 feet from the soloist and mix it with the other microphones.

Many recording companies prefer to use several spot mics and a multitrack recorder when taping classical music. Such a method gives more control of balance and definition and is necessary in many situations. If you use spot or accent mics on various instruments or instrumental sections, mix them at a low level relative to the main pair—just loud enough to add definition, but not loud enough to destroy depth. Operate the spot-mic faders subtly or leave them untouched. Otherwise the close-miked instruments may "jump forward" when you bring up the fader, and then "fall back in" when you bring down the fader.

Spot mics sound more natural if you delay their signals according to this formula:

Delay in seconds = 1.25 × Miking distance in feet/1130 feet per second.

For example, if the main stereo mics are 12 feet from the ensemble, the spot-mic delay should be about $1.25 \times 12/1130 = 13$ milliseconds.

Solo the main stereo pair, and note the image locations of instruments that you are spot-miking. Pan the spot mics so their image locations coincide with those of the main mic pair.

Recording

Now that the mics are positioned properly, you're ready to record. At a live concert, you might want to set your recording levels to read about −15 dBFS with the opening applause. This procedure should result in approximately correct recording levels when the musicians start playing. Or set the record-level controls where they were at previous sessions.

Start recording a few seconds before the music starts. Once the recording is in progress, let the recording-level meters peak at −3 dB maximum. This allows a little headroom for surprises. Leave the recording level alone as much as possible. If you must adjust the level, do so slowly and try to follow the dynamics of the music.

If there is applause at the end of a musical piece, you can fade it out over 3 to 5 seconds by slowly turning down the recording-level controls or the mixer master volume control. Or leave it alone and do the fade-out while editing the program.

Most classical recording sessions are done in several takes. Write down the recorder counter times for these takes, and slate each one. Mark the keeper takes for later editing.

After the concert, pack the mics away first; otherwise, they may be stolen or damaged.

Editing

Once you have your tapes home, you may want to edit them to make a tight presentation. Editing is done with a Digital Audio Workstation (DAW), as described in Chapter 13.

Put about 3 or 4 seconds of silence between each selection. Or you may want to insert an interval of recorded "room sound" ("room tone") especially between movements of a symphony.

Some on-location recordings have a fair amount of background noise. If you insert silence between pieces, there will be an abrupt cutoff of this noise at the end of each piece. To prevent that, either fade the noise down to silence, or replace the silence with room tone.

If you recorded a live concert with applause, fade down the applause over several seconds. At the point where the applause fades to silence, edit that to a point a few seconds before the start of the next piece, and fade up there.

Note the start time of each piece. Finally, launch your CD burning software and enter Start IDs as described in Chapter 15.

Congratulations! You now have your finished product: a realistic, professional recording of a classical-music ensemble.

19

SURROUND SOUND: TECHNIQUES AND MEDIA

So far we've covered techniques that result in a 2-channel stereo recording. With stereo, you hear all the instruments and reverb in front of you, in the area between the two loudspeakers. But with surround sound, you hear audio images in every direction. For example, the musical ensemble could be up front, while the hall ambience envelops you from the sides and rear.

Stereo uses two channels feeding two loudspeakers in front of you. Surround sound uses multiple channels feeding multiple speakers placed all around you. A disadvantage of stereo is that you must sit in a tiny "sweet spot" to hear correct localization. In contrast, surround sound can be heard correctly in a wide area between the speakers.

Surround gives a wonderfully spacious effect. It puts you inside the concert hall with the musicians. You and the music occupy the same space—you're part of the performance. For this reason, surround is more musically involving, more emotionally intense, than regular stereo. There's a sense of envelopment. Surround mixing is fast becoming a valuable new tool to offer your customers. Because many listeners have home theater systems with multiple speakers, they are already set up to play surround audio recordings.

A magazine devoted to multichannel sound production is *Surround Professional* found at www.surroundpro.com.

Surround Speaker Arrangement

Inherited from the film industry, surround sound uses six channels feeding six speakers placed around the listener. This forms a 5.1 surround system, where the "point 1" is the subwoofer or low-frequency effects (LFE) channel. The LFE channel is band-limited to 125 Hz and below, while the other channels are full bandwidth (20 Hz to 20 kHz). "5.1" is a channel format—a surround-sound standard that states the number of channels, their frequency response, and the speaker placement.

The six speakers are:

- Left-front
- Center
- Right-front
- Left-surround
- Right-surround
- Subwoofer

Figure 19.1 shows the recommended placement of monitor speakers for 5.1 surround sound. It is the standard setup proposed by the International Telecommunication Union (ITU). From the center speaker, the left and right speakers should be placed at ±30 degrees, and the surrounds at ±110 degrees. (Some engineers prefer 120 to 125 degrees for the surround speakers.) If you are mixing movie soundtracks, use dipole speakers for the surrounds, and place them to the sides (±90 degrees).

The left-front and right-front speakers provide regular stereo. The surrounds provide a sense of envelopment due to room ambience. They also allow sound images to appear behind the listener. Deep bass is filled in by the subwoofer. Because we do not localize low frequencies below about 120 Hz, the sub can be placed almost anywhere without degrading localization.

Originally developed for theaters, the center-channel speaker is mounted directly in front of the listener. In a home-theater system, it is placed just above the TV screen, or just below and in front of the TV screen. This speaker plays center-channel information in mono, such as dialog.

Figure 19.1 Recommended placement of monitor speakers for 5.1 surround sound.

Why use a center speaker, when two stereo speakers create a phantom center image? If you use only two speakers and you sit off-center, the phantom image shifts toward the side on which you're sitting. But a center-channel speaker produces a real image, which does not shift as you move around the listening area. The center speaker keeps the actors' dialog on-screen, regardless of where the listener sits.

Also, the phantom center image does not have a flat frequency response, but a center speaker does. Why is this? Remember that a center image results when you feed identical signals to both stereo speakers. The right-speaker signal reaches your right ear, but so does the left-speaker signal after a delay around your head. The same thing occurs symmetrically at your left ear. Each ear receives direct and delayed signals, which interfere and cause phase cancellations at 2 kHz and above. A center-channel speaker does not have this response anomaly.

With a phantom center image, the response is weak at 2 kHz because of the phase cancellation just mentioned. To compensate, recording engineers often choose mics with a presence peak in the upper midrange for vocal recording. The center-channel speaker does not need this compensation.

441

For sharpest imaging and continuity of the soundfield, all the speakers should be:

- The same distance from the listener
- The same model (except the sub)
- The same polarity
- Direct-radiator types
- Driven with identical power amps
- Matched in sound pressure level with pink noise (the sub is 10 dB louder)

If you are mixing movie soundtracks, the surround speakers should be dipole designs rather than direct-radiator types. The dipole speakers project sound forward and backward to create a diffuse effect.

Typically the speakers are 4 to 8 feet from the listener and 4 feet high. Use a length of string to place the monitors the same distance from your head. The sub can go along the front wall on the floor. Be sure that all the speakers sound the same so there is no change in tonal balance as you pan images around.

A DSP algorithm to reproduce surround sound on headphones has been developed by Lake Technology. It is licensed to Dolby Labs under the name "Dolby Headphone." This feature is beginning to appear in consumer receivers and surround processors.

Setting Up a Surround Monitoring System

Working in surround, of course, requires more equipment than working in stereo. You'll need:

- Five "satellite" monitors and a subwoofer, and six channels of power amplification. Or five powered monitors and a powered sub.
- A sub/satellite crossover (a bass management system). This is built into some surround monitor packages.
- A six-channel volume control (hardware or software).

Be sure to include a subwoofer in your monitor system. If you don't, you might not hear low-frequency noises that a home listener with a sub will hear. These noises include breath pops, mic-stand thumps, air-conditioning rumble, and excessively heavy deep-bass notes.

You could use a home-theater surround receiver for power amplification and bass management. It has a single volume control that simultaneously adjusts the level of all the tracks. Most home-theater receivers have five amp channels and a line output that feeds a powered subwoofer. The sub's power amp should be at least 100 watts, and the receiver should have six analog inputs for your surround mix.

A feature to look for in surround sound receivers is Dolby Digital and Digital Theater Systems (DTS) decoders (explained later under the heading "Surround Encoding for DVD"). Receivers labeled "5.1 ready" or "Dolby Digital ready" are not Dolby Digital compatible. The receiver must have Dolby Digital and DTS decoders to play those formats. However, you don't need those decoders to do surround mixes.

Bass Management

In the surround receiver is a "bass management" circuit. Bass management is nothing more than a subwoofer/satellite crossover filter. It sends the deep lows to the sub, and sends other frequencies to the five "satellite" speakers that surround you.

A bass management circuit routes frequencies above about 100 Hz to the five full-range speakers, and routes frequencies below about 100 Hz from all six channels to the subwoofer. In other words, the bass management circuit routes low-frequency signals from all the five channels—and the LFE channel—to your sub. By keeping the deep lows out of the full-range speakers, bass management reduces their low-frequency distortion and lets them be made relatively small for home use. Note that bass management affects only what you monitor, not what you record.

Bass management can be done by a surround-receiver circuit, a standalone box, a special circuit in a subwoofer, or a software plug-in.

You set the bass management crossover frequency as low as possible to remove the directionality of the bass, and to extend the headroom of the sub. Typically you'd set the crossover frequency to the frequency where your full-range speakers are down 3 dB on the low end. If your five main speakers extend down to 20 or 30 Hz, you don't need bass management. In some receivers, the bass management crossover frequency is adjustable among 120/100/80 Hz—whatever your system needs for flattest response.

The crossover frequency is 120 Hz with the Dolby Digital standard and 80 Hz with the DTS standard. However, the Dolby Digital, DTS, and THX standard crossover frequencies are irrelevant to the frequency of the bass-management filter for the mixing work in your studio. And the

443

upper frequency limit of the LFE track (125 Hz) has nothing to do with the crossover frequencies of the bass management (120 to 40 Hz). As mentioned above, you set the crossover frequency according to the frequency response of your monitor speakers.

LFE Channel Filtering

The LFE channel lowpass filter used in bass management is a very steep filter (48 dB per octave) so it causes a lot of phase shift. But you can turn off the LFE filter in the encoder, and use your own gentler filter instead. Try 80 Hz, 24 dB per octave. Use any filter you like, as long as it removes all the energy at 125 Hz. Insert this filter between your console's LFE channel output and the input of the mixdown recorder's LFE track. In other words, record the LFE track pre-filtered, and turn off filtering in the encoder.

Surround Mixing Equipment

Here's what you need to mix in surround:

- Multitrack recordings in any format: analog tape, digital tape, hard disk, etc.
- If you're using a mixing console, it needs to have at least six output channels (also called busses or subgroups). Most digital consoles have a surround matrix, a section set up for mixing and monitoring in surround.
- If you're using a console, you need an 8-track recorder (MDM or hard-disk) to record the surround mix. Tracks 1 through 6 record the surround channels, while tracks 7 and 8 record a separate stereo mix.
- If you're using a Digital Audio Workstation (DAW) for surround mixing, you need an audio interface with at least 8 outputs. Set up the DAW to feed the 5.1 output channels to six of the interface outputs, then connect those outputs to your five satellites and subwoofer. M-Audio's Sonic Theater is a low-cost audio interface and software that accepts six audio channels from your computer (via USB) and sends them to your powered surround monitor system. It can replace the surround receiver mentioned earlier.
- If you're using a DAW, you need surround mixing software. For example, MOTU's Digital Performer and Cakewalk's Sonar Producer 4.0 include surround mixing. You could mix to six tracks on

your hard drive, and copy the six resulting wave files to a CD or DVD.

Connections

Now that you know the necessary equipment for surround mixing (including bass management), you can wire the system together. Two methods are described below. Method 1 uses an external multitrack recorder to record the surround mix tracks. Method 2 uses the DAW to record the surround mix tracks.

Method 1 using an external multitrack recorder: Basically you connect line-level signals from six busses to the associated tracks on your mixdown recorder. To monitor those six tracks, connect that recorder's outputs either to a surround receiver, or to a bass-management filter that feeds six channels of power amps. The receiver or amplifiers drive the five small speakers and the sub.

Figure 19.2 shows the connections. Patch the console's bus outputs or surround matrix outputs to the inputs of the 8-track mixdown recorder. On the back of this recorder, connect track outputs 1 through 6 to the inputs of the surround receiver or power amp inputs. On the receiver or power amps, connect the speaker outputs to the speakers. If your subwoofer is self-powered, connect the LFE channel line output to the sub

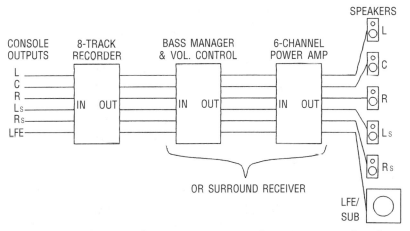

Figure 19.2 Mixdown and monitoring system for 5.1 surround with an external multitrack recorder.

445

line input. If your speakers are self-powered, connect them to outputs 1 through 6 of your bass manager.

When you connect your mixer or DAW to the 8-track mixdown recorder, which signal goes on which track? The most common track assignment (the Dolby Digital, ITU, and SMPTE standard) is given below:

1. Left-front
2. Right-front
3. Center
4. LFE or subwoofer channel
5. Left-surround
6. Right-surround
7. Stereo mix left
8. Stereo mix right

Be sure to label your tapes or CDs with the track assignment.

Method 2 using the DAW to record the surround mix tracks: Connections for powered monitors are shown in Figure 19.3, top. Set the overall listening level with your computer surround-monitoring software, and trim individual speaker levels at each speaker. Connections for

Figure 19.3 Mixdown and monitoring system for 5.1 surround, using a DAW as the multitrack recorder. Top: with powered monitors. Bottom: with passive monitors.

passive monitors are shown in Figure 19.3, bottom. Set the overall listening level with the volume control on the surround receiver. The bass-management filter is built into the powered sub or into the surround receiver.

Calibration

It's essential to calibrate your speakers so that their levels match. You need a sound level meter (a low-cost one from Radio Shack is adequate). Here's a suggested procedure:

1. Turn on the LFE channel and disable bass management.
2. Set each power amp channel to the same volume setting.
3. Get a pink noise signal from a console noise generator, receiver, or CD. Connect the signal to a console input. Other options are a surround calibration DVD, such as the Audio DVD Toolkit from www.goldline.com, or surround calibration software and an audio interface, such as M-Audio's Sonica Theater.
4. Assign that input channel to one surround channel (output bus or group fader). For example, start with the front-left channel.
5. Set the console's or DAW's bus faders and master faders to unity gain (the shaded portion of fader travel).
6. Adjust the input fader so the level on the mixer's meter is –20 dBFS (decibel Full Scale), the standard level for surround mixes. This level allows for some headroom.
7. Hold the sound level meter at your mixing position and set it to C-weighting, slow response. Aim the meter's mic at the speaker. Set the level of the power-amp channel (or the powered monitor) so you read 85 dBSPL on the meter.
8. Repeat steps 4 through 7 for each channel and its speaker, one at a time, so that all speakers are putting out the same level. (In most home-theater receivers is a menu that lets you set the level of each channel.) If you are mixing for theater sound with dipole speakers, set their level to 82 dBSPL instead.
9. Assign pink noise to just the LFE channel so that you hear the noise coming from the subwoofer. The playback level of the LFE channel is set 10 dB higher than the main speaker levels because that's done in consumer surround systems. Using the receiver's LFE gain control, or the sub's level control, set the sub's SPL to 95 dBSPL if

you're using an RTA, or to 89 dBSPL if you're using a sound level meter.

Why the difference? Unlike an RTA, an SPL meter integrates the energy of all the octaves it is measuring to come up with a single reading. The main speakers produce 8 to 10 octaves, but the sub produces 2 octaves. The sub's 2-octave energy reads lower on the SPL meter than the main speakers' 8- to 10-octave energy. So, instead of setting the sub's SPL to 95 dB, you set it to 89 dB to allow for this energy difference.

It's important to phase-align the subwoofer's signal with the full-range speakers' signals. Improper alignment can cause a dip in the system's frequency response at the crossover frequency. If your sub has a phase-matching control, send a sine wave to all speakers at the crossover frequency. Adjust the phase control to get the loudest sound at the mix position. If the sub has a polarity switch, try it both ways, and use the position that gives the loudest bass when all the speakers are playing.

10. After calibration, record your mixes at 0 dBFS maximum, and set your master monitor level as desired.

Here's another calibration method developed by Lorr Kramer of DTS and surround-sound expert Mike Sokol: Use bandwidth-limited pink noise (20 to 80 Hz) for the woofer and 500 to 2000 Hz for the main speakers. Set each speaker to the same SPL, one at a time. Using 500 Hz to 2 kHz pink noise for the main speakers keeps the signal out of the LFE channel and avoids exciting room modes. And limiting the top end to 2 kHz helps avoid directionality effects of the mic in the SPL meter.

Now that your monitors are calibrated, do the same for your mixing console or DAW recording software.

1. Feed pink noise into one channel input module.
2. Route or patch that module's signal equally to all six surround output busses.
3. Set each output bus fader to unity gain.
4. At the input module, set the trim to make the output level −20 dB on the console's meters.
5. Trim the console's ADAT or TDIF outputs (if used) so that all six tracks read −20 dBFS. (The LFE channel does not get 10 dB extra gain in the recording path, only in the monitoring path.)

6. On your surround receiver, set the small/large speaker switch to small, and enable bass management.

Recording and Mixing Pop Music for Surround

Doing a multitrack recording for later mixdown in surround is almost identical to recording for mixdown to stereo. Some engineers add a few ambience mics to pick up the studio acoustics. These ambience tracks are panned to the rear speakers during mixdown.

Many producers like to monitor the original recording session in stereo, not surround, to reduce the technical issues that can slow down the session. They wait until mixdown to monitor in surround.

Earlier in this book we showed some ways to mix multitrack recordings to 2-track stereo. Now let's look at mixing those same recordings to 5.1 surround.

Panning

To place the image of each track in space around you, some consoles include a pair of pan controls (left/right and front/rear) for each channel. Others use a trackball, mouse, or joystick. Some digital audio editing programs, or plug-ins for those programs, permit surround panning as well. For example, SmartPan Pro by Kind of Loud Technologies is a surround panner plug-in for Digidesign's Pro Tools. Minnetonka's SurCode is a program to mix 5.1 surround sound, including a "Build your own mixer" GUI and automated surround panners.

Suppose your console or recording software has 8 output busses, and you want to pan or move a track between front-left and rear-right. Assign the track to the front-left and rear-right busses. They should be odd-even numbered. When you turn the pan knob from left to right, the sound image should move as desired.

Where should you pan tracks in a surround mix? There are no set rules. Some producers like to put all the musicians up front, and put the hall ambience and applause in the rear. This works especially well for live concerts and recordings of classical music. This ambience can be created artificially with a digital reverb, or recorded with mics in the concert hall. Stereo reverb returns are typically sent to the rear channels. Several multichannel digital reverbs are available, both as hardware and as software.

Some producers use the rear speakers for background vocals, percussion, horn stabs, or strings, leaving the lead vocal and rhythm

449

instruments in front. A group of five singers could be panned to place one singer in each speaker, so that the listener is in the middle of the group. You might put the lead vocal in the center speaker, and also partly in the surrounds to pull the vocal out toward the listener. Spreading instruments between left-front and left-surround, and right-front and right-surround, gives a greater sense of envelopment.

You can place instruments in fixed positions (static panning), or move them around in space (dynamic panning). Some consoles let you move a sound by drawing its path in space on your monitor screen. You might move a sound along a circle, arc, or line. Then save these pans as part of your automated mix.

Using the Center Speaker

Some engineers prefer to send center tracks equally to the left and right speakers—not the center speaker. This creates a phantom center image. However, sending a track to the center speaker gives a better tonal balance and more stable imaging than the center phantom image.

It's not recommended to send a track only to the center speaker. Some listeners don't have a center speaker. So if you send, say, a vocal to just the center speaker, those listeners without a center speaker won't hear it. Also, home listeners can solo the center channel and hear punch-ins on that channel. So many producers route center tracks mainly to the left and right speakers, and also feed a little signal to the center speaker (maybe 6 dB down). Some producers don't use the center speaker at all.

Using the LFE Channel

With music mixes you seldom need to send anything to the LFE sub-woofer track—it gets deep bass from the bass management circuit in the listener's playback system, and at the proper level. Because the five main channels of 5.1 surround encoders reproduce from 20 Hz to 20 kHz, there's no need to put anything musical in the LFE channel. The consumer's bass management filter will redirect any sub-80-Hz energy to their own subwoofer anyway, whether or not you put this bass signal in the LFE track. On the other hand, suppose the end listener has a sub-woofer/satellite system with six channels, but no bass management system. In that case, the sound they hear will be thin in the bass if there is no signal in the LFE channel.

450

With film mixes, you typically send low-frequency sound effects (explosions, tornadoes, crashes) to the LFE channel. The LFE track also helps with DJ mixes and synth mixes that have lots of lows. Even if you mix only music, you need a subwoofer to hear the very deep noises (breath pops, room rumble) that a home listener will hear with a sub.

Downmixing

Downmixing is making a stereo mix from a 5.1 surround mix. It is done in the consumer's home theater receiver. In the downmixing circuit, the left and right surround channels are blended with the left and right front channels. The center channel is blended equally with the left and right channels. The LFE channel is either mixed with the front signals or not used.

Downmixes made this way seldom create a well-balanced stereo mix, so be sure to check your 5.1 mix for stereo compatibility. Surround monitoring systems should have a downmix button so you can hear how your surround mixes will sound when downmixed to stereo by consumer receivers.

It's best to do a separate stereo mix and record it on tracks 7 and 8 of your 8-track mixdown deck. This stereo mix can be put on DVD-Audio discs or Super Audio CDs (explained later) along with the surround mix.

Surround Mix Delivery Format

You will deliver your final 6-track recording to a mastering facility that will encode it onto a DVD. Supply them either the multitrack mixdown recording, or a DVD or CD-R with AIFF or WAV files, one for each output channel. Be sure to label the recording with the track contents. Identify each track with a recorded slate (front-left, center, sub, etc.). Print a 30-second test tone at 1 kHz at −20 dBFS on all tracks. Also include this information: LFE channel filtered or not, LFE channel filter frequency, mixdown listening level in dBSPL, sampling rate, bit resolution, SMPTE format (if included), media formats, program length, intended final audio sampling rate and bit depth, and a note about any glitches on your master.

Surround-Sound Mic Techniques

So far we've talked about mixing multitrack recordings to surround—a method intended mainly for pop music. But for classical music, you can

451

record in surround using five microphones, which capture the spatial character of the concert hall in which the musical ensemble is playing. Each mic feeds a separate track of a multitrack recorder.

Surround mic techniques are somewhat different from stereo mic techniques. In addition to the usual front-left and front-right mics, you need two surround mics to pick up the hall ambience, and sometimes a center mic to feed to the center channel. Note that listening in surround reduces the stereo separation (stage width) because of the center speaker, but mic techniques for surround are optimized to counteract this effect.

A number of mic techniques have been developed for recording in surround. Let's take a look at them.

Soundfield 5.1 Microphone System

This system is a single, multiple-capsule microphone (Soundfield ST250 or MKV) and Soundfield Surround Decoder for recording in surround. The decoder translates the mic's B-format signals (X, Y, Z, and W) into L, C, R, LR, RR, and mono subwoofer outputs.

Delos VR² Surround Miking Method

John Eargle, Delos' director of recording, developed their VR² (Virtual Reality Recording) format. Recordings made with this method offer discrete surround. They also are claimed to sound good in stereo and very good with "steered" analog decoding, such as Dolby Pro Logic.

In making these recordings, Eargle typically uses the mic placement shown in Figure 19.4. This method employs an ORTF pair in the center, flanked by two spaced omnis typically 12 feet apart. Two house mics (to pick up hall reverb) are placed about 23 to 52 feet behind the main pair. These house mics are cardioids aiming at the upper rear corners of the hall, spaced about 12 feet apart and about 30 feet high. Spot mics (accent mics) are placed within the orchestra to add definition to certain instruments.

The mics are assigned to various tracks of a digital multitrack recorder:

- 1 and 2: A mix of the coincident-pair mics, flanking mics, house mics, and spot mics
- 3 and 4: Coincident-pair stereo mic

Figure 19.4 A Delos surround miking method.

- 5 and 6: Flanking mics
- 7 and 8: House mics (surround mics)

NHK Method

The Japanese NHK Broadcast Center has worked out another surround miking method. They found that, for surround recording, cardioid mics record a more natural amount of reverb than omni mics. The mics are placed as described below:

- A center-aiming mic feeds the center channel.
- A near-coincident pair feeds front-left and front-right.
- Widely spaced flanking mics add expansiveness.
- Up to three pairs of ambience mics aim toward the rear.

453

Figure 19.5 An NHK surround-sound miking method.

Figure 19.5 shows the mic placement. The front mics are placed at the critical distance from the orchestra, where the direct-sound level matches the reverberant-sound level. Typically, this point is 12 to 15 feet from the front of the musical ensemble and 15 feet above the floor.

NHK engineers make this recommendation: When you're monitoring the surround program, the reverb volume in stereo listening should match the reverb volume in multichannel listening. That is, when you fold down or collapse the monitoring from 5.1 to stereo, the direct/reverb ratio should stay the same.

The KFM 360 Surround System

Jerry Bruck of Posthorn Recordings developed this elegant surround-miking method. It is a form of the mid-side (MS) stereo technique.

Bruck starts with a modified Schoeps KFM 6U stereo microphone, which is a pair of omni mics mounted on opposite sides of a 7-inch hard sphere. Next to those mics, nearly touching, are two figure-8 mics, one on each side of the sphere, each aiming front and back (Figure 19.6). This array creates two MS mic arrays aimed sideways in opposite directions. The mics do not seriously degrade each other's frequency response.

Figure 19.6 The KFM 360 Surround Miking System.

In the left channel, the omni and figure-8 mic signals are summed to give a front-facing cardioid pattern. They are also differenced to give a rear-facing cardioid pattern. The same thing happens symmetrically in the right channel. The sphere, acting as a boundary and a baffle, "steers" the cardioid patterns off to either side of center, and makes their polar patterns irregular.

By adjusting the relative levels of the front and back signals, the user can control the front/back separation. As a result, the mic sounds like it is moving closer to or farther from the musical ensemble.

According to Bruck:

> The system is revelatory in its ability to recreate a live event. Perhaps most remarkable is the freedom a listener has to move around and select a favored position, as one might move around in a concert hall to select a preferred seat. The image remains stable, without a discernible "sweet spot." The reproduction is unobtrusively natural and convincing in its sense of "being there.

The four mic signals can be recorded on a 4-track recorder for later matrixing. The figure-8 mics need some equalization to compensate for their low-frequency rolloff and loss in the extreme highs. To maintain a good signal-to-noise ratio, this EQ can be applied after the signal is digitized.

Five-Channel Microphone Array with Binaural Head

This method was developed by John Klepko of McGill University. It combines an array of three directional mics with a 2-channel dummy head (Figure 19.7):

Figure 19.7 The Klepko surround-sound miking method.

- For the front left and right channels: identical supercardioid mics
- For the center channel: a cardioid mic
- For the surround channels: a dummy head with two pressure-type omni mics fitted into the ear molds

The mics are shock mounted and have equal sensitivity and equal gains. Supercardioids are used for the front left/right pair to reduce center-channel buildup. Although the dummy head's diffraction causes peaks and dips in the response, it can be equalized to compensate. During playback, the listener's head reduces the acoustical crosstalk that would normally occur between the surround speakers.

According to Klepko:

The walkaround tests form an image of a complete circle of points surrounding the listening position. Of particular interest is the imaging between ±30 degrees and ±90 degrees. The array produces continuous, clear images here where other (surround) techniques fail.

The proposed approach is downward compatible to stereo, although there will be no surround effect. However, stereo headphone reproduction will resolve a full surround effect due to the included binaural head-related signals. Downsizing to matrix multichannel (5-2-4 in this case) is feasible except that it will not properly reproduce binaural signals to the rear because of the mono surrounds. As well, some of the spatial detail recorded by the dummy-head microphone would be lost due to the usual bandpass filtering scheme (100 Hz to 7 kHz) of the surround channel in such matrix systems.

DMP Method

DMP engineer Tom Jung has recorded in surround using a Decca Tree stereo array for the band and a rear-aiming stereo pair for the surround ambience (Figure 19.8). Spot mics in the band complete the miking. The Decca Tree uses three mics spaced a few feet apart, with the center mic placed slightly closer to the performers. It feeds the center channel in the 5.1 system.

Figure 19.8   A DMP surround miking method.

The rear-aiming mics are either a coincident stereo mic, another Decca tree, or a spaced pair whose spacing matches that of the Decca tree outer pair. Jung tries to aim the rear mics at irregular surfaces to pick up diffuse sound reflections.

Woszcyk Technique (PZM Wedge plus Opposite-Polarity, 180-Degree Coincident-Cardioid Surround Mics)

A recording instructor at McGill University, Wieslaw Woszcyk, developed an effective method for recording in surround that also works well in stereo. The orchestra is picked up by a PZM wedge made of two 18 × 29 inch hard baffle boards angled 45 degrees. A mini omni mic is mounted on or flush with each board. At least 20 feet behind the wedge are the surround mics: two coincident cardioids angled 180 degrees apart, aiming left and right, and in opposite polarity (Figure 19.9).

According to Woszcyk, his method has several advantages:

• Imaging is very sharp and accurate, and spaciousness is excellent due to strong pickup of lateral reflections.

Figure 19.9 Woszcyk surround miking method.

- The out-of-phase impression of the surround pair disappears when a center coherent signal is added.
- The system is compatible in surround, stereo, and mono. In other words, the surround signals do not phase-interfere with the front-pair signals. That is because (1) the surround signals are delayed more than 20 msec, (2) the two mic pairs operate in separate sound fields, and (3) the surround mics form a bidirectional pattern in mono, with its null aiming at the sound source.

If a PZM wedge is not acceptable because of its size and weight, other arrays with wide stereo separation may be substituted.

Williams Five Cardioid Mic Array

Michael Williams, an independent audio consultant, worked out the math to determine the best cardioid microphone arrangement for realistic reproduction of surround-sound fields. His method is shown in Figure 19.10.

Double MS Technique

Developed by Curt Wittig and Neil Muncy, the double MS technique uses a front-facing mid-side mic pair for direct sound pickup and a rear MS pair facing away from the front (Figure 19.11). The rear pair is placed at or just beyond the critical distance of the room-where the reverberant sound level equals the direct sound level. The matrixed outputs feed front-left, front-right, rear-left, and rear-right speakers. No center channel

Figure 19.10 Williams five cardioid mic array (multichannel microphone array or MMA).

Figure 19.11 Double MS technique.

mic is specified, but you could use the front-facing cardioid mic of the front MS pair for this purpose.

Surround Ambience Microphone Array

The surround ambience microphone (SAM) array was developed by Gunther Theile of the Institute für Rundfunktechnik (IRT). Four cardioid mics are placed 90 degrees to each other and 21 to 25 cm apart. No center channel is described.

Spider Microphone Array

This system uses a special mic mount with five arms that radiate out from a center point, like a star. At the end of each arm is a condenser mic aiming outward from the center. Two examples: The Microtech Gefell INA 5 uses five M930 mics in shock mounts. In the SPL Atmos 5.1/ASM 5 Surround Recording System, five Brauner condenser mics feed a five-channel mixing console, which adjusts the mic polar patterns and offers panning, bass management, and surround monitoring. SPL's Web site is www.spl-usa.com. Both systems use the Ideal Cardioid Arrangement (ICA 5, ITU-775 specification, Figure 19.12) developed by Volker Henkels and Ulf Herrmann.

Figure 19.12 Ideal cardioid arrangement (ICA) used in Brauner SPL Atmos 5.1/ASM5 Adjustable Surround Microphone.

The Holophone H2-PRO Surround Mic

This is a surround microphone using several omni mic capsules flush-mounted in a football-shaped surface. It captures up to eight channels of discrete surround sound and has eight XLR connectors. See the Web site: www.holophone.com.

Mike Sokol's FLuRB Array

This array uses four coincident cardioid mics at 90 degrees to each other aiming to the front, left, right, and back (Figure 19.13). The four mic signals feed a matrix processor that delivers the correct signals for 5.1 surround and up to 8.1 surround. The array is compact, relatively low cost, and convenient to use. Plus, it will sum to stereo or mono without phase cancellations.

Stereo Pair plus Surround Pair

In this method, the center-channel mic is omitted. You use a standard stereo pair of your choice to pick up the musical ensemble, plus another stereo pair of your choice to pick up the hall ambience. The hall mics feed the left- and right-surround channels. For example, two Crown SASS-P MKII microphones can be placed back-to-back, separated by several feet.

461

Figure 19.13 Mike Sokol's FLRB Array.

You might try a hybrid approach for a pop-music concert: Feed the front speakers a mix of multiple close-up mics on stage, and feed the rear speakers the signals from a rear-aiming stereo mic.

Surround Media

So far we've discussed surround recordings, and their creation either by multitrack mixdown or by surround mic techniques. Now let's turn our attention to the media that will play those surround recordings. They can be distributed to the public on CD, Super Audio CD (SACD), or DVD. Two DVD formats for surround sound are DVD-Audio and DVD-Video. Let's examine all these formats in detail.

Compact Disc

The compact disc is a 2-channel format that uses linear pulse code modulation (PCM) encoding. The CD's bit depth is fixed at 16 bits, and the sampling rate is fixed at 44.1 kHz. Storage capacity is 640 MB, or about 74 minutes of stereo audio. To fit the six channels of surround into the CD's two channels, two data-reduction encoding schemes are used: Dolby Digital or DTS. We'll explain them later under the heading "Surround Encoding for CD."

DVD

Another medium for playing surround recordings is the Digital Versatile Disc or DVD. It is a high-capacity optical storage medium the size of a compact disc (4.73 inches diameter). DVD can store digital data in three formats: audio, video, and computer.

DVD Compatibility

The DVD player reads:

- CD
- CD-R (in some units)
- CD-ROM
- DVD-Video disc (video plus audio)

A DVD-Video player can play a DVD-Audio disc if the latter has a Dolby digital version of the audio in the DVD-Video zone on the disc.

DVD Capacity

Compared to a CD, DVD has much greater capacity due to its smaller pits and closer tracks. The scanning laser has a shorter wavelength than in a CD player, which lets it read DVD's denser data stream. In addition, MPEG-2 data compression of the video data increases the data density on the DVD.

Some DVDs have a single layer of pits; others have a dual layer at different depths. The laser automatically focuses on the required layer. The Blu-Ray Disc records and plays up to 27 GB of data on a single-sided, single-layer CD-sized disc using a 405 nanometer blue-violet laser. It uses MPEG-2 video-recording format and AC3 and MPEG-1 Layer 2 audio-recording format.

Listed below are the storage capacities of various media:

- A compact disc holds 640 MB.
- A single-sided, single-layer DVD holds 4.7 GB (7 times CD capacity).
- A single-sided, dual-layer DVD holds 8.5 GB.
- A double-sided, single-layer DVD (DVD-RAM) holds 9.4 GB.

- A double-sided, dual-layer DVD holds 17 GB.
- A single-sided, single-layer Blu-Ray Disc holds up to 27 GB.

A DVD with 4.7-GB capacity is enough to hold a 2-hour and 10-minute movie plus subtitles.

DVD Players

Most DVD players have these outputs:

- 2-channel analog outputs playing a Dolby-Surround-encoded stereo signal
- A digital output that can be connected to an external decoder for 5.1-channel surround sound (either Dolby Digital or DTS)
- 6 analog outputs fed from built-in Dolby Digital or DTS decoders (in some units)

DVD Data Rate

The DVD data transfer rate is 10.1 megabits per second (mps). A 6-channel surround track, with data compression, consumes 384 or 448 kilobits per second (Kbps). Linear PCM stereo, at 16 bits and 48-kHz sampling rate, takes 1.5 mps. A 96-kHz sample rate doubles the transfer rate to 3 mps. A 24-bit resolution increases the rate to 4.5 mps.

Two forms of DVD are DVD-Video and DVD-Audio, and we'll look at them below.

DVD-Video

DVD-Video is a DVD format for videos: movies with a surround-audio soundtrack. The audio tracks can be Dolby Digital (AC-3) surround; 5.1 channels, with MPEG-1 compressed audio; or they can be 2 channels, 16- to 24-bit, 48- or 96-kHz, linear PCM audio. In other words, you can have surround sound with compromised fidelity, or 2-channel sound with excellent fidelity. A DVD disc can be encoded with both formats, and listeners can choose which one they want to hear.

Although the spec provides for multichannel PCM audio, current players have only 2 channels of PCM audio. All players support Dolby Digital (AC-3) surround.

Many software packages for DVD-Video authoring are easy to use and cost under $700. Some titles are Sonic Solutions' DVDit! and MyDVD, Roxio Toast 5 Titanium, MedioStream neoDVD Standard, Arcsoft Showbiz, Veritas PrimoDVDFormac Devideo, Cyberlink PowerDVD, and

RecordNow. Several manufacturers of DVD recorders offer bundled DVD authoring software.

DVD-Audio

DVD-Audio is a DVD format that features audio programs. It also can have optional still pictures (slide shows), Internet links, visual interactive menus, on-screen text and lyrics, and about 15 minutes of video clips.

DVD-Audio discs use PCM encoding. This encoding can be linear, as on CDs, or "packed" using Meridian Lossless Packing (MLP). MLP reduces the data rate up to 50%, but without any data loss: The reproduced signal is identical, bit for bit, with the original signal. While a standard CD is a 2-channel format fixed at 16-bit/44.1 kHz resolution, DVD-Audio is a multichannel surround format with higher resolutions. Compared to a standard CD, DVD-Audio permits:

- Better fidelity due to higher bit depth and higher sampling rate (up to 24-bit, 96 kHz with six channels, or 176.4 and 192 kHz with two channels).
- Longer playing time in some formats (up to 6 hours of 16-bit, 44.1-K stereo).
- More channels (up to six in the 5.1 scheme, allowing surround sound).
- Optional formats in addition to PCM, such as MPEG-2 BC, DTS, Dolby Digital, or Direct Stream Digital (DSD; the format used in the Super Audio CD).
- Option for different sampling rate and bit depth on different channels. DVD-Audio allows a wide range of sampling rates, number of channels, and quantizations. For example, a 4.7-GB, single-layer disc can hold a 75-minute program in which the left-center-right signals are 24-bit/88.2 kHz and the two surround channels are 20-bit/44.1 kHz.
- Option for different formats on different tracks. One track might be 16-bit/96 kHz surround, while another could be 24-bit/192 kHz stereo.
- Other optional media formats besides audio. For example, you can get video content from computer graphics, still photos, tracking over still photos, and mini-DV camera shots of concerts.

DVD-Audio content is stored in a separate DVD-Audio zone on the disc (the AUDIO__TS directory). The program can be copy protected by a digital signature signal and digital watermarking.

Most DVD-Audio players will also play DVD-Videos and CDs, and some will play Super Audio CDs. A DVD-Video player can play a DVD-Audio disc if the latter has a Dolby digital version of the audio in the DVD-Video zone on the disc. The digital outputs on a DVD-Audio player include PCM and Dolby Digital. Some units have DTS and DSD outputs, and all have multichannel analog outputs. Future players might have FireWire (IEEE 1394) connections.

You compile and record a DVD-Audio disc as you do a CD-R disc: with software, a hardware disc recorder, and blank discs. Specifically, you need DVD-Audio authoring software, a DVD-R recorder or DLT tape backup drive, and blank DVD discs. The write-once discs have dye on one side like a CD-R. Handling up to 4.7 GB on a single-sided disc, they can record DVD-Video, DVD-Audio, or DVD-ROM files.

Two major titles of authoring software are Sonic Solutions Studio HD and DVD Creator for Macintosh G3 or G4. If you want to make pure audio DVD-Audio discs, Sonic HD will do the job. Just click the mouse to assemble a playlist of surround-sound tracks. If you want to include video, etc., Sonic DVD Creator can handle it. In that case, compiling the DVD-Audio disc can become complex, with a variety of sampling rates, bit depths, channels, video, interactive menus, and so on. The Web site is www.sonic.com.

Another major DVD-authoring package is discWelder Bronze by Minnetonka software (www.minnetonkaaudio.com, www.discwelder.com). This low-cost program lets you record PCM stereo files up to 24-bit, 192 kHz and PCM 5.1 files up to 24-bit, 48 kHz to DVD-Audio, using a computer DVD recorder.

Super Audio CD

An alternative to DVD-Audio is the Super Audio CD developed by Sony and Philips. It uses the Direct Stream Digital (DSD) process, which encodes a digital signal in a 1-bit (bitstream) format at a 2.8224 MHz sampling rate. This system offers a frequency response from DC to 100 kHz with 120-dB dynamic range.

The Super Audio CD has two layers that are read from one side of the disc. On one layer is a 2-channel stereo DSD program followed by the same program in six channels for surround sound. On the other layer is a Red-Book 16-bit/44.1 K program, which makes dual inventories unnecessary. The 16-bit/44.1 kHz signal is derived from Sony's Super Bit Mapping Direct processing, which downconverts the bitstream signal with minimal loss of DSD quality. Standard CD players can play the 16-bit/44.1 kHz layer, and future players would be able to play the 2- or 6-channel DSD layer for even better sound quality.

The TASCAM DV-RA1000 records DSD audio or high-resolution DVD audio—up to 24-bit/192 kHz—to DVD blanks. This standalone unit also records standard CD-DA, wave, and DSDIFF files to CD and DVD discs.

Encoding Surround for Release on Various Formats

As we said, surround mixes can be released on CD, DVD-Video, DVD-Audio, or Super Audio CD (SACD). We need to consider how to get six channels of surround audio onto those formats.

In the past, DVD-Video discs and videocassettes used Dolby Surround—a matrix encoding system that combined four channels (left, center, right, surround) into two channels. The Dolby Pro Logic decoder in the consumer playback system unfolded the two channels back into four. The surround channel, which was mono and limited bandwidth, was reproduced over left and right surround speakers. Unfortunately, matrixing reduced the separation between channels because it mixed the channels together.

Recent encoding schemes let us encode six channels of surround audio onto disc, while keeping the channels discrete (not matrixed together). Let's look at how this is done for CD, DVD-Video, and DVD-Audio.

Surround Encoding for CD

A CD has only enough capacity for two channels of audio. So to get six channels of surround audio onto CD, data reduction is needed. Two data-reduction schemes are Dolby Digital and DTS.

Dolby Digital and DTS

Dolby Digital and DTS each use a different lossy encoding process to reduce the bit rate needed to transmit the six channels via a 2-channel

467

bitstream. "Lossy" means that the reproduced signal is missing some data that appeared in the original signal, so there is a slight loss in sound quality.

Why is this data reduction necessary? Audio at 44.1-kHz sampling frequency and 16-bit word length flows at a rate of about 700 Kbps for each channel. The rate for six channels is 4.2 Mbps. In order to handle this huge data flow, data reduction is needed. The data-reduction encoder in Dolby Digital is AC-3. It can be used on any number of channels.

First used in movie theaters, Dolby Digital and DTS are perceptual coding methods. They use data reduction (data compression) to remove sounds deemed inaudible due to masking. Dolby Digital's AC-3 encoder compresses the data about 12:1, while DTS compresses about 3:1. Both formats offer six discrete channels of digital surround sound. DTS resolution is 20-bit, while Dolby Digital is 16, 18, or 20 bits (perhaps 24 bits in the future). Dolby's AC-3 encoder accepts data at sample rates of 32, 44.1, or 48 kHz; DTS accepts 44.1 kHz. Compared to Dolby Digital, DTS uses less data compression and needs more data storage space and bandwidth. But some listeners say that DTS sounds more transparent—more like the original discrete tracks that were fed into the DTS encoder. Still, Dolby is the most common format.

To play a Dolby Digital or DTS recording, you need a newer DVD player or a CD player with a digital output. Plug the digital output into a decoder, which extracts the six channels of digital audio and converts them to analog. The decoder has a DSP chip that can decode both the DTS and Dolby Digital formats. Some surround receivers have DTS and Dolby Digital decoders built in.

Surround Encoding for DVD-Video

In a DVD-Video disc, much of the disc's capacity is taken up by video signals. So this format also can handle only two channels of audio. DVD-Video uses Dolby Digital's AC-3 scheme to encode (data-reduce) the six channels of surround into two channels. During playback, the two channels are decoded to recover the six channels. As we said, those six channels are separate or discrete—not matrixed.

Surround Encoding for DVD-Audio

A DVD-Audio disc has lots of capacity for audio signals. The six channels can be encoded without any data loss by using MLP.

Summary of Media Formats

Let's summarize the surround encoding schemes used in different media:

- CD uses DTS or Dolby Digital. Both are lossy systems.
- DVD-Video uses Dolby Digital, a lossy system.
- DVD-Audio uses MLP, a lossless system.

However, a DVD-Audio disc also can include Dolby Digital mixes so that it will play on CD players and DVD-Video players.

Table 19.1 summarizes the sample rate, the resolution (bit depth), and the stereo or surround systems associated with various media formats. Figure 19.14 summarizes the types of data handled by various formats of surround media and players.

Encoding Hardware and Software for CD and DVD-Video

With the CD and DVD-Video formats, the surround encoding can be done either by you or by the mastering house that manufactures the disc. To do it yourself you need a hardware or software encoder. Some examples of software encoders are Minnetonka Audio's SurCode CD Pro DTS and SurCode Dolby Digital, Sony Media Software's 5.1 Surround Plug-in Pack, Sonic Foundry's SoftEncode 5.1 Channel, or Kind of Loud's Smart-Code Pro/DTS plug-in for the Pro Tools platform. If you're using a hardware encoder, connect its inputs to the outputs of your 8-track mixdown deck, in parallel with the connections to the monitor system.

Surround-sound expert Mike Sokol suggested a way to put Dolby Digital or DTS encoded surround mixes on a CD-R. Load the six audio

Table 19.1 Parameters of Surround Media Formats

Format	Sample rate	Resolution	Stereo/surround systems
CD	44.1 kHz	16-bit	Stereo PCM. 16-, 20-, or 24-bit DTS (6 channels)
SACD	2.82 MHz	1-bit	DSD (2 or 6 channels)
DVD-V	48 kHz	16- or 24-bit	Stereo PCM, Stereo MPEG-2, Dolby Digital, or DTS (2 to 6 channels)
DVD-A	48 kHz	16-bit	Dolby Digital (6 channels)
	96 kHz	24-bit	PCM encoded with MLP (6 channels)
	192 kHz	24-bit	PCM encoded with MLP (stereo)

SURROUND MEDIA FORMATS

6 digital tracks in

ENCODERS

| Dolby Digital AC-3 Encoder (hardware or software) | OR | DTS Encoder (hardware or software) | OR | 6 channels of digital audio to SACD recorder |

6 channels of digital audio to DVD-A recorder

2 encoded surround channels out to CD or DVD recorder

MEDIA

CD disc
16-bit/44.1 kHz stereo PCM

Dolby Digital (AC-3) (not for commercial release) OR DTS

DVD-Audio disc
(PCM or MLP)

Up to 6 ch. 24/96 PCM OR Dolby Digital or DTS 6 ch. encoded to 2 ch. with data reduction

Graphics   DSD

DVD-Video disc

Dolby Digital 6 ch. encoded to 2 ch. with data reduction OR 2-channel PCM with no data reduction

SACD disc
DSD format (2.8 MHz, 1-bit)

2 ch. and 6 ch. DSD on one layer. 2 ch. stereo PCM (16-bit/44.1k) on 2nd layer.

PLAYERS

CD player with digital output

Digital out

Dolby Digital / DTS decoder

Six channels of analog out

DVD-A or DVD-V player

PCM 2 channel 16-24 bits 48 or 96 kHz

Dolby Digital (or DTS in DVD-A only)

Internal Dolby Digital decoder (or DTS decoder in DVD-A only)

SACD

DSD out

Digital out

2 ch. of analog out

6 ch. of analog out

External Dolby Digital or DTS decoder

SACD player

6 discrete DSD surround channels

2 stereo DSD channels

2 stereo PCM 16-bit/44.1k channels

6 ch. of analog out

2 ch. of analog out

2 ch. of analog out

Figure 19.14 Surround media formats.

tracks as WAV files into your computer and feed them to the encoding software. Select an AC-3 WAV file output or DTS WAV file output. Encode the tracks. Finally, burn the resulting WAV file onto a CD-R. To hear the surround mix, connect the CD player's digital output to a surround decoder, such as the one in a surround receiver. Commercially released CDs must use only DTS. Details are in Mike's upcoming book entitled *Surround Sound Production*.

DVD Pre-Mastering Formats

You don't necessarily have to buy an encoder; the encoding can be done by a mastering house or DVD manufacturer. Simply send them your 8-track tape. But you might want to put an encoder in your monitor system in order to hear the effect of the encoder as the end listener will hear it.

Here are acceptable media formats to send to a DVD manufacturer. For 2-channel stereo, use one of these formats:

* A removable hard drive with SMPTE time code
* DAT with SMPTE
* The digital audio tracks on the video master tape (D-1, D-5, Digital Betacam, DCT, or D-2)

For 5.1 channel surround, you can use one of these formats:

* 20-bit Alesis ADAT with time code.
* Genex magneto-optical 8-track 20-bit machine.
* DLT, DVD-R, or Sonic Solutions 24- or 20-bit Exabyte.
* Tascam DTRS (DA-38/88/78/98). Record the six audio channels on tracks 1 through 6. Put SMPTE on track 8, or put a stereo mix on tracks 7 and 8. For Dolby Digital, track assignments are 1, left; 2, center; 3, right; 4, LFE (low-frequency effects or subwoofer); 5, left surround; 6, right surround; 7, data channel; 8, data channel. For DTS the track assignment is 1, left; 2, right; 3, left surround; 4, right surround; 5, center; 6, LFE; 7, data; 8, data.
* Or use all eight Tascam DA-88 tracks to simulate six tracks with 20-bit resolution. This is done with a Prism MR-2024T processor.
* Pro Tools 24 with Jaz drives.
* Nagra-D is fine for four-channel surround.

If you want to create the LFE or 5.1 channel, you can mix the five other channels and feed them through a 120-Hz lowpass filter. Note that most music-only surround mixes don't use the LFE channel.

You might have video slide-show data that accompanies the audio. Store the video on a removable hard drive or send the file by modem on the Internet. TIFF and bitmap (BMP) formats are acceptable.

Dolby Units for DVD Mastering

This information is for those who produce the actual DVD. DVD mastering requires MPEG-2 video and Dolby Digital encoders. You can use either one for multichannel audio on DVD titles. Dolby Digital soundtracks are compatible with mono, stereo, or Dolby Pro Logic systems. Dolby encoders and decoders are described on Dolby's Web site at www.dolby.com/professional/index.html.

Special thanks to Mike Sokol for his generous technical help with this chapter.

PUTTING YOUR MUSIC ON THE WEB

You've recorded a new song and you can't wait for people to hear it. You might want to distribute it on the World Wide Web, in addition to (or instead of) CDs.

Why put your music online? Web excerpts of your songs can entice listeners to buy your CDs. Making songs available on the Web costs less than duplicating CDs and selling them in stores. Also, distributing online can be done much more quickly than distributing CDs. Keep in mind, though, that it's hard to get your songs noticed among the millions of other titles that are online.

Streaming versus Downloading

There are two basic ways to transmit music on the Internet: streaming files and downloading files. A streaming file plays as soon as you click on its title. A downloaded file doesn't play until you copy the entire file to your hard disk. Streaming audio is heard almost instantly (after the buffer memory fills), but usually sounds muffled and can be interrupted by net congestion.

The sound quality of streaming audio depends on the speed of the modem and ranges from funky AM radio (with a 56K modem) to near-CD quality (with a cable modem). With downloaded audio you must wait up to several minutes to download the song. But when it plays, the sound is high-fidelity and continuous.

Data Compression

Audio files that you record on your computer are usually WAV or AIFF files. They have no data reduction (data compression), so they take up lots of memory. A 3-minute song recorded at 16-bits, 44.1 kHz consumes about 32 MB. Downloading a 32-MB WAV file on the Web would take several hours using a 56K modem. So, audio files intended for Web download are generally data-reduced or data-compressed. For example, if you compressed a 3-minute song by 10:1 (as in an MP3 file), it would consume about 3.2 MB and would download 10 times faster than the equivalent wave file.

Most types of data compression tend to degrade sound quality. The sound quality of a compressed-data format depends on its bitrate, measured in kilobits per second (kbps). The higher the bitrate, the better the sound, but the greater the file size. A bitrate of 128 kbps for stereo MP3 files is considered "near-CD" quality. At low bitrates (below 128 kbps), you can hear artifacts such as "swirly" cymbals, smeared transients such as drum hits, a general "phasey" effect, and less treble. At higher bitrates above 200 kbps, the artifacts start to disappear, and to most listeners the sound is CD quality.

A stereo MP3 file encoded at 128 kbps is 64 kbps per channel. A mono MP3 file encoded at 64 k is equal in quality to a stereo MP3 file encoded at 128 k. Mono files consume half the file space of stereo files if both are the same bitrate.

Table 20.1 relates MP3 stereo bitrate to sound quality.

Table 20.1 Data Compression Ratio and Sound Quality of Various Bitrates

Bitrate	Compression	Sound Quality
8 kbps	176:1	CB radio quality
64 kbps	20:1	AM radio quality
128 kbps	10:1	Near-CD quality (22-kHz bandwidth)
320 kbps	4.4:1	CD quality

Web-Related Audio Files

You can put audio on the Web in several file formats. Let's look at some of the file types in current use.

- WAV: A standard PC format for audio files. It encodes sound without any data reduction by using pulse code modulation. Audio CD resolution is 16-bit, 44.1 kHz. Because wave files consume a lot of memory (about 32 MB for a 3-minute song in stereo), they are seldom used on the Internet for songs because they take so long to download. This may change when we start using the much faster Internet 2, which employs fiber-optic cables.

- Audio Interchange File Format (AIFF): A standard Mac format for audio files. Like wave files, AIFF files are not data-compressed.

- MP3 (MPEG Level-1 Layer-3): A format that stores audio in a small space with high quality. In an MP3 file (.mp3), the data has been compressed or reduced to one-tenth of its original size or less. Compressed files take up less memory, so they download faster. You download MP3 files to your hard drive, then listen to them. MP3 audio quality at a 128-kbps rate is nearly the same as that of CDs (depending on the source material).

- MP3Pro: An improvement over MP3. Songs encoded at 64 kbps with MP3Pro are said to sound as good as songs encoded at 128 kbps with MP3. MP3Pro offers faster downloads and nearly doubles the amount of music you can put on a Flash memory player. MP3 and MP3Pro files are compatible with each others' players, but an MP3Pro player is needed to hear MP3Pro's improvement in sound quality.

- MP3 with VBR: MP3 encoding done with variable bitrate, which depends on the requirements of the signal from one instant to the next. VBR reduces file size but keeps the audio quality high.

- RealAudio: A highly compressed audio file format used for streaming audio. Generally, RealAudio has lower fidelity (less treble) than MP3, but the fidelity depends on modem speed and the current Internet bandwidth. RealAudio files (.ra or. rm) are often used as short excerpts or previews of songs.

- Windows Media Audio (WMA): Another popular compressed audio file format for streaming audio and for downloads. Windows Media

9 promises performance similar to MP3Pro: near-CD quality at 48 kbps and CD quality at 64 kbps. For more information see www.microsoft.com/windows/windowsmedia/default.aspx.

- OGG (Ogg Vorbis): A new compressed audio format, Ogg Vorbis is license-free open software. For a given file size, Vorbis sounds better than MP3. Vorbis takes up less file size than MP3 files of equal quality. For more information see www.xiph.org and www.vorbis.com.

- MPEG Advanced Audio Coding (AAC): A relatively new compressed audio file format, AAC is intended to replace MP3. AAC offers better sound quality than MP3 at the same bitrate. Many listeners claim that AAC files made at 128 kbps sound like the original uncompressed audio source. What's more, AAC supports multi-channel audio and a wide range of sample rates and bit depths. It's also used with Digital Rights Management technology by helping to control the copying and distributing of music.

- MID (MIDI, or Musical Instrument Digital Interface): A string of numbers that represent performance gestures, such as note on/off, which note is played, key velocity, and so on. See Chapter 16. Because MIDI files do not include audio, they are very compact.

- Rich Music Format (RMF): A MIDI file that might have digital audio embedded. It's used with the Beatnik browser plug-in, which plays RMF files using sampled instruments or custom samples triggered by the MIDI data. Beatnik features interactive control of the music.

There are many other formats—some of which include video—but MP3, Windows Media, and RealAudio are currently the most popular audio formats on the Internet. For more information on using Windows Media, see www.microsoft.com/windows/windowsmedia/.

Currently, Windows Media and RealAudio are the most popular streaming types, and MP3 and Windows Media are the most popular download types. *CD track 41 demonstrates MP3 and WMA data compression.*

What You Need

To put your music on the Web, you need to download a few pieces of software, which are low-cost or free:

- Audio editing software: A program that takes an audio signal through a sound card and records the audio on your hard drive. Most programs let you edit the audio on-screen with a mouse, then save the edited program as a WAV or AIFF file. Examples are Windows Sound Recorder, MusicMatch Jukebox from www.musicmatch.com, Adobe Audition, Cakewalk Home Studio and Sonar, emagic Logic, Steinberg Cubase, MOTU Digital Performer, SAW Studio, and Pro Tools.

- Ripper: A program that converts audio from a CD or CD-R to a WAV file. Examples are MusicMatch Jukebox from www.musicmatch.com, Xing Audiocatalyst from www.real.com/accessories, and Windows Media Player from www.microsoft.com/windows/windowsmedia. Not all CD-ROM drives support ripping.

- MP3 Encoder: A program that converts a WAV or AIFF file to an MP3 file. Examples are AltoMP3 from www.Yuansoft.com, SoloH mpeg Encoder from www.mp3proclub.com, and dBPower Amp from www.dbpoweramp.com. Some other sites for encoders are www.team-mp3.com and www.liquidaudio.com. Some audio editing software includes an MP3 encoder.

- WMA Encoder: A program that converts a WAV or AIFF file to a WMA file. It's a free download from Microsoft at www.microsoft.com/windows/windowsmedia/download/default.asp. Select Windows Media Encoder. Windows Media Technologies has Digital Rights Management, which limits who can hear your music. Another source is www.team-mp3.com.

- Ogg Vorbis Encoder (optional) from www.vorbis.com. Converts CD tracks or wave files to Vorbis format. For a given file size, Vorbis sounds better than MP3. Vorbis takes up less file size than MP3 files of equal quality.

- RealAudio Encoder (optional): A program that converts a WAV or AIFF file to a RealAudio file for streaming. Examples are RealProducer or RealProducer Plus from www.real.com/products/tools/producer/index.html.

To listen to your music that you put on the Web, you need:

- MP3 Player: A program or a device that plays MP3 files. Some software player examples are MusicMatch Jukebox from www.musicmatch.com, Winamp from www.winamp.com, Windows Media

Player from www.microsoft.com/windows/mediaplayer, RealPlayer from www.real.com, MacAMP from www.macamp.net, and QuickTime-4 in Apple's OS. Portable MP3 players play MP3 files downloaded from the Web.

- RealAudio Player: A free program that plays RealAudio streaming files. Examples are RealNetwork's RealPlayer from www.real.com or Windows Media Player (below).

- Windows Media Player: Although intended to play mainly Windows Media files, this free program plays many file types: .avi, .asf, .asc, .rmf, .wav, .wma, .wax, .mpg, .mpeg, .m1v,.mp2, .mp3, .mpa, .mpe, .ra, .mid, .rmi, .qt, .aif, .aifc, .aiff, .mov, .au, .snd, .vod. This player is a free download from www.microsoft.com/windows/windowsmedia/download/default.asp. Another source is www.winamp.com.

- Ogg Vorbis player (optional) such as Winamp.

All-in-one software combines several of these functions in one program. RealJukebox Plus from www.real.com and MusicMatch Jukebox from www.musicmatch.com are MP3 player/ripper/ encoders. Xing-Tech's Audio-Catalyst from www.real.com\accessories is a ripper/encoder that converts CD data directly to MP3. Another source for players, rippers, and encoders is www.team-mp3.com. Some digital audio editing programs can output MP3, WMA, and RealAudio files.

How to Upload Compressed Audio Files

Got everything you need? Let's go. Basically, here's what you will be doing:

1. Convert your song from a cassette, DAT, or CD to a WAV file by using audio editing software. Or convert your song from a CD to a WAV file by using a ripper.

2. Edit and process the WAV file to optimize it for the Internet.

3. Then use an encoder to convert the WAV file to an MP3, RealAudio, or WMA file.

4. Finally, send (upload) the converted file to a Web site that features music in the MP3, RealAudio, or WMA format.

Let's go over the procedure in more detail. You can substitute WMA or RealAudio for MP3 in the procedure below.

1. Start with a cassette, DAT, or CD recording of a song.
2. If you have a cassette, plug your cassette-deck line outputs into your sound card's line input. If you have a DAT, plug your DAT recorder analog outputs into your sound card's line input. If your sound card has a digital input, connect it to your DAT recorder's digital output.
3. If your source is a cassette or DAT, use the editing software to copy the recording to your hard disk. If the source is a CD or CD-R, put it in your CD-ROM drive. Turn off all other programs. Then use a ripper to convert the CD audio file to a WAV file on your hard drive. (Not all CD-ROM drives support ripping.)
4. Edit the start and stop points of the song. You might want to edit a 30-second excerpt of a song to use as a preview of your music. If so, add a fade-in and fade-out to the preview.
5. Next, you might want to process the audio so that it will sound louder and clearer when played on the Web. To do this, reduce the bandwidth: apply low cut below 40 Hz and high cut above 15 kHz. You might want to try a higher frequency low cut and a lower frequency high cut. Apply compression or multiband compression, apply peak limiting, then normalize the song to maximize its level.
6. Save the edited, processed song as a WAV file (PC) or AIFF file (Mac) on your hard drive.
7. Use an MP3 encoder to convert the song's WAV or AIFF file to an MP3 file on your hard drive. You might want to use a bitrate of 128 kbps, which many Web sites require. It's a good compromise between file size and sound quality. A 3-minute MP3 song encoded at 128 kbps typically is 3 MB in size. Higher bitrates give better sound but longer download times. A file encoded at 256 kbps sounds the same as CD, but is one-sixth the file size.

You might want to convert the excerpt's WAV or AIFF file to a RealAudio file by using a RealAudio encoder. **Note**: Some Web sites—such as www.iuma.com—automatically create a RealAudio file from your uploaded MP3 file.

8. Next you will upload your MP3 files to an MP3 server—a Web site that accepts MP3 files for distribution on the Web. Examples are www.iuma.com, www.peoplesound.com, or www.AWAL.com. (Also see the MP3 Links page at www.team-mp3.com.) Some sites offer free downloads of their music files; others charge the listener so that you make some income.

Point your Web browser to the desired site. Then click on the button labeled "Artists Only," "Submit your music," or something similar. This is where you upload your MP3 files.

9. Once you sign up and fill out some forms, click on "Upload" (or whatever). With a 56K modem, an upload of one 3-minute song might take up to 30 minutes. You can also upload scanned photos of your band, your album cover, and text describing your band and its music. Some sites take a few days to approve your songs.

Congratulations—you're on the Web!

Putting Your Music on Your Web Site

Does your band have a personal Web site? You'll want to put samples of your music there. So far I've covered how to upload your songs to an MP3 server such as www.iuma.com. Now I'll explain how to put your songs on your own Web site. People visiting your site can click on a song title to hear a streaming song preview, or to download an entire song.

Let's start by creating a Web-page link to each MP3 file. It's easy. You can create the link with Web-page design software or with one line of HTML code.

Putting MP3 or WMA Files on Your Site

Again, you can substitute WMA for MP3 in the procedure below.

1. Suppose you have a song on CD called "Blues Bash." Using ripper software, transfer that song to your hard drive as a WAV file (PC) or AIFF file (Mac). In this example, we'll call the file BLUES.WAV. If you have a ripper/encoder, convert that song to an MP3 or WMA file and go to Step 3.

2. Use an MP3 encoder to convert the WAV file to an MP3 file. I recommend a setting of 128 kbps bitrate, stereo for MP3 files. This setting gives hi-fi sound with a relatively small file, so it's fairly quick to download. You now have an MP3 file called BLUES.MP3.

3. In your computer, move BLUES.MP3 to the same directory on your hard drive that your Web page is in. Use Web-page design software to open your band's Web page. Here we'll call it PAGE1.HTM. On

that page, where you want the song title to appear, type in the title of the song. In this example it's "Blues Bash." Link that title to BLUES.MP3.

When you left-click on the song title, BLUES.MP3 should load and play. When you right-click on the song title, you can select "Save Target As" to copy the MP3 file to a directory on your hard drive.
If you want to use HTML code:

1. Open your Web page with a browser such as Windows Explorer.
2. Select View > Source. You'll see the HTML code for that page in plain text format.
3. Find a spot on the page where you want the song title to appear. Type in the title of the song and the link to its MP3 file. In HTML it would look something like this:

```
<a href = "blues.mp3">Blues Bash</a>
```

For a WMA file it would be

```
<a href = "blues.wma">Blues Bash</a>
```

4. Save and close the text file, go to Windows Explorer, and select View > Refresh. You should see the link to the MP3 or WMA song. When you left-click on the link, you should hear the song. When you right-click on it, you can select "Save Target As" to copy the MP3 file to a directory on your hard drive.

Now that your song links are working correctly on your computer, its time to upload them.

1. Upload (send) PAGE1.HTM and BLUES.MP3 to your Web server. Make sure that both files go to the same directory. Some Web servers have an upload page for this purpose. If yours doesn't, you can upload the files using file transfer protocol (ftp) client software. Do a Web search for ftp freeware, such as www.freshmeat.net/search/?fourohfour=1&q=ftp.html. **Note**: Some Web servers do not allow MP3 files. Find the page on the Web server site that tells what files they permit.
2. Now start your Web browser and go to PAGE1.HTM on your site. Click "Refresh" in your browser so that you see the PAGE1.HTM that you just uploaded.

3. Left-click the song title (in this case, "Blues Bash"), and it should play after the buffer fills. Or right-click the song title and select "Save Target As." The song should download over a few minutes. Then you can play it with your MP3 player or WMA player in glorious hi-fi stereo.

Putting RealAudio Files on Your Site

Now let's consider streaming audio. If you are running your own Web site from your own computer, you need RealAudio server software, such as RealAudio Basic Server at www.real.com/products. Setting up your own server is beyond the scope of this book.

If your Web site is hosted on someone else's computer, you don't need server software. For example, if you have a free Web site at www.Geocities.com, they already have a RealAudio server, so you don't need one.

Download the free RealProducer software from www.real.com. RealProducer Plus is another option. Also download the latest free RealAudio player (RealPlayer) from www.real.com.

Now decide what music you want to stream. Generally, you'll want your streaming audio files to be short samples or "clips" of your music, but they can be entire songs. Let's start by creating a short sample.

1. Suppose you have a song on CD called "Blues Bash." Using ripper software, transfer your song onto your hard drive as a WAV file (PC) or AIFF file (Mac). In this example, we'll call the file BLUES.WAV.

2. Load BLUES.WAV into your digital editing software. Trim it to about 30 seconds long. Fade out the music toward the end. In this example, let's save the edited song as BLUES2.WAV.

3. Use RealProducer to convert BLUES2.WAV into BLUES2.RM, which is a RealMedia file.

4. Next, use a word processor or text editor (such as Notepad) to create a small text file that points to the location of your RealMedia file; for example; www.webserver.com/mywebsite/blues2.rm.

Fill in your own Web server and the URL of your Web site. Save this line in your Web page directory on your hard drive as a plain ASCII text file with a.RAM extension. In this case, you'd call it BLUES2.RAM. This type of file is called a metafile.

5. Using Web-page design software, open your Web page on your hard drive. Here we'll call it PAGE1.HTM. Find a spot where you want the song title to appear. Type the song title, such as "Blues Bash." Link that title to BLUES2.RAM. In HTML it would look something like this:

```
<a href = "blues2.ram">Blues Bash</a>
```

In your Web page, left-click on the song link. It should play.

Important: Don't link your song title directly to the sound file BLUES2.RM. Instead, link it to the text file (metafile) called BLUES2.RAM, which you just created. This metafile makes the RealPlayer pop up and play the streaming audio. If you incorrectly link your song to BLUES2.RM, and a listener clicks on "Blues Bash," that file will download instead of streaming.

Now that your song links are working correctly in your computer, it's time to upload them.

1. Upload (send) PAGE1.HTM, BLUES2.RAM, and BLUES2.RM to your Web site. Be sure they all go to the same directory (your Web site directory). As we said before, some Web servers have an upload page. If yours doesn't, try uploading the files with ftp software. You might want to search for ftp freeware on the Web.

2. Now start your Web browser and go to your site. Open PAGE1.HTM. Then click "Refresh" in your browser so that you see the current page, not the one in your cache.

3. Click on the song title. In a few seconds RealPlayer should pop up, fill the buffer memory, and play the tune. If not, recheck everything you typed. Then try www.real.com for tech support.

An important part of this process is providing the right bitrate for your target audience. When you use RealPublisher, you might create one file suited for a 28K modem, and another file for a 56K modem. Why use both formats? The faster the modem speed, the better the sound quality (more extended highs). So a file optimized for a 56K modem sounds better than a file made for a 28K modem. However, a 28K modem can't play 56K modem files without skipping, so you need both formats. If you are working in RealPublisher's Advanced mode, select "mono" because it gives brighter sound than "stereo" for the same bitrate.

Play low-fi "Tango" sample (28K modem)

Play low-fi "Tango" sample (56K modem)

Right-click to download hi-fi "Tango" sample in MP3 format

Right-click to download hi-fi "Tango" sample in WMA format

Figure 20.1 Sample Web-page song links with four song formats: RealAudio for 28K modem, RealAudio for 56K modem, MP3, and WMA.

Examples of Web-page Song Links

Figure 20.1 shows some examples of Web-page song links with four formats: RealAudio for 28K modem, RealAudio for 56K modem, MP3, and WMA. Below Figure 20.1 is the HTML code for that page.

The methods described here will provide streaming audio and downloadable audio from any HTTP server.

```
<a href = "tango28.ram">Play low-fi "Tango" sample (28K modem)</a> <br>

<a href = "tango56.ram">Play low-fi "Tango" sample (56K modem)</a> <br>

<a href = "tango.mp3">Right-click to download hi-fi "Tango" sample in MP3 format</a> <br>

<a href = "tango.wma">Right-click to download hi-fi "Tango" sample in WMA format</a>
```

The code <br> above means a line break.

Streaming Audio from a RealServer Site

Now we'll look at ways to stream audio from a RealServer (RTSP) site, which has some advantages.

In your Web surfing, have you ever clicked on a song title to hear the music, only to have it interrupted? You try to play a streaming audio file, but it skips instead of playing continuously. One cause is too much traffic on the Internet. Your connection speed to the Net changes continuously depending on how many folks are using it the same time you are. If your connection speed is too slow to keep up with the streaming file's bitrate, the sound skips.

Is there a way to play streaming files that adapts to changing connection speeds? There is, and it's called SureStream. Developed by RealNetworks, SureStream is a file format for streaming audio files that varies the bitrate to match what your modem can handle at the moment. The sound quality varies with the bitrate—sometimes bright, sometimes dull. But at least the stream is continuous. No skipping—unless traffic is really bad. With SureStream, you don't need to offer several audio files at different bitrates to handle different modem speeds. One SureStream file can work for any modem.

If you'd like to have SureStream ability on your Web site, have your Web page hosted by a server who has RealServer software. Geocities.com is one example. If you are a member of Geocities, and you are a member of their Geomedia group, you can access their RealServer and play SureStream files from your Web site.

There are two ways to stream RealAudio files: via Hyper Text Transfer Protocol (HTTP) or Real Time Streaming Protocol (RTSP). Any standard Web server is HTTP format. Earlier in this chapter we looked at ways to stream your music from any Web-site host using HTTP. Now we'll cover streaming from a Web-site host that has RealServer.

Setting Up RealServer Files with RealProducer

Imagine that your band's Web page is on a RealServer host, and you want to add SureStream RealAudio files to your page. Here's how to go about it.

First, download the free RealProducer software from www.realnetworks.com/products/producer/index.html. RealProducer Plus is another option. Also download the free RealAudio player from www.real.com.

Note: Your Web server might require a special version of RealProducer that you download from the server. For example, Geocities says that you must download and use the Geomedia-optimized RealProducer; otherwise your files won't play correctly. You might need to become a RealMedia member in your Web server. For instance, the Geocities Web server has a RealMedia option called Geomedia. Once you've subscribed, you can use their RealServer to play SureStream files.

Now decide what music you want to stream. Generally, you'll want your streaming audio files to be short samples or "clips" of your music, but they can be entire songs. Let's start by creating a short sample.

1. Suppose you have a song called "Blues Bash" recorded on your hard drive as BLUES.WAV. Load the file BLUES.WAV into your digital

485

editing software. Trim it to about 30 seconds long. Fade out the music toward the end. In this example, let's save the edited song as BLUES2.WAV.

2. Use RealPublisher to convert BLUES2.WAV into BLUES2.RM, which is a RealMedia file. In this process, select "SureStream" file rather than "Fixed Rate" file. If you are working in RealPublisher's Advanced mode, select "mono" because it gives brighter sound than "stereo" for the same bitrate.

3. If you wish, use RealPublisher's Web Page Wizard to create a Web page, such as PAGE1.HTM. This page includes the song title, such as "Blues Bash." You can add more text and graphics to the page. The Web Wizard will also create a short text file called a metafile. Its filename is BLUES2.RAM (for a pop-up player) or BLUES2.RPM (for an embedded player). A pop-up player appears only when you click on a song title. An embedded player is part of a Web page. In working with the Web Page Wizard, try to use the default settings. Be sure that BLUES2.RM, BLUES2.RAM, and PAGE1.HTM are all saved in the same directory on your hard drive.

4. Use RealPublisher's Publishing Wizard to upload those three files to your Web site. If for some reason it doesn't work, go to your Web site and upload those three files from there. They should all go to the same directory (your Web page directory).

5. At your Web site, open PAGE1.HTM (or whatever). Click "Refresh" so that you see the current version, not the one in your cache.

6. Click on the song title. The RealPlayer should pop up and play the song. If you selected "Embedded Player" when you created the song file, you should see a RealPlayer on your Web page. Click the PLAY button to hear the song. **Note**: If Internet traffic is high, you might get an error message saying "Requested file not found" even though the file is there. You might have to wait for a less busy time.

Setting Up RealServer Files Without RealPublisher

You can do all the above without RealPublisher by following the procedure below:

1. Using a word processor or text editor (such as Notepad), create a small text file that points to the location of your RealMedia file. For example:

486

```
rtsp://real.webserver.com/your Web site address/blues2.rm.
```

Replace "webserver.com" with your own Web server; for example, geoc-ities.com. Save this line as a plain ASCII text file with a .RAM extension. In this case, you'd call it BLUES2.RAM. This type of file is called a metafile. Save it in your Web site's directory on your hard drive.

The URL above has three parts:

* The protocol
* The server
* The address

The protocol part is rtsp:, which stands for Real Time Streaming Proto-col—used by the RealNetworks server. The server part is real.web-server.com, which points to the Web server's address. The "real" at the beginning directs your request to the RealNetworks servers. The address part includes your entire Web server and Web site address, along with the audio filename.

2. Now use Web-page design software to open your Web page on your hard drive. Here we'll call it PAGE1.HTM.

3. Find a place on the page where you want the song title to go. Type the title of the song, such as "Blues Bash."

4. Link that title to the metafile you just wrote and saved (in this case, link it to BLUES2.RAM). In HTML it would look something like this:

```
<a href = "blues2.ram">Blues Bash</a>
```

Important: Don't link your song title directly to the sound file BLUES2.RM. Instead, link it to the text file (metafile) called BLUES2.RAM that you created earlier. This metafile makes the RealPlayer pop up and play the streaming audio. If you incorrectly linked your song to BLUES2.RM, and a listener clicks on "Blues Bash," that file will down-load instead of streaming.

Now that your song links are working correctly on your computer, it's time to upload them.

1. Upload (send) PAGE1.HTM, BLUES2.RAM, and BLUES2.RM to your Web site. Be sure they all go to the same directory (your Web site directory).

2. Now start your Web browser and go to your site. Open PAGE1.HTM. Then click "Refresh" in your browser so that you see the current page, not the one in your cache.

487

3. Click on the song title. In a few seconds RealPlayer should pop up, fill the buffer memory, and play the tune. If not, re-check everything you typed. Also, net congestion or Web-server glitches can cause false error messages. Try again when the Net is not so busy. If all else fails, contact tech support at your Web server and at www.real.com.

The methods described here will provide streaming audio from any Web host that offers RealServer. I hope to hear your music on the Web!

Liquid Audio

An online music service you should know about is Liquid Audio (www.liquidaudio.com). This server company provides music retailers and record labels a secure way to sell and distribute high-quality audio tracks on the Internet, with Digital Rights Management. Once you open an account with Liquid Audio, you have access to software that encodes your music to the high-quality AAC format and uploads it to the Liquid Audio server. The encoder adds to the file an inaudible digital "watermark," which tracks the song owner and controls copies.

To hear your music, listeners download the free Liquid Player. It plays free preview clips and handles a variety of file types. It also lets the user purchase full-length tracks online. In addition, the player displays graphics, liner notes, lyrics, and promotional material, and includes CD burning software.

dB or not dB

In the studio, you need to know how to set and measure signal levels and match equipment levels. You also need to evaluate microphones by their sensitivity specs. To learn these skills, you must understand the decibel—the unit of measurement of audio level.

Definitions

In a recording studio, level originally meant power, and amplitude referred to voltage. Today, many audio people also define "level" in terms of voltage or sound pressure, even though this terminology is not strictly correct. You should know both definitions in order to communicate.

Audio level is measured in decibels (dB). One dB is the smallest change in level that most people can hear—the just-noticeable difference. Actually, the just-noticeable difference varies from 0.1 dB to about 5 dB, depending on bandwidth, frequency, program material, and the individual. But 1 dB is generally accepted as the smallest change in level that most people can detect. A 6- to 10-dB increase in level is considered by most listeners to be "twice as loud." Sound pressure level, signal level, and change in signal level all are measured in dB. *Play CD track 42 to hear the effects of various decibel changes on loudness. Track 43 finishes the CD.*

Sound Pressure Level

Sound pressure level (SPL) is the pressure of sound vibration measured at a point. It's usually measured with a sound level meter in dB SPL (decibels of sound pressure level).

The higher the sound pressure level, the louder the sound (Figure A.1). The quietest sound you can hear, the threshold of hearing, is 0 dBSPL. Average conversation at 1 foot is 70 dBSPL. Average home-stereo listening level is around 85 dBSPL. The threshold of pain—so loud that the ears hurt and can be damaged—is 125 to 130 dBSPL.

Sound pressure level in decibels is 20 times the logarithm of the ratio of two sound pressures:

$$dBSPL = 20 \log P/P_{ref}$$

where P is the measured sound pressure in dynes/cm^2, and P_{ref} is a reference sound pressure: 0.0002 dyne/cm^2 (the threshold of hearing).

Signal Level

Signal level also is measured in dB. The level in decibels is 10 times the logarithm of the ratio of two power levels:

Figure A.1 A chart of sound pressure levels.

$$dB = 10 \log P/P_{ref}$$

where P is the measured power in watts, and Pref is a reference power in watts.

Recently it's become common to use the decibel to refer to voltage ratios as well:

$$dB = 20 \log V/V_{ref}$$

where V is the measured voltage, and V_{ref} is a reference voltage.

This expression is mathematically equivalent to the previous one, because power equals the square of the voltage divided by the circuit resistance:

$$dB = 10 \log P_1/P_2 = 10 \log((V_1^2/R)/(V_2^2/R))$$
$$= 10 \log(V_1^2/V_2^2) = 20 \log(V_1/V_2)$$

The resistance R (or impedance) in this equation is assumed to be the same for both measurements, and thus divides out.

Signal level in decibels can be expressed in various ways, using various units of measurement:

- dBm: decibels referenced to 1 milliwatt
- dBu or dBv: decibels referenced to 0.775 volt (dBu is preferred)
- dBV: decibels referenced to 1 volt

dBm

If you're measuring signal power, the decibel unit to use is dBm, expressed in the equation

$$dBm = 10 \log P/P_{ref}$$

where P is the measured power, and P_{ref} is the reference power (1 milliwatt).

For an example of signal power, use this equation to convert 0.01 watt to dBm:

$$dBm = 10 \log (P/P_{ref})$$
$$= 10 \log (0.01/0.01)$$
$$= 10$$

491

So, 0.01 watt is 10 dBm (10 decibels above 1 milliwatt).
Now convert 0.001 watt (1 milliwatt) into dBm:

$$dBm = 10 \log (P/P_{ref})$$
$$= 10 \log (0.001/0.001)$$
$$= 0$$

So, 0 dBm = 1 milliwatt. This has a bearing on voltage measurement as well. Any voltage across any resistance that results in 1 milliwatt is 0 dBm. This relationship can be expressed in the equation

$$0\,dBm = V^2/R = 1 \text{ milliwatt}$$

where V = the voltage in volts, and R is the circuit resistance in ohms.
For example, 0.775 volt across 600 ohms is 0 dBm. One volt across 1000 ohms is 0 dBm. Each results in 1 milliwatt.
Some voltmeters are calibrated in dBm. The meter reading in dBm is accurate only when you're measuring across 600 ohms. For an accurate dBm measurement, measure the voltage and circuit resistance, then calculate:

$$dBm = 10 \log ((V^2/R)/0.001)$$

dBv or dBu

Another unit of measurement expressing the relationship of decibels to voltage is dBv or dBu. This means decibels referenced to 0.775 volt. This figure comes from 0 dBm, which equals 0.775 volt across 600 ohms (because 600 ohms used to be a standard impedance for audio connections):

$$dBu = 20 \log V/V_{ref}$$

where V_{ref} is 0.775 volt.

dBV

Signal level also is measured in dBV (with a capital V), or decibels referenced to 1 volt:

$$dBV = 20 \log (V/V_{ref})$$

where V_{ref} is 1 volt. For example, use this equation to convert 1 millivolt (0.001 volt) to dBV:

$$dBV = 20 \log (V/V_{ref})$$
$$= 20 \log (0.001/1)$$
$$= -60$$

So, 1 millivolt = -60 dBV (60 decibels below 1 volt). Now convert 1 volt to dBV:

$$dBV = 20 \log (1/1)$$
$$= 0$$

So, 1 volt = 0 dBV.

To convert dBV to voltage, use the formula

$$Volts = 10^{(dBV/20)}$$

Change in Signal Level

Decibels also are used to measure the change in power or voltage across a fixed resistance. The formula is

$$dB = 10 \log (P_1/P_2)$$

or

$$dB = 20 \log (V_1/V_2)$$

where P_1 is the new power level, P_2 is the old power level, V_1 is the new voltage level, and V_2 is the old voltage level.

For example, if the voltage across a resistor is 0.01 volt, and it changes to 1 volt, the change in dB is

$$dB = 20 \log (V_1/V_2)$$
$$= 20 \log (1/0.01)$$
$$= 40 \, dB$$

Doubling the power results in an increase of 3 dB; doubling the voltage results in an increase of 6 dB.

The VU Meter, Zero VU, and Peak Indicators

A VU meter is a voltmeter of specified transient response, calibrated in volume units or VU. It shows approximately the relative volume or loudness of the measured audio signal. VU meters are used on analog tape recorders, broadcast consoles, some live-sound mixing consoles, and older recording consoles.

The VU-meter scale is divided into volume units, which are not necessarily the same as dB. The volume unit corresponds to the decibel only when measuring a steady sine-wave tone. In other words, a change of 1 VU is the same as a change of 1 dB only when a steady tone is applied.

Most recording engineers use 0 VU to define a convenient "zero reference level" on the VU meter. When the meter on your mixer or recorder reads "0" on a steady tone, your equipment is producing a certain level at its output. Different types of equipment produce different levels when the meter reads 0 (Figure A.2). Zero VU corresponds to:

- +8 dBm in older broadcast and telephone equipment
- +4 dBm in balanced recording equipment
- −10 dBV in unbalanced recording equipment

A 0 VU recording level (0 on the record level meter) is the normal operating level of an analog tape recorder; it produces the desired recorded flux on tape. A "0 VU recording level" does not mean a "0 VU signal level."

Figure A.2 VU-meter scale.

With a VU meter, 0 VU corresponds to a recording level 8 dB below the level that produces 3% third-harmonic distortion on tape at 400 Hz. Distortion at 0 VU typically is below 1%.

The response of a VU meter is not fast enough to track rapid transients accurately. In addition, when a complex waveform is applied to a VU meter, the meter reads less than the peak voltage of the waveform. (This means you must allow for undisplayed peaks above 0 VU that use up headroom.)

In contrast, a peak indicator responds quickly to peak program levels, making it a more accurate indicator of recording levels. One type of peak indicator is an LED that flashes on peak overloads. Another is the LED bar graph meter commonly seen on digital recorders and some mixers. Yet another is the PPM (peak program meter). It is calibrated in dB, rather than VU. Unlike the VU meter reading, the PPM reading does not correlate with perceived volume.

In a digital recorder, the meter is an LED or LCD bar graph meter that reads up to 0 dBFS (Full Scale). In a 16-bit digital recorder, 0 dBFS means all 16 bits are ON. In a 24-bit digital recorder, 0 dBFS means all 24 bits are ON. The OVER indication means that the input level exceeded the voltage needed to produce 0 dBFS, and there is some short-duration clipping of the output analog waveform. Some manufacturers calibrate their meters so that 0 dBFS is less than 16 bits or 24 bits ON; this allows a little headroom.

Balanced versus Unbalanced Equipment Levels

Generally, audio equipment with balanced (3-pin) connectors works at a higher nominal line level than equipment with unbalanced (phono) connectors. There's nothing inherent in balanced or unbalanced connections that makes them operate at different levels; they're just standardized at different levels.

These are the nominal (normal) input and output levels for the two types of equipment:

- Balanced: +4 dBu (1.23 volts)
- Unbalanced: −10 dBV (0.316 volt)

In other words, when a balanced-output recorder reads 0 VU on its meter with a steady tone, it is producing 1.23 volts at its output connector. This voltage is called +4 dBu when referenced to 1 milliwatt. When an unbalanced-output recorder reads 0 on its meter with a steady tone, it is usually producing 0.316 volt at its output connector. This voltage is called −10 dBV when referenced to 1 volt.

Interfacing Balanced and Unbalanced Equipment

There's a difference of 11.8 dB between +4 dBu and −10 dBV. To find this, convert both levels to voltages:

$$dB = 20 \log (1.23/0.316)$$
$$= 11.8$$

So, +4 dBu is 11.8 dB higher in voltage than −10 dBV (assuming the resistances are the same).

A cable carrying a nominal +4 dBu signal has a signal-to-noise ratio (S/N) 11.8 dB better than the same cable carrying a −10 dBV signal. This is an advantage in environments with strong radio-frequency or hum fields, such as in a computer. But in most studios with short cables, the difference is negligible.

Connecting a +4 dBu output to a −10 dBV input might cause distortion if the signal peaks of the +4 equipment exceed the headroom of the −10 equipment. If this happens, use a pad to attenuate the level 12 dB (Figure A.3, top). The wiring shown converts from balanced to unbalanced as well as reducing the level 12 dB. If you hear hum with this connection, add an isolation transformer as shown in Figure A.3 (bottom). All the leads should be twisted.

You don't always need that pad. Many pieces of equipment have a +4/−10 level switch. Set the switch to the nominal level of the connected equipment. If there is no such switch on either device, connect between them a +4/−10 converter box such as the Ebtech Line Level Shifter (www.ebtechaudio.com) or use the wiring shown here.

To connect an unbalanced −10 dBV output to a balanced +4 dBu input, use the wiring shown in Figure A.4. All the exposed leads should be twisted.

Microphone Sensitivity

Decibels are an important concern in another area: microphone sensitivity. A high-sensitivity mic puts out a stronger signal (higher voltage) than

Figure A.3 Top: Wiring balanced out to unbalanced in. The resistors form a 12-dB pad to match the balanced +4 dBu output to the unbalanced −10 dBV input. Bottom: Same, with an isolation transformer added to reduce hum.

Figure A.4 Top: Wiring unbalanced out to balanced in. Bottom: Same, with an isolation transformer added to reduce hum.

497

a low-sensitivity mic when both are exposed to the same sound pressure level.

A **microphone-sensitivity specification** tells how much output (in volts) a microphone produces for a certain input (in SPL). The standard is millivolts per pascal, where one pascal (Pa) is 94 dB SPL.

A typical "open-circuit sensitivity" spec is 5.5 mV/Pa for a condenser mic and 1.8 mV/Pa for a dynamic mic. "Open-circuit" means that the mic is unloaded (not connected to a load, or connected to a mic preamp with a very high input impedance). If the spec is 5.5 mV/Pa, that means the mic produces 5.5 mV when the SPL at the mic is 94 dB SPL.

If you put a microphone in a 20 dB louder soundfield, it produces 20 dB more signal voltage. For example, if 74 dBSPL in gives 0.18 mV out (−75 dBV), then 94 dBSPL in gives 1.8 mV out (−55 dBV); 150 dBSPL in gives 1.1 volt out (+1 dBV), which is approximately line level! That's why you need so much input padding when you record a kick drum or other loud source.

B

OPTIMIZING YOUR COMPUTER FOR DIGITAL AUDIO

Once you've chosen some recording software and installed it, you'll want to make your computer run as fast and glitch-free as possible. Speed becomes critical as you use more tracks, more effects, high bit rates and high sampling rates.

The data flow of multitrack digital audio places high demands on computer speed. Recording and playback of digital audio requires long, continuous periods of streaming audio data. The more tracks in use, and the higher the bit depth and sampling rate, the faster the data flow has to be. And the more soft synths and effects plug-ins that are in use, the greater the load on the CPU.

Clearly, you need a fast computer for multitrack recording. But you also need to optimize its settings for best results. I will cover some ways to speed up the data flow and to reduce the CPU usage. If you follow these tips, you will have a faster system that handles more plug-ins and more tracks at once. Also, when you play tracks or burn CDs, clicks and drop-outs in the audio will be less likely.

Most of these tips apply to a PC computer. Mac advice is offered at the end of this appendix in the section Optimizing MacIntosh for Audio.

Disclaimer: Backup your system and data files first, then proceed at your own risk. Neither I nor Focal Press will be responsible for damage to your computer system or files from making these changes. If you are not comfortable doing a particular tweak, don't do it. However, I have done all these adjustments with no problems, and they are reversible.

Speeding Up Your Hard Drive

The hard drive should have a fast average access time (under 9 msec) and a high internal sustained transfer rate. Some recommended hard drives are the Seagate Barracuda, Maxtor Diamond Max, and Western Digital models.

What transfer rate is needed? The data rate of 24 tracks at 24-bits/96 kHz is 6.6 MB/sec. That is a practical minimum.

The best current hard drives can transfer about 40 MB/sec continuously with a PCI bus and an Ultra ATA/66 or Ultra ATA/100 interface (for EIDE drives), or a SCSI interface for SCSI drives. Do not use an ISA bus interface—its rate tops out at 2 MB/sec. SCSI drives tend to be faster than IDE, but SCSI costs more, and many IDE drives are fast enough.

To use ATA66 or higher, your motherboard and BIOS must support it. You need an 80-pin ribbon cable that is 18 inches maximum. Plug the blue connector into the motherboard, the black one into the Master device, and the gray one into the Slave device.

Hard drives with high rotational speed or spindle rate (7200 rpm or greater) tend to have faster transfer rate. They are recommended for 24-bit multitrack productions. The typical sustained transfer rate of a 7200 rpm drive is 30 to 40 MB/sec. This provides up to 160 16-bit/44.1 kHz tracks or 50 24/96 tracks, depending on file fragmentation.

For fastest speed, use one hard drive for applications and Windows files, and another for audio data. That way the audio drive head does not waste time looking for system files. Put the Windows drive on the Primary IDE channel as a master, and put the audio drive on the Secondary IDE channel as a master. To set a drive to master or slave, change the jumpers on the back of the drives. A laptop computer can connect to an external USB or FireWire hard drive that is used just for audio files.

You can put the two hard drives on the same IDE cable (both on the Primary channel) as master and slave. Usually, but not always, they will run a little slower that way. If you have a second CD-ROM drive, put it on the secondary channel as a slave.

If you have only one hard drive, use one disk partition for audio files and another partition for programs. That way you can defragment or reformat the audio data partition frequently.

Defragment the audio file drive often. Defragging reorganizes files into contiguous areas on the disk so that each audio file can be accessed with minimum head movements. Before you defrag, disable programs in the System Tray, then select Start > Run > Scandisk.exe (Win98) or Chkdsk.exe (Win2000 and XP). Then select Start > Run > Defrag.exe.

When you are finished using all the files on the audio data drive, reformat it to zero-out the clusters. This is necessary even if you deleted all the files.

Your system will run faster if you delete unneeded applications. Select Start > Settings > Control panel > Add/Remove programs. You must uninstall a program, rather than deleting its folders, to notice an improvement.

Install lots of RAM, at least 256 MB for audio, or at least 512 MB for soft synths. How does more RAM help? When your DAW application is streaming audio, it continuously feeds data into a RAM buffer for each track. If the RAM buffer is big, it needs to be filled less often, so the disk-drive heads need to search less often. A larger RAM buffer lets the disk heads work more efficiently by getting bigger continuous chunks of data into RAM. Also, having lots of RAM prevents frequent access of the swap file (described later). The goal is to have your programs running in RAM, rather than swapping data to your hard drive.

Important: Use drivers that allow bus mastering. If necessary, install bus-mastering drivers for your hard drives. To check for bus mastering in Windows, select Start > Settings > Control panel > System > Hardware > Device manager > Hard disk controllers. Look for Bus master controller.

Important: Check that Direct Memory Access (DMA) is activated on hard drives and CD ROMs. DMA mode transfers data directly from the hard drive to RAM, bypassing the CPU. With DMA, your CPU overhead (while accessing the drive) will fall from about 50 to 5 percent or less. Here is the procedure:

- In Windows 2000 and XP, select Start > Settings > Control Panel > System > Hardware > Device Manager > IDE ATA/ATAPI Controllers (or hard disk controller) > Right-click Primary IDE channel > Properties > Advanced Settings Tab > Transfer Mode: "DMA if available." Repeat for the Secondary IDE channel. Reboot.

501

- In Windows 98, select Start > Settings > Control Panel > System > Device Manager > Disk drives > Right-click your hard drive > Properties > Settings > Check "DMA" if it isn't already. Reboot.

Two more tweaks: Select Start > Settings > Control panel > System > Performance > File system > Hard disk. Set Read-ahead optimization to minimum. Set Typical Role to Network Server because it gives higher priority to disk use, and will provide more RAM to open programs faster.

You can also speed the data transfer of CD-ROM drives and CD burners. Go to Start > Settings > Control panel > System > Device manager, and double-click CD-ROM. Select your model of CD-ROM > Properties > Settings. Check Disconnect, Sync data transfer, and DMA.

Disable write-behind caching on your hard drives: Write-behind caching waits until the system is idle before writing data to the hard drive, causing a delay. Here's how to turn it off:

- In Win2000: Select My Computer > right-click a hard drive > Properties > Hardware > Highlight a hard drive > Properties > Disk properties > Uncheck "Write cache enabled" > OK. Do this for all the IDE drives listed. Reboot.

- In XP: Select Start > Settings > Control Panel > System > Hardware > Device Manager > Disk drives > Double click on your audio data drive > Policies > Uncheck "Enable Write Caching" > OK. Reboot.

- In Win98: Select Start > Settings > Control Panel > System > Performance > File System > Troubleshooting > Check "Disable write-behind caching for all drives" > OK. Reboot.

Disable read-ahead optimization: In Win 98, select Start > Settings > Control Panel > System > Performance > File System > Hard disk > Move the read-ahead slider far left. In Win2000, select My Computer > Right-click the drive you want to alter > Properties > Hardware > Properties > Disk Properties > Move the read-ahead slider far left.

Increasing Processing Speed

A computer has a central processing unit (CPU) that does most of the calculations needed to run software. For example, in some DAWs the CPU performs all the signal processing for real-time effects. Get a computer or motherboard with the fastest CPU that you can afford—a clock speed of 2 Gigahertz or greater. Consider getting a computer or motherboard with a multiprocessing CPU such as the AMD Athlon MP.

502

The following tips reduce the amount of data that the CPU has to handle:

Do not enable hyperthreading in BIOS unless your software requires it.

If you hear drop-outs or glitches, maybe too many real-time effects are running. Select a track that has real-time effects, and bounce (export) that track with effects to an open track. That way the effects are recorded or embedded in the bounced track, rather than running in real time. This reduces CPU loading. Then archive (save) the original track and delete it from the project. Muted audio tracks also put a load on the CPU. Archive tracks that are muted, then delete those tracks from the project.

Consider converting MIDI soft-synth tracks into audio tracks.

Do not insert the same delay effects (echo, reverb, chorus, flanging) in several tracks. Instead, set up an aux bus that has the desired effect. Use aux send controls on tracks to adjust the amount of effects on each track. This reduces the number of effects processes that are running and reduces CPU loading.

In extreme cases, consider installing a Digital Signal Processing (DSP) card, such as the TC Works PowerCore. It handles the processing for the plug-ins. The plug-ins access the card rather than the CPU. You can install multiple cards to expand the system. Pro Tools includes its own DSP cards.

Before recording, turn off the waveform drawing function of your sound clips or regions.

Reduce the number of colors: Right-click the Desktop > Properties > Settings > set Colors to 16 bit or 256 colors. A higher setting reduces audio performance when the level meters redraw in your audio application.

Reduce video acceleration: Select Start > Settings > Control panel > System > Performance > Graphics > Advanced. Reduce Hardware Acceleration as much as you can without degrading the display of your audio program.

In Win2000 and XP, adjust visual effects for best performance: Select Start > Settings > Control Panel > System > Advanced > Performance Settings > Visual Effects Tab > Adjust for best performance.

Change the Performance mode for the operating system: Right-click My Computer > Properties > Advanced > Performance > Settings > Select Background Services. This setting allocates more processor time to background activities, such as streaming audio.

Preventing Interruptions

Unnecessary background programs can rob CPU cycles, interrupt the audio program, and interrupt the hard-drive head from playing audio. The following tweaks to your PC can increase the number of tracks and effects you can play without drop-outs or clicks by disabling various background programs or increasing free memory.

Don't multitask (run other programs) while the audio is streaming. Each additional task slows down the system. Press Ctrl+Alt+Del to see what's running, highlight anything you don't need, and select End Task. Leave Explorer and Systray running. Don't scroll while recording or playing back.

Turn off autosave except when working just on MIDI files.

Use a PS/2 mouse instead of a USB mouse, which can cause audio instability.

Switch off your Anti-Virus program when using your programs. Do this by right clicking on the icon on the bottom right of your screen, then click "Disable." You need Anti Virus for downloading files from outside sources but not for DAW work.

Check Interrupts (IRQs):

- Right click My Computer > Properties > Device Manager" > Double-click "Computer." You will see a complete list of all the IRQs in use. Make sure that SCSI controllers do not share interrupts with anything. Make sure the sound card has its own IRQ number. Do not share audio cards/devices IRQs with USB, SCSI, or graphics cards. Do not share Network/Modems IRQs with USB/SCSI/Video. Do not assign IRQ #9 to anything; that is the cascaded IRQ from #2. To reset IRQ's, you might need to remove ALL cards except for the video card and add one card at a time in different slots.

- Place your audio card in a non-IRQ-shared PCI slot. Look in the PCI table of your motherboard manual for the highest-priority slot that does not share with another device or another slot. Consider using a shareware utility called PowerStrip, which lets you adjust PCI latency timers. Set the latencies high enough to get the needed performance, and no higher.

- If you are not using any devices on your USB, serial or parallel ports, go to the BIOS and disable them. This will reduce the number of IRQs and may prevent conflicts. Also disable or disconnect other devices that you are not using; this frees up interrupts.

Disable CD-ROM autoplay (turn off Auto Insert Notification): In Win XP, select Start > Run > type regedit > HKEY__LOCAL__MACHINE > System > CurrentControlSet > Services > Cdrom. Set autorun to 0 instead of 1 at the right end of the displayed number. In Win 98, Right-click My Computer and select Device Manager. Click on the + sign next to the CD-ROM. Double-click a device. Then click Settings and uncheck Auto Insert Notification. Repeat for other CD-ROMs.

Disable Background image: Right-click Desktop > Properties > Background > None.

Disable Screen Saver: Right-click Desktop > Properties > Screen Saver > None.

Switch off power schemes: Start > Settings > Control Panel > Power Options > Set "Turn off hard discs" to "Never."

Switch off hibernation: Start > Settings > Control Panel > Power Options > Hibernate > Uncheck "Enable hibernate support."

Remove programs that load on startup: In Win98 and XP, select Start > Run > Type in MSCONFIG and press OK. Select the "Startup" tab and deselect everything but the system tray and the speaker volume control. Press Apply, then OK. Reboot. For faster startup, archive fonts you don't use. They are in Windows\fonts or WinNT\fonts. You might want to remove autoexec.bat and config.sys from the boot sequence. This removes most DOS-compatibility mode drivers that can sap resources. For Win2000, get "Startup Control Center" from www.download.com.

Disable virtual memory (also called a swap file or paging file): When your system has used all its RAM, it saves and loads data to your hard drive in a swap file.

This process can interrupt audio streaming. It's best to install more memory and disable the swap file. Here's what to do:

- In Windows 2000 and XP: Select Start > Settings > Control Panel > System > Advanced > Performance options > Advanced > Virtual Memory > Change > Disable Virtual Memory > OK. Close the windows and reboot.

- In Windows 98: Select Start > Control Panel > System > Performance > Virtual memory > Disable Virtual Memory > OK. Close the windows and reboot.

Disable System Sounds: This prevents interruptions caused by Windows' bells and chimes. It also may prevent audio conflicts due to Windows sounds being at low sample rates and bit rates. Select Start > Settings > Control Panel > Sounds and Audio Devices (or Sounds and

505

Multimedia) > Sounds tab > Set "Sound Scheme" to "No sounds . . ." Press No to "Do you want to save the previous scheme?" Hit Apply.

In WinXP, disable unnecessary services: A service is a background program. Some are necessary, but many are not. They consume memory and resources. Select Start > Run > Type services.msc > Services. Double-click each service to read about it and set its status to "automatic," "manual," or "disabled." For advice on setting Services, go to www.blkviper.com. When done, exit Services and reboot.

Setting the Buffer Size

When you play or record several tracks at once with real-time effects, or record at a high sample rate, you might hear drop-outs or clicks in the audio. Playback might stop. This is caused by too much data flow for the CPU to handle, or by buffer memory being emptied faster than it can be refilled.

In your recording software, increase the size of the buffer a little at a time until the drop-outs or glitches stop. Ideally you'd do this just during playback or mixdown, because enlarging the software buffer increases latency (monitoring delay). If a small buffer setting causes problems during overdubs, temporarily disable any plug-ins.

In Windows XP, reduce your audio interface buffer size to 128 MB in its software control panel. This reduces latency. Pro Tools might need a buffer size of 256 or 512.

Other Tips

- Check out www.tweakXP.com, www.Tweak3D.net, and www.extremetech.com.
- More information on computer recording can be found at www.pcrecording.com.
- Three vendors of audio-optimized computers are www.dawbox.com, www.sweetwater.com (Creation Station), and www.liquiddaw.com.
- Microsoft has a patch that fixes audio stuttering with USB devices in XP. Go to www.microsoft.com and look for 307271.
- Go to the manufacturer Web sites of your sound card (or audio inter-face) and audio editing software. Check out the support sections for

advice on optimizing your computer and on troubleshooting sound problems.

- Keep your sound card(s) away from the AGP graphics card to prevent hum.

- Don't install much software (it can corrupt the registry).

- On the Web, download the latest drivers for your motherboard, video card, IDE controller, CD burner, recording software, and sound card or audio interface. Also download the Windows updates. However, Windows XP Service Pack 2 is not recommended for most audio software.

- Consider using an AGP video card rather than a PCI video card to reduce the load on the PCI bus.

- Free up disk space by uninstalling Windows components that you don't use. Select Start > Settings > Control Panel > Add or remove programs > Add/Remove Windows Components. For Win98, select Start > Settings > Control Panel > Add or remove programs > Windows Setup.

- For Win2000 and XP, disable Disc Indexing Service on NTFS formatted drives if you do NOT do frequent searches for filenames. Double-click My Computer > Right-click each hard drive > Properties > Uncheck "Allow Indexing Service to index this disc for fast file searching" > Apply. Choose all files and subfolders within the drive. This action takes a while as all the files are scanned. If you get a message that says Access Denied . . . , press "Ignore All."

- Vcache is an amount of RAM that is set aside by Windows to store data recently read from your hard drive that may be required again. Go to Start > Run > and type Sysedit to open System.ini in a text editor. Find [vcache]. Under [vcache], make sure MinFileCache and MaxFileCache are both set to the same value. Use 4096 with 32MB of RAM installed; use 8192 with 64MB, and use 16,000 with 128MB or more. Save System.ini and reboot. These settings can prevent clicks and skipping audio.

- If you have a utility that measures disk thruput (sustained disk transfer rate) you might want to measure your system's speed before and after each hard-drive change mentioned above. That way you will know which changes were effective. A good utility is DskBench, available free from www.prorec.com and other sources.

507

Optimizing MacIntosh for Audio

Many of the PC tips apply to Mac: Use a second, fast hard drive running at 7200 rpm or higher for audio files. Use a minimum of non-audio software. Disable energy-saver or sleep software. Update drivers. Reduce monitor resolution. Disable video acceleration software and cards. Check online user groups for advice.

- Check the Memory panel and turn off virtual memory.
- Check for enough application memory. Select the program's icon and press Command-I. In the info window that appears, try increasing the amount of allotted memory.
- Minimize extensions. Extensions are small programs that extend the functionality of the operating system when they are loaded. But they use up CPU clock cycles and slightly reduce stability. You might disable all extensions but these: sound card, Appearance, Quick-Time, SystemAV, and video drivers. Consider loading the CD-ROM driver only when you need it. If the recording application freezes after boot-up, check the application's documentation for known extension conflicts.
- Upgrade to OS 8 or higher.
- Use a G4 or better, not an iMAC.
- Use HFS on the audio drive—not HFS+ (extended). HFS+ can make the cluster sizes too small for efficient streaming of audio.
- Buy more RAM: 256 MB or more is recommended.
- Delete outtakes and unused files to prevent a "too many files open" message.
- If you can't play enough tracks, increase the buffer size.
- If you hear pops or clicks, check your sound card's clock settings. When recording an analog source, set it to Internal Clock or Sync. If your source is digital, set it to external S/PDIF sync or the format you are using. Be sure sample rates match. Move audio and digital cables away from USB cables and power lines. Try turning off USB hubs, modems and printers.
- If your system crashes frequently, try removing downloaded software. Delete your application's preferences file if it has become corrupt.
- If the application stops in the middle of a recording, check to see whether you have set a maximum recording time in advance.

INTRODUCTION TO SMPTE TIME CODE

SMPTE time code is a special signal recorded on tape or hard disk that can sync together two multitrack recorders so that they operate as one. Time code can synchronize an audio recorder with a video recorder, or sync audio clips with a video program in a window on a computer screen.

SMPTE stands for the Society of Motion Picture and Television Engineers. The SMPTE standardized the time-code signal for use in video production, and you can use it in audio recording as well. SMPTE time code is something like a digital tape counter, where the counter time is recorded as a signal on tape or hard disk.

How the Time Code Works

A time-code generator creates the time-code signal (a 1200-Hz modulated square wave). You record—or stripe—this signal onto one track of both recorders. A time-code reader reads the code off the two recorders. Then a time-code synchronizer compares the codes from the two transports and locks them together in time by varying the motor speed of one of the transports.

The counter time is recorded as a signal on tape or hard disk. Pictures on a video screen are updated approximately 30 frames per second, where a frame is a still picture made of 525 lines on the screen. 525 lines/30 frames is the format used only in the United States. SMPTE time code assigns a unique number (address) to each video frame—8 digits that specify hours:minutes:seconds:frames.

Each video frame is identified with its own time-code address; for example, 01:26:13:07 means "1 hour, 26 minutes, 13 seconds, and 7 frames." These addresses are recorded sequentially: for each successive video frame, the time-code number increases by one frame count. There are approximately 30 frames per second in the American TV system, so the time code counts frames from 0 to 29 each second.

Time-Code Signal Details

The SMPTE time code is a data stream that is divided into code words. Each code word includes 80 binary digits (or bits) that identify each video frame (Figure C.1).

The 80-bit time-code word is synchronized to the start of each video frame. The code uses binary 1's and 0's. During each half-cycle of the square wave, the voltage may be constant (signifying a 0) or changing (signifying a 1). That is, a voltage transition in the middle of a half-cycle of the square wave equals a 1. No transition signifies a 0. This is called biphase modulation (Figure C.2). It can be read forward or reverse, at almost any tape speed. A time-code reader detects the binary 1's and 0's and converts them to decimal numbers to form the time-code addresses.

SMPTE words also can include user information. There are 32 multipurpose bits (8 digits or 4 characters) reserved for the user's data—for example, the take number.

The last 16 bits in the word are a fixed number of 1's and 0's called sync bits. These bits indicate the end of the time-code word, so that the time-code reader can tell whether the code is being read forward or in reverse.

Drop-Frame Mode

SMPTE code can run in various modes depending on the application. One of these, Drop-Frame mode, is needed for specific reasons.

Black-and-white video runs at 30 frames per second. A time-code signal also running at 30 frames per second will agree with the clock on the wall. Color video, on the other hand, runs at 29.97 frames per second.

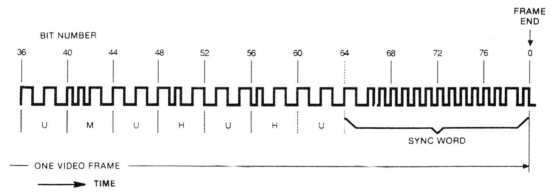

Figure C.1   An 80-bit time-code word.

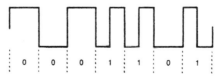

Figure C.2   Biphase modulation used in SMPTE time code.

If a color program is clocked at 30 frames per second for 1 hour, the actual show length will run 3.6 seconds (108 frames) longer than an hour.

The Drop-Frame mode causes the time code to count at a rate to match the clock on the wall. Each minute, frame numbers 00 and 01 are dropped, except every 10th minute. (Instead of seeing frames . . . 27, 28, 29, 00 on the counter; you see frames . . . 27, 28, 29, 02.) This speeds up the time-code counter to match the rate of the video frames. The video

511

frames still progress at 29.97 frames per second, and the time code progresses at 30 frames per second, but it drops every few frames—so the effective time-code frame rate is 29.97 frames per second.

You program the time-code generator to operate in Drop (Drop-Frame) or Non-Drop (Non-Drop-Frame) mode. Non-Drop can be used for audio-only synchronizing, but Drop mode should be used if the audio will be synced to a video tape later on.

Setting Up a Time-Code System

To use the SMPTE time code, you need a time-code generator, reader, and synchronizer. These may be all-in-one or separate units. Figure C.3 shows a typical system hookup, in which the generator, reader, and synchronizer are combined in one unit.

Set the generator to Time-of-Day code, or any other convenient starting time. If you are syncing to video, feed the generator a sync signal from the video source being recorded. This will lock the generator together in time with the video source. For audio-only applications, use the internal crystal sync.

Select Drop-Frame or Non-Drop-Frame mode, and stay with it for the entire production. Use Drop-Frame mode if you anticipate syncing audio to video in the future.

Figure C.3 Typical hookup for synchronizing two tape transports with SMPTE time code.

Next, set the frame rate: 29.97, 30, 24, or 25 frames per second, using the following guidelines:

- Color video productions require 29.97 frames per second.
- Black-and-white video or audio-only productions use 30 frames per second.
- Film usually runs at 24 frames per second.
- European TV-using EBU (European Broadcast Union) time code-requires 25 frames per second.

The time-code signal appears at the generator output, which is a standard 3-pin audio connector. Signal level is +4 dBm. The signal is fed through a standard 2-conductor shielded audio cable. To avoid crosstalk of time code into audio channels, separate the time-code cables from audio cables. Patch the time-code signal into an outside track of the recorders you want to lock together. Then, patch the outputs of those time-code tracks to the inputs of the time-code reader.

The reader decodes the information recorded on tape or hard disk and, in some models, displays the time-code data in the hours:minutes:seconds frames format. Some readers have an error bypass feature that corrects for missing data.

The time-code synchronizer matches bits between two time-code signals to synchronize them. The synchronizer compares tape direction, address, and phase to synchronize two SMPTE tracks via servo control of the transport motors. The two tape machines to be synced are called "master" and "slave." The synchronizer controls the slave by making its tape position and speed follow that of the master.

Connect the shielded multipin interface cable between the synchronizer and slave machine to control the slave's tape transport and motors. This interface cable has channels for controlling the capstan motor, tape direction, shuttle modes, and tachometer (more on the tach later).

Because the time-code signal becomes very high in frequency when the tape is shuttled rapidly, special playback amplifier cards with extended high-frequency response may be needed to reproduce the SMPTE signal accurately. These cards are available from the recorder manufacturer.

Unfortunately, when the tape is in shuttle mode (fast forward or rewind), the tape usually is lifted from the heads—losing the SMPTE signal. In this case, the recorders are synchronized using tach pulses from the recorders as a replacement for the SMPTE time code. Some synchronizers are fed tach pulses from the slave only.

If chase mode is available, the slave follows the shuttle motions of the master. If the master is put in fast-forward, the slave goes into fast-forward, and so on. Without chase mode, the synchronizer notes the address of the master tape when it is stopped and cues the slave to match that location. Chase mode is useful for repetitive overdubs.

How to Use SMPTE Time Code

Suppose you want to synchronize two multitrack recorders. Follow this procedure:

1. If you're working with analog tape recorders, clean the heads and the tape path.
2. Record the SMPTE time code on an outside track of both analog recorders at −5 to −10 VU, leaving the adjacent track blank, if possible, to avoid time-code crosstalk. Don't put high-transient sounds (such as drums) on that adjacent track; they can cause sync problems in analog tape machines. You might record the time code on two tracks, which reduces the potential for dropouts.
3. Start recording, or striping, the code about 20 seconds before the music starts, and continue nonstop with no breaks in the signal. Stripe the two tapes simultaneously. If that is not possible, you need a time-code editor to correct or insert an offset.
4. During playback, manually cue the slave to approximately the same point as the master tape, using time-code address information as a reference.
5. Engage the synchronizer in Lock and Chase mode, and enable it.
6. Put both recorders in Play mode.
7. Adjust the slave's tape speed to gradually reduce the error between transports to less than one time-code frame.

With some synchronizers, this operation is automatic. You set the slave tape to approximately the same point as the master tape. Then put the master in Play. When the synchronizer detects master time code, it sets the slave machine in Play mode and, in a few seconds, adjusts the slave's speed to synchronize the two recorders. This condition is called locked up.

514

When you record on two synchronized transports, try not to split stereo pairs between two tapes. The slight time differences between machines can degrade stereo imaging. Keep all stereo pairs on the same tape, copying them if necessary onto the other tape.

When synching audio to video, get a word-clock signal derived from video sync, and also get time code from the video system.

Restriping Defective Code

You may encounter degraded or erased sections on a time-code track. This lost code must be replaced with good code in proper sequence. If you need to re-record (restripe) a defective SMPTE track, use the Jam Sync mode on the time-code generator. This feature produces new code that matches the original addresses and frame count.

For example, suppose the slave tape needs to be restriped. Follow this procedure:

1. Patch the slave's time-code track into the generator set to Jam Sync mode.
2. Patch the generator output into the time-code track input on the slave machine (or into another track).
3. Play the tape. The time-code reader built into the generator detects a section of good code and initializes the generator with that information.
4. Start recording the new, regenerated code over the bad data (or on a new track).

Jam Sync also should be used when you copy a tape containing time code. With Jam Sync in operation, the code is regenerated to create a clean copy. This procedure is preferable to copying the time-code track directly because each generation can distort the code signal.

Audio-for-Video SMPTE Applications

With the advent of music videos and other audio/video combinations, there's a widespread need to sync audio to video. Studios doing sound-track work for film or video can use SMPTE time code to synchronize sound and picture for overdubbing narration, dialog, lip-sync, music, environmental sounds, or sound effects.

515

Synchronizing to Video

Running audio and video recordings in synchronization for TV audio editing is a typical postproduction method. You can edit the audio and video portions of a program independently even though they are locked together in time.

When you sync audio and video, select Longitudinal Time Code (LTC) or Vertical Interval Time Code (VITC). Longitudinal code records along the length of an audio track on the video tape. Vertical Interval code is combined with the video signal and is placed in the vertical blanking interval—the black bar seen over the TV picture when it is rolling vertically. VITC frees up an audio track for other purposes.

If you record the time-code signal on an audio or cue track of the video recording, do not use automatic level control because it may distort the SMPTE waveform. Instead, adjust the time-code signal level manually.

Some time-code systems include a character inserter that displays the address on the video monitor. If desired, these addresses can be recorded with (burned into) the picture—a feature called window dub.

The Audio-Tape Synchronization Procedure

At a typical on-location video shoot, the video from the camera(s) is recorded on a videocassette recorder, while the audio from the microphones is recorded on a separate high-quality tape recorder (such as a Nagra) or a portable DAT with SMPTE time-code capability. Both video and audio tapes are prestriped with SMPTE time code so that they can be synchronized later in postproduction.

Back at the studio, you connect the audio and video decks for SMPTE sync as described earlier. When you play the video tape, the SMPTE time code locks the picture and sound together. You can equalize the audio tape or change levels, and then lay it back (copy it) to the video cassette.

If you sync video to a multitrack tape recorder, you can run the video over and over as you refine the mix. Update your mix moves with an automated mixer or automated mixing program. Finally, when the mix is satisfactory, record the mixer output signal onto the video tape.

This procedure eliminates the dubbing step when transferring the audio soundtrack to video tape. That is, you can mix the multitrack tape master directly to the video tape (keeping sync), rather than mixing down to 2-track and dubbing that to video tape.

Using SMPTE with a Digital Audio Workstation

SMPTE can be used with a Digital Audio Workstation (DAW). The recording/editing software automates the playback of music and sound-effects cues for motion-picture and video post-productions. Using SMPTE time code, you can synchronize audio events, such as sound effects, to film or video tape. Some popular programs for audio/video work are Cakewalk Sonar Producer 4.0, BIAS Deck, MOTU Digital Performer, Sonic Foundry's Vegas Video, Avid Xpress Studio, and Adobe Audition.

Either SMPTE or MIDI time code (MTC) keeps the soundtrack and video in sync. MTC is a SMPTE signal sent on a MIDI cable. Using MIDI time code, you can cue MIDI devices to play music and sound effects at various SMPTE times relative to the video program.

You create a cue list (also known as an edit decision list or EDL) of audio events, each with its own time-code address. These events are played by MIDI instruments, or are played from recordings on a computer hard disk.

You record the video and audio onto hard disk. The video shows up in a window on your computer screen. While watching the video, you note the SMPTE times where you want sound effects or music to occur. In this way you create the sound track and sync it to video, all in your computer. Then dump the edited soundtrack to a prestriped video-cassette or DVD.

This process can go through various steps.

1. First, you're handed a work tape or DVD, which is a video recording of the program you're working on. It has SMPTE time code already striped on the DVD, and the SMPTE time code appears in a window on-screen called a window dub. As stated before, some programs let you copy the video onto your hard disk.

2. Watch the picture. Using a MIDI keyboard workstation or your DAW, compose and record musical parts related to the video scenes and their SMPTE start/stop times. These times indicate how long the music needs to be for each musical segment.

517

3. Dialog and wild sound are often recorded on a portable 2-track recorder with a center track for time code. (Wild sound is ambient noise recorded on the set while shooting, not synced with the video.) Sync the dialog tape to the video.

4. Missing or poorly recorded dialog is replaced during a process called looping or automatic dialog replacement (ADR). Have actors watch the video and lip-sync their lines using the DAW.

5. Record sound effects so that they line up with corresponding events in the video. Audition several sound effects from CD libraries, pick the ones you like, and import them into tracks in the digital audio workstation.

6. Using slow motion or freeze frame, go through the video and note the SMPTE times where each sound effect should occur. You also can nudge audio clips in time to align with video events.

DAW Features

The following are some of the many tasks that are offered in some DAWs:

- Spot and lay back sound effects (turn them on at the proper times and record them onto the video tape).
- Do an automated mix.
- Back time events.
- Repeat events.
- Name events.
- Print cue lists, libraries, and recording logs.
- Map keyboards (show each effect's location on a piano-style keyboard).
- Cut/copy/paste, insert, and delete events.
- Enter cue locations by tapping on the space bar as the program progresses.
- Display sound-effect cues graphically along with the music.
- Convert sequencer files to a cue sheet (hit list).
- Indicate both SMPTE time code and bar/beat for each event.
- Expand or compress the duration of a soundtrack.
- Enter subtle tempo variations, or introduce time offsets, to make cues fall exactly on the beat.

- Write standard MIDI files with meter, beat, tempo, and event data for use in a sequencer.
- Lock to MTC and SMPTE.

You also can perform various other tasks found in sequencer programs, such as quantization, tap-in tempo, and so on.

Other Time-Code Applications

SMPTE time code allows video editing under computer control. In editing a video program, you copy program segments from two or more video recordings onto a third recorder. On a computer you specify the edit points (time-code addresses) where you want to switch from one video source to another. You can rehearse edits as often as required.

Time code is used also as an index for locating cue points in a recording. During a mixdown, you can use these cue points to indicate where to make changes in the mix.

Time code also can be used as a reference for console automation and MIDI instruments. With this latter application, MIDI synthesizers can be cued to any point within a sequence, rather than having to start at the beginning.

By using SMPTE time code to lock together audio or video programs, you can greatly expand your operating flexibility.

FURTHER EDUCATION

The following books, magazines, and literature are recommended to anyone who wants more education in recording technology.

Books and Videos

The Library

Go to your local library and search their database for books on recording, home recording, music recording, audio recording, MIDI, or sound production. Read everything you can on the subject.

Pro Audio Books

This company offers an online catalog of all sorts of audio and recording books, videos and courses at www.proaudiobooks.com.

Music Books Plus

This is a catalog with a huge number of audio and recording books. The Web site is www.musicbooksplus.com.

Focal Press

This is another major audio/recording books catalog and it is found at http://books.elsevier.com/focalbooks/. For a detailed explanation of stereo theory and stereo mic techniques, search the catalog for *On Location Recording Techniques* by Bruce Bartlett, published by Focal Press.

Amazon.com

This online bookstore has a great search engine. In the search field, type in whatever audio subject you are interested in: DVD, MIDI, digital audio, recording, mixing, microphones, etc.

Recording Magazines

Recording (home and project studio recording), www.recording-mag.com.

Mix (pro recording and concert sound), www.mixonline.com.

EQ (home, project, and pro recording), www.eqmag.com.

EQ also offers a recording and sound buyer's guide.

Electronic Musician (home and project studio recording), www.emusician.com.

Electronic Musician offers a supplement called "Desktop Music Production Guide."

Tape Op (home and project studio creative recording), www.tapeop.com.

Sound On Sound (Britain's premier recording publication), www.soundonsound.com.

In those magazines are ads related to recording products and services.

Pro Audio Magazines

Journal of the Audio Engineering Society (JAES; pro audio engineering. Scholarly. With its journal, conventions, workshops and local chapters, the Audio Engineering Society is a tremendous resource, www.aes.org.

Pro Audio Review (reviews of pro audio equipment), www.proaudioreview.com.

Live Sound (concert sound reinforcement), www.livesoundint.com.

Church Production (audio for houses of worship), www.churchproduction.com.

Technologies for Worship (audio for houses of worship), www.tfwm.com.

Consumer Audio Magazines

Sound and Vision (consumer audio and video), www.soundandvisionmag.com.

Stereophile (high-end audio), www.stereophile.com.

The Absolute Sound (high-end audio), www.theabsolutesound.com.

Guides, Brochures, and Other Literature

Microphone application guides are available from

Crown International, www.crownaudio.com.

Shure Inc., www.shure.com.

Countryman Associates Inc., www.countryman.com.

AKG Acoustics Inc., www.akg.com.

Sennheiser Electronic Corp., www.sennheiserusa.com.

Audio-Technica U.S. Inc., www.audiotechnica.com.

Neumann/USA, www.neumannusa.com.

Schoeps Mikrofone, www.schoeps.de.

You can find valuable information in user manuals and free sales literature provided by manufacturers of recording equipment. Ask your equipment dealer for manufacturers' phone numbers and Web site URLs.

Guides to Recording Schools

The Audio Engineering Society offers a directory of educational programs at www.aes.org/education/directory.cfm.

Each July issue of *Mix* magazine contains a comprehensive directory of recording schools, seminars, and programs. Universities and colleges in most major cities have recording-engineering courses.

An index of recording schools is at www.modrec.com/schools/

The Internet

A great place to ask questions, besides magazines, is on the Internet. In Google, select Groups, then type in rec.audio.pro. You can ask questions and get answers from pro engineers. You may get conflicting answers, because often there are many ways to do the same thing. Also, some who reply are more expert than others. But you'll often find stimulating debates.

Some other valuable Web sites are www.homerecording.com, www.prorec.com, www.kvraudio.com, www.recording.org, www.vintagesynth.com, www.allmusic.com, www.microphones.com, www.digido.com, www.prosoundweb.com, www.digifreq.com, www.cdrfaq.org/, www.harmony-central.com/recording, www.musicplayer.com (forums), www.recordingconnection.com, www.popeye-x.com/tech/xbert.htm, www.audio-recording-center.com, and www.josephson.com/mic-faq.html.

Also see pcrecording.com, audiogrid.com, digitalprosound.com, dvdforum.org, superaudiocd.com/, artofthemix.org, recordproduction.com/, audiodirectory.nl/, binaural.com, artistpro.com, sweetwater.com (check out Publications and Forums), tweakheads.com/ how_to_research_studio_gear.html, http://dir.yahoo.com/Entertainment/Music/Recording/, and www.solorb.com/dat-heads/.

If you lack good multitrack recordings with which to practice mixing, go to www.raw-tracks.com. There you can download individual tracks in wav or mp3 format, or purchase a CD of raw tracks.

Some online MIDI resources are: midi.com, midiworld.com, ultimatemidi.com/, harmony-central.com/midi, borg.com/~jglatt/tutr/miditutr.htm, cakewalk.com/tips/desktop.asp, and atarimagazines.com/v7n2/MIDIResources.html.

The Web sites of audio equipment manufacturers have support sections, online discussion groups, and FAQs with lots of information. Search for the name and model number of equipment you have, or are interested in.

Do a Google search with the keywords audio, recording, recording vocals, audio recording links, DAT, digital audio, and so on. Also search for FAQs on various audio topics. You'll discover hundreds of audio-related websites and links. In Google, you can search Groups as well as Web sites.

Recording Equipment Catalogs

Here are but a few catalogs from which to order recording gear:

American Musical Supply, www.americanmusical.com.

BSW, www.bswusa.com.

Guitar Center, www.guitarcenter.com.

Musician's Friend, www.musiciansfriend.com.

Sam Ash, www.samash.com.

Sweetwater Sound, www.sweetwater.com.

The Woodwind and the Brasswind, www.wwandbw.com.

Experience

It's the best teacher. Record all you can with any equipment you have. You can buy a 4-track cassette recorder on ebay for $45, or download free multitrack recording software. Then buy a cheap mic, and practice mic techniques and overdubbing.

IMPEDANCE

Impedance is one of audio's more confusing concepts. To clarify this topic, I'll present a few questions and answers about impedance.

What Is Impedance?

Impedance (Z) is the resistance of a circuit to alternating current, such as an audio signal. Technically, impedance is the total opposition (including resistance and reactance) that a circuit has to passing alternating current.

A high-impedance circuit tends to have high voltage and low current. A low-impedance circuit tends to have relatively low voltage and high current.

I'm Connecting Two Audio Devices. Is It Important to Match Their Impedances? What If I Don't?

First some definitions. When you connect two devices, one is the source and one is the load. The source is the device that puts out a signal. The load is the device you are feeding the signal into. The source has a certain output impedance, and the load has a certain input impedance.

A few decades ago in the vacuum-tube era, it was important to match the output impedance of the source to the input impedance of the load. Usually the source and load impedances were both 600 ohms. If the source impedance equals the load impedance, this is called "matching" impedances. It results in maximum power transfer from the source to the load.

In contrast, suppose the source is low Z and the load is high Z. If the load impedance is 10 times or more the source impedance, it is called a "bridging" impedance. Bridging results in maximum voltage transfer from the source to the load. Today, nearly all devices are connected bridging—low-Z out to high-Z in—because we want the most voltage transferred between components.

If you connect a low-Z source to a high-Z load, there is no distortion or frequency-response change caused by this connection. But if you connect a high-Z source to a low-Z load, you might get distortion or altered response. For example, suppose you connect an electric bass guitar (a high-Z device) into an XLR-type mic input (a low-Z load). The low frequencies in the signal will roll off, so the bass will sound thin. And the highs might roll off, making the sound dull.

We want the bass guitar to be loaded by a high impedance, and we want the mic input to be fed by a low-impedance signal. A direct box or impedance-matching adapter does this (Figure E.1). Such adapters are available from Radio Shack.

The adapter is a tube with a phone jack on one end and a male XLR connector on the other. Inside the tube is a transformer. Its primary winding is high Z, wired to the phone jack. The transformer's secondary winding is low Z, wired to the XLR. You plug the guitar cord into the phone jack, and plug the XLR into a mic input in a snake or mixer. Use it with a bass guitar, electric guitar, or synth.

This impedance-matching adapter works, but is not ideal. The load it presents to the bass guitar might be 12 kilohms, which will slightly load down the high-Z guitar pickup, causing thin bass.

HI-TO-LOW IMPEDANCE MATCHING ADAPTER

Figure E.1 High-to-low impedance matching adapter.

An active direct box solves this problem. In place of a transformer, the active DI usually has a field effect transistor (FET). The FET has a very high input impedance that does not load down the bass guitar.

What About Microphone Impedance?

Recording-quality mics have XLR (3-pin) connectors and are low Z (150 to 300 ohms). A low-Z mic can be used with hundreds of feet of cable without picking up hum or losing high frequencies.

I'm Connecting a Mic to a Mixer. Is Impedance a Consideration?

Yes. If your mixer has phone-jack inputs, they are probably high Z. But most mics are low Z. When you plug a low-Z mic into a high-Z input you get a weak signal. That's because a high-Z mic input is designed to receive a relatively high voltage from a high-Z mic, and so the input is designed to have low gain. So you don't get much signal amplification.

If you can't get enough level when you plug a mic into a phone-jack input, here's a solution: Between the mic cable and the input jack, connect an impedance matching adapter (Figure E.2). It steps up the voltage of the mic, giving it a stronger signal.

The adapter is a tube with a female XLR input and a phone-plug output. Inside the tube is a transformer. Its primary winding is low Z, wired to the XLR. It secondary winding is high Z, wired to the phone plug. Connect the mic to the XLR; connect the phone plug to the mixer's phone jack. Then the mixer will receive a strong signal from the mic.

If you're using a phantom-powered condenser mic, the connections are different. First, turn off any phantom power in your recorder-mixer.

Figure E.2   Low-to-high impedance matching adapter.

Connect your mic to a standalone phantom-power supply, and connect the supply output to the impedance-matching adapter.

If your mixer has XLR inputs, they are low-Z balanced. In this case, simply connect the mic to the mixer using a mic cable with a female XLR on the mic end and a male XLR on the mixer end. A low-Z mic input is typically about 1500 ohms, so it provides a bridging load to a mic that is 150 to 300 ohms.

Should I Consider Impedance When I Connect Two Line-Level Devices?

This is seldom a problem. In most audio devices, the impedance of the line output is low—about 100 to 1000 ohms. The impedance of the line input is high—about 10 kilohms to 1 megohm. So every connection is bridging, and you get maximum voltage transfer. Some audio devices, such as passive equalizers, require a terminating resistor at the input or output for best performance.

Can I Connect One Source to Two or More Loads?

Usually yes. You can connect several devices in parallel across one line output. Suppose you connect a mixer output simultaneously to a recorder input, an amplifier input, and another mixer's input in parallel. The combined input impedance of those three loads might be 4000 ohms, which still presents a bridging load to the mixer's 100-ohm output impedance.

Mics are a different story. If you connect one mic to two or more mixers with a Y cable, the combined input impedance will be about 700 ohms or less. This can load down some microphones, reducing the bass in dynamic mics or causing distortion in condenser mics. One solution is to use a transformer mic splitter.

Can I Connect Two or More Sources to One Input?

Not recommended. If you combine two or more sources into a single load, the low-output impedance of one source will load down the output of the other source, and vice versa. This can cause level loss and distortion.

If you want to combine the signals from two devices into one input, you need to put a series resistor in line with each device before combining them. That prevents each device from loading down the other. A minimum resistor value might be 470 ohms per source. If the source is balanced, use one resistor on pin 2 and one on pin 3—two resistors per source.

Summary

- Impedance (Z) is the opposition to alternating current, measured in ohms.
- Microphones and line outputs are usually low Z.
- Electric guitars, synthesizers, and line inputs are usually high Z.
- XLR mic inputs are low impedance; phone jack mic inputs are high impedance.
- Speakers are usually 4 to 8 ohms.
- Equal impedances in parallel result in half the impedance.
- Equal impedances in series result in twice the impedance.
- Connect low-Z sources to low-Z inputs. (A low-Z input is usually 7 to 10 times the source impedance, but it's still called a low-Z input.)
- Connect high-Z sources to high-Z inputs.
- Connect a low-Z source to a high-Z input through a step-up transformer (impedance matching adapter).
- Connect a high-Z source to a low-Z input through a step-down transformer (impedance matching adapter, or direct box).

GLOSSARY

A WEIGHTING   *See* Weighted.

AAC (Advanced Audio Codec)   A relatively new compressed audio file format, AAC offers better sound quality than MP3, but with less storage space and bandwidth. It's also used with Digital Rights Management technology to help control the copying and distributing of music.

A-B   A listening comparison between two audio programs, or between two components playing the same program, performed by switching immediately from one to the other. The levels of the two signals are matched. *See* also Spaced-Pair.

AC-3   Same as Dolby Digital, a perceptual encoding scheme that data-reduces the six surround channels of a 5.1 system to two channels.

ACCENT MICROPHONE   *See* Spot Microphone.

ACCESS JACKS   *See* Insert Jacks.

ACTIVE COMBINING NETWORK   A combining network with gain. *See* Combining Network.

AES   Audio Engineering Society.

AES/EBU   Also called IEC 988 Type 1. An interface format for digital signals, using a balanced 110 ohm mic cable terminated with XLR-type connectors. *See* also S/PDIF.

AIFF (Audio Interchange File Format)   A standard Mac format for uncompressed digital audio files.

ALIGNMENT   The adjustment of tape-head azimuth and of tape-recorder circuitry to achieve optimum performance from the particular type of tape being used.

ALIGNMENT TAPE A prerecorded tape with calibrated tones for alignment of a tape recorder.

AMBIENCE Room acoustics, early reflections, and reverberation. Also, the audible sense of a room or environment surrounding a recorded instrument.

AMBIENCE MICROPHONE A microphone placed relatively far from its sound source to pick up ambience.

AMPLITUDE, PEAK On a graph of a sound wave, the sound pressure of the waveform peak. On a graph of an electrical signal, the voltage of the waveform peak. The amplitude of a sound wave or signal as measured on a meter is 0.707 times the peak amplitude.

ANALOG-TO-DIGITAL (A/D) CONVERTER A circuit that converts an analog audio signal into a stream of digital data (bitstream).

ANTI-ALIAS FILTER In an A/D converter, a lowpass filter that removes all frequencies above 20 kHz before sampling in order to prevent audio artifacts called aliasing.

ANTI-IMAGING FILTER In a D/A converter, a lowpass filter that smooths the voltage steps in the analog signal that was generated by translating digital numbers into analog voltages. The anti-imaging filter recovers the waveform of the original analog signal.

ASIO (Audio Stream Input/Output) Steinberg's computer audio driver spec for Mac and Windows. ASIO has low latency because it interfaces directly between a sound card and the audio application software.

ASSIGN To route or send an audio signal to one or more selected channels.

ATRAC (Adaptive Transform Acoustic Coding) A data compression scheme (used in the MiniDisc) that reduces by 5:1 the storage needed for digital audio. ATRAC is a perceptual coding method that omits data deemed inaudible due to masking.

ATTACK The beginning of a note. The first portion of a note's envelope in which a note rises from silence to its maximum volume.

ATTACK TIME In a compressor, the time it takes for gain reduction to occur in response to a musical attack.

ATTENUATE To reduce the level of a signal.

534

ATTENUATOR   In a mixer (or mixing console) input module, an adjustable resistive network that reduces the microphone signal level to prevent overloading of the input transformer and mic preamplifier.

AUDIO INTERFACE   A device that connects to a computer and converts an audio signal into computer data for storage in memory or on hard disk. The interface also converts computer data into an audio signal. *See* Breakout Box, I/O Box, and Sound Card.

AUTOLOCATE   A recorder function that makes the tape or disk go to a program address (counter time) at the press of a button.

AUTOMATED MIXING   A system of mixing in which a computer remembers and updates mixer control settings and moves. With this system, a mix can be performed and refined in several stages and played back at a later date exactly as set up previously.

AUXILIARY BUS (AUX BUS)   A bus or channel that contains a mix of the aux-send signals of the input modules in a mixer. An aux bus is used to send signals to an effects unit or monitor system. *See* Effects Bus.

AUXILIARY SEND (AUX-SEND)   A control in a mixer's input module used to send that module's signal to an aux bus. The aux-send level adjusts the amount of effects heard on an instrument, or adjusts the loudness of that instrument in the monitor system.

A/V DRIVE   A hard-disk drive meant for audio/video use. It postpones thermal recalibration until the disk is inactive, preventing data errors.

AZIMUTH   In a tape recorder, the angular relationship between the head gap and the tape path.

AZIMUTH ALIGNMENT   The mechanical adjustment of the record or playback head to bring it into proper alignment (90 degrees) with the tape path.

BACK-TIMING   A technique of cueing up the musical background or a sound effect to a narration track so that the music or effect ends simultaneously with the narration.

BAFFLED-OMNI   A stereo miking arrangement that uses two ear-spaced omnidirectional microphones separated by a hard padded baffle.

BALANCE   The relative volume levels of various tracks or instruments.

535

BALANCED AC POWER   AC power from a center-tapped power transformer. Instead of one 120 V line and one 0 V line, it has two 60 V lines. They are in phase with each other and sum to 120 V. But they are connected to the center-tap ground out of phase (one is +60 V; the other is −60 V). Any hum and noise on the grounding system cancel out.

BALANCED LINE   A cable with two conductors surrounded by a shield, in which each conductor is at equal impedance to ground. With respect to ground, the conductors are at equal potential but opposite polarity; the signal flows through both conductors.

BANDPASS FILTER   In a crossover, a filter that passes a band or range of frequencies but sharply attenuates or rejects frequencies outside the band.

BASIC TRACKS   Recorded tracks of rhythm instruments (bass, guitar, drums, and sometimes keyboard).

BASS MANAGEMENT   A subwoofer/satellite crossover, usually part of a surround receiver. Bass management routes frequencies above about 100 Hz to the five full-range speakers, and routes frequencies below about 100 Hz from all six channels to the subwoofer. It affects only what you monitor, not what goes on tape. Bass management can be done by a surround-receiver circuit, a standalone box, a special circuit in a sub-woofer, or a software plug-in.

BASS TRAP   An assembly that absorbs low-frequency sound waves in the studio.

BI-AMPLIFICATION (BI-AMPING)   Driving a woofer and tweeter with separate power amplifiers. An active crossover is connected ahead of these power amplifiers.

BIAS   In tape-recorder electronics, an ultrasonic signal that drives the erase head. This signal is also mixed with the audio signal applied to the record head to reduce distortion.

BIDIRECTIONAL MICROPHONE   A microphone that is most sensitive to sounds arriving from two directions-in front of and behind the microphone. It rejects sounds approaching either side of the microphone. Sometimes called a cosine or figure-eight microphone because of the shape of its polar pattern.

BINAURAL RECORDING   A 2-channel recording made with an omni-directional microphone mounted near each ear of a human or a dummy head, for playback over headphones. The object is to duplicate the acoustic signal appearing at each ear.

BIT DEPTH (word length)   The number of bits (ones and zeros) making up a word in a digital signal (such as 16- or 24-bits). Each word is a binary number that is the value of each sample. A sample is a measurement of an analog waveform that is done several thousand times a second during analog-to-digital conversion. High bit depth = long word length = high resolution of the analog signal amplitude = high sound quality.

BLUMLEIN ARRAY   A stereo microphone technique in which two coincident bidirectional microphones are angled 90 degrees apart (45 degrees to the left and right of center).

BOARD   *See* Mixing Console.

BOUNCING TRACKS   A process in which two or more tracks are mixed, and the mixed tracks are recorded on an unused track or tracks. Then the original tracks can be erased, which frees them up for recording more instruments.

BOUNDARY MICROPHONE   A microphone designed to be used on a boundary (a hard reflective surface). The microphone capsule is mounted very close to the boundary so that direct and reflected sounds arrive at the microphone diaphragm in phase (or nearly so) for all frequencies in the audible band.

BREAKOUT BOX (I/O BOX)   A group of audio input and output connectors in a chassis, which is wired to a sound card, a USB port, or a FireWire port in a computer. Used to interface analog audio signals (and often MIDI and digital signals) with a computer.

BREATHING   The unwanted audible rise and fall of background noise that may occur with a compressor. Also called pumping.

BULK TAPE ERASER   A large electromagnet used to erase a whole reel of recording tape at once.

BUS   A common connection of many different signals. An output of a mixer or submixer. A channel that feeds a tape track, signal processor, or power amplifier.

BUS IN   An input to a program bus, usually used for effects returns.

BUS MASTER   In the output section of a mixing console, a potentiometer (fader or volume control) that controls the output level of a bus. Also called Group Fader.

BUS OUT   The output connector of a bus.

BUS TRIM   A control in the output section of a mixing console that provides variable gain control of a bus, used in addition to the bus master for fine adjustment.

BUZZ   An unwanted edgy tone that sometimes accompanies audio, containing high harmonics of 60 Hz.

CALIBRATION   *See* Alignment.

CAPACITOR   An electronic component that stores an electric charge. It is formed of two conductive plates separated by an insulator called a dielectric. A capacitor passes AC but blocks DC.

CAPACITOR MICROPHONE   *See* Condenser Microphone.

CAPSTAN   In a tape-recorder transport, a rotating post that contacts the tape (along with the pinch roller) and pulls the tape past the heads at a constant speed during recording and playback.

CARDIOID MICROPHONE   A unidirectional microphone with side attenuation of 6 dB and maximum rejection of sound at the rear of the microphone (180 degrees off-axis). A microphone with a heart-shaped directional pattern.

CD   *See* Compact Disc.

CD-R (CD-Recordable)   A recordable compact disc that cannot be rewritten. Once recorded, it cannot be erased and reused.

CD-ROM   A computer disk drive that plays computer data from CD-ROM disks. The disks are read optically by a laser, as in a compact disc player.

CD-RW (CD-Rewritable)   A recordable compact disc that can be rewritten. Once recorded it can be erased and reused.

CHANNEL   A single path of an audio signal. Usually, each channel contains a different signal.

CHANNEL ASSIGN   *See* Assign.

CHORUS   1. A special effect in which a signal is delayed by 15 to 35 msec, the delayed signal is combined with the original signal, and the delay is varied randomly or periodically. This creates a wavy, shimmering effect. 2. The main portion of a song that is repeated several times throughout the song with the same lyrics.

CLEAN   Free of noise, distortion, overhang, leakage. Not muddy.

CLEAR   Easy to hear, easy to differentiate. Reproduced with sufficient high frequencies.

CLIP   *See* Region.

COINCIDENT-PAIR   A stereo microphone, or two separate microphones, placed so that the microphone diaphragms occupy approximately the same point in space. They are angled apart and mounted one directly above the other.

COMB-FILTER EFFECT   The frequency response caused by combining a sound with its delayed replica. The frequency response has a series of peaks and dips caused by phase interference. The peaks and dips resemble the teeth of a comb.

COMBINING AMPLIFIER   An amplifier at which the outputs of two or more signal paths are mixed together to feed a single track of a tape recorder.

COMBINING NETWORK   A resistive network at which the outputs of two or more signal paths are mixed together to feed a single track of a tape recorder.

COMPACT DISC (CD)   A read-only optical disc medium for storing digital audio programs up to 74 minutes long. The compact disc stores data in the form of a spiral groove of microscopic pits and is read optically by a laser. CD's digital audio format is 44.1 kHz sampling rate and 16-bit word length.

COMPLEX WAVE   A wave with more than one frequency component.

COMPING   Recording several musical performances (takes) of a single instrument or vocal on different tracks, and selecting the best parts of each take to be played in order on a composite track during mixdown.

COMPOSITE TRACK   A track containing the best parts of several takes of a musical-instrument or vocal performance.

COMPRESSION   1. The portion of a sound wave in which air molecules are pushed together, forming a region with higher-than-normal atmospheric pressure. 2. In signal processing, the reduction in dynamic range or gain caused by a compressor. 3. Data compression or data reduction is an encoding scheme to reduce the size of a data file by throwing away audio data deemed inaudible because of masking. ATRAC, MP3, MLP, AAC, RealAudio, OGG, and Microsoft Media are examples of compressed data formats.

COMPRESSION RATIO (SLOPE)   In a compressor, the ratio of the change in input level (in dB) to the change in output level (in dB). For example, a 2:1 ratio means that for every 2 dB change in input level, the output level changes 1 dB.

COMPRESSOR   A signal processor that reduces dynamic range or gain by means of automatic volume control. An amplifier whose gain decreases as the input signal level increases above a preset point.

CONDENSER MICROPHONE   A microphone that works on the principle of variable capacitance to generate an electrical signal. The microphone diaphragm and an adjacent metallic disk (called a backplate) are charged to form two plates of a capacitor. Incoming sound waves vibrate the diaphragm, varying its spacing to the backplate, which varies the capacitance, which in turn varies the voltage between the diaphragm and backplate.

CONNECTOR   A device that makes electrical contact between a signal-carrying cable and an electronic device, or between two cables. A device used to connect or hold together a cable and an electronic component so that a signal can flow from one to the other.

CONSOLE   *See* Mixing Console.

CONTACT PICKUP   A transducer that contacts a musical instrument and converts its mechanical vibrations into a corresponding electrical signal.

CONTROL ROOM   The room in which the engineer controls and monitors the recording. It houses most of the recording hardware.

CONTROLLER SURFACE   A chassis with faders (and sometimes buttons and knobs) that resembles a mixer, used to adjust virtual controls that appear on-screen in computer editing software. Connected to the computer by USB or FireWire, the controller surface might include analog and digital input/output connectors and MIDI connectors.

CONVOLUTION REVERB (SAMPLING REVERB) A reverberation device or plug-in which creates the reverb from impulse-response samples (wave files) of real acoustic spaces, rather than from algorithms. The resulting sound quality is very natural.

CROSSOVER An electronic network that divides an incoming signal into two or more frequency bands.

CROSSOVER, ACTIVE (ELECTRONIC CROSSOVER) A crossover with amplifying components, used ahead of the power amplifiers in a bi-amped or tri-amped speaker system.

CROSSOVER FREQUENCY The single frequency at which both filters of a crossover network are down 3 dB.

CROSSOVER, PASSIVE A crossover with passive (nonamplifying) components, used after the power amplifier.

CROSSTALK The unwanted transfer of a signal from one channel to another. Crosstalk often occurs between adjacent tracks within a record or playback head in a tape recorder, or between input modules in a console.

CUE, CUE SEND In a mixing-console input module, a control that adjusts the level of the signal feeding the cue mixer that feeds a signal to headphones in the studio.

CUE LIST *See* Edit Decision List.

CUE MIXER A submixer in a mixing console that takes signals from cue sends as inputs and mixes them into a composite signal that drives headphones in the studio.

CUE SHEET Used during mixdown, a chronological list of mixing-console control adjustments required at various points in the recorded song. These points may be indicated by counter or ABS-time readings.

CUE SYSTEM A monitor system that allows musicians to hear themselves and previously recorded tracks through headphones.

DAMPING FACTOR The ability of a power amplifier to control or damp loudspeaker vibrations. The lower the amplifier's output impedance, the higher the damping factor.

541

DAT (R-DAT)   A digital audio tape recorder that uses a rotating head to record digital audio on tape.

DATA COMPRESSION   A data encoding scheme for reducing the amount of data storage on a medium. Same as Data Reduction. *See* Compression, ATRAC, and MP3.

DAW   Abbreviation for digital audio workstation.

dB   Abbreviation for decibel.

DEAD   Having very little or no reverberation.

DECAY   The portion of the envelope of a note in which the envelope goes from maximum to some midrange level. Also, the decline in level of reverberation over time.

DECAY TIME   *See* Reverberation Time.

DECIBEL   The unit of measurement of audio level. Ten times the logarithm of the ratio of two power levels. Twenty times the logarithm of the ratio of two voltages. dBV is decibels relative to 1 volt. dBu is decibels relative to 0.775 volt. dBm is decibels relative to 1 milliwatt. dBA is decibels, A weighted. *See* Weighted.

DECODED TAPE   A program on analog recording tape that is expanded after being compressed by a noise-reduction system. Such a program has normal dynamic range.

DE-ESSER   A signal processor or plug-in that removes excessive sibilance ("s" and "sh" sounds) by compressing high frequencies around 5 to 10 kHz.

DELAY   The time interval between a signal and its repetition. A digital delay or a delay line is a signal processor that delays a signal for a short time.

DELAY COMPENSATION   Adjusting the timing of a track that is processed by a plug-in, to make it in sync with non-processed tracks. Plug-ins add latency (delay) to a track.

DEMAGNETIZER (DEGAUSSER)   An electromagnet with a probe tip that is touched to elements of an analog recorder tape path (such as tape heads and tape guides) to remove residual magnetism.

DEPTH   The audible sense of nearness and farness of various instruments. Instruments recorded with a high ratio of direct-to-reverberant sound are perceived as being close; instruments recorded with a low ratio of direct-to-reverberant sound are perceived as being distant.

DESIGN CENTER The portion of fader travel (usually shaded), about 10 to 15 dB from the top, in which console gain is distributed for optimum headroom and signal-to-noise ratio. During normal operation, each fader in use should be placed at or near design center.

DESIGNATION STRIP A strip of paper taped near console faders to designate the instrument that each fader controls. Also called a Scribble Strip.

DESK The British term for mixing console.

DESTRUCTIVE EDITING In a digital audio workstation, editing that rewrites the data on disk. A destructive edit cannot be undone unless a copy of the original data is saved before the edit is done.

DI Short for direct injection, recording with a direct box.

DIFFUSION An even distribution of sound in a room.

DIGITAL AUDIO An encoding of an analog audio signal in the form of binary digits (ones and zeroes).

DIGITAL AUDIO WORKSTATION (DAW) A computer, audio interface, and recording software that allows you to record, edit, and mix audio programs entirely in digital form. A standalone DAW is a digital multitrack recorder-mixer. Standalone DAWs include real mixer controls; computer DAWS have virtual controls on-screen.

DIGITAL RECORDING A recording system in which the audio signal is stored as binary digits (ones and zeroes).

DIGITAL-TO-ANALOG (D/A) CONVERTER A circuit that converts a digital audio signal into an analog audio signal.

DIM To reduce the monitor volume temporarily by a preset amount so that you can carry on a conversation.

DIRECT BOX A device used for connecting an amplified instrument directly to a mixer mic input. The direct box converts a high-impedance unbalanced audio signal into a low-impedance balanced audio signal.

DIRECT INJECTION (DI) Recording with a direct box.

543

DIRECTIONAL MICROPHONE A microphone that has different sensitivity in different directions. A unidirectional or bidirectional microphone.

DIRECT OUTPUT, DIRECT OUT An output connector following a mic preamplifier, fader, and equalizer, used to feed the signal of one instrument to one track of a multitrack recorder.

DIRECT SOUND Sound traveling directly from the sound source to the microphone (or to the listener) without reflections. Also, DirectSound is an audio driver for the Windows operating system, intended as an enhancement to MultiMedia Extensions (MME).

DIRECTX (DX) A package of Windows audio, video, and game-controller drivers, also called multimedia application programming interfaces (APIs).

DISTORTION An unwanted change in the audio waveform, causing a raspy or gritty sound quality. The appearance of frequencies in a device's output signal that were not in the input signal. Distortion is caused by recording at too high a level, using improper mixer settings, components failing, or vacuum tubes distorting. (Distortion can be desirable, e.g., for an electric guitar.) In digital recording, distortion called quantization error can also occur at very low signal levels, where there are not enough bits to record the signal accurately.

DITHER Low-level noise added to a digital signal to reduce quantization distortion caused by truncating (removing) bits in a digital word. It's a good idea to add dither to a 24-bit program just before converting it to 16 bits for CD release.

DOLBY DIGITAL A perceptual coding method using AC-3 data compression, offering 6 discrete channels of digital surround sound. Dolby Digital uses a lossy encoding process to reduce the bit rate needed to transmit the six channels via a 2-channel bitstream. The standard audio format for DVD-Video.

DOLBY PRO LOGIC A surround decoder that decodes the two channels of Dolby-Surround-encoded programs back into four channels (left, center, right, surround).

DOLBY SURROUND A matrix encoding system that combines four channels (left, center, right, surround) into two channels. The Dolby Pro Logic decoder unfolds the two channels back into four. The surround channel, which is mono and limited bandwidth, is reproduced over left and right surround speakers.

DOLBY TONE   A reference tone recorded at the beginning of a Dolby-encoded analog tape for alignment purposes.

DOUBLING   A special effect in which a signal is combined with its 15- to 35-msec-delayed replica. This process mimics the sound of two identical voices or instruments playing in unison. In another type of doubling, two identical performances are recorded and played back to thicken the sound.

DROP-FRAME   For color video production, a mode of SMPTE time code that causes the time code to match the clock on the wall. Once every minute, frame numbers 00 and 01 are dropped, except every 10th minute.

DROP-OUT   During playback of a tape recording, a momentary loss of signal caused by separation of the tape from the playback head by dust, tape-oxide irregularity, etc. During playback of a hard-disk recording, a momentary loss of signal caused by a buffer memory being emptied. Increasing buffer size usually prevents drop-outs.

DRUM MACHINE   A device—hardware or software—that plays samples of real drums and includes a sequencer to record rhythm patterns.

DRY   Having no echo or reverberation. Referring to a close-sounding signal that has not yet been processed by a reverberation or delay device or plug-in.

DSD (Direct Stream Digital)   A process that encodes a digital signal in a 1-bit (bitstream) format at a 2.8224 MHz sampling rate. Offers state-of-the-art sound quality; used in the Super Audio CD.

DSP (Digital Signal Processing)   Modifying a signal in digital form by doing calculations on the numbers. DSP is used for level changes, EQ, effects, and so on.

DTS (Digital Theater System)   A perceptual coding method using data compression, offering 6 discrete channels of digital surround sound. DTS uses a lossy encoding process to reduce the bitrate needed to transmit the 6 channels via a 2-channel bitstream.

DVD (Digital Versatile Disc)   A storage medium the size of a compact disc that holds much more data. The DVD stores video, audio, or computer data. DVD-RAM and DVD-R are recordable; DVD-RW is rewritable, and DVD-A is audio only.

DVD-AUDIO   A DVD intended mainly for audio programs. It can use Dolby Digital or DTS encoded programs.

DX   *See* DirectX.

DXi   DirectX Instruments, Cakewalk's standard format for plug-in software synthesizers based on Microsoft DirectX technology.

DYNAMIC MICROPHONE   A microphone that generates electricity when sound waves cause a conductor to vibrate in a stationary magnetic field. The two types of dynamic microphones are moving coil and ribbon. A moving-coil microphone is usually called a dynamic microphone.

DYNAMIC RANGE   The range of volume levels in a program from softest to loudest.

EARTH GROUND   A connection to moist dirt (the ground we walk on). This connection is usually done via a long copper rod or an all-metal cold-water pipe.

EASI (Enhanced Audio Streaming Interface)   Emagic's sound card driver spec. EASI achieves low latency by interfacing the sound card directly with the audio application software.

ECHO   A delayed repetition of a signal or sound. A sound delayed 50 msec or more that is combined with the original sound.

ECHO CHAMBER   A hard-surfaced room containing a widely separated loudspeaker and microphone, once used for creating reverberation.

EDIT DECISION LIST (EDL)   A list of program events in order, plus their starting and ending times.

EDITING   The cutting and rejoining of magnetic tape to delete unwanted material, to insert silent spaces, or to rearrange recorded material into the desired sequence. Also, the same actions performed on a digital recording with a DAW, hard-disk recorder, or MiniDisc recorder-mixer.

EDITING BLOCK   A metal block that holds magnetic tape during the editing/splicing procedure.

EFFECTS   Interesting sound phenomena created by signal processors, such as reverberation, echo, flanging, doubling, compression, or chorus. *See* Sound Effects.

EFFECTS BUS   The bus that feeds effects devices (signal processors).

EFFECTS LOOP   A set of connectors in a mixer for connecting an external effects unit, such as a reverb or delay device. The effects loop includes a send section and a receive section. *See* Effects Send, Effects Return.

EFFECTS MIXER   A submixer in a mixing console that combines signals from effects sends, and then feeds the mixed signal to the input of an effects device, such as a reverberation unit.

EFFECTS RETURN (AUX RETURN)   In the output section of a mixing console, a control that adjusts the amount of signal received from an effects unit. Also, the connectors in a mixer to which you connect the effects-unit output signal. They might be labeled "bus in" instead. The effects-return signal is mixed with the program bus signal.

EFFECTS SEND (AUX SEND)   In an input module of a mixing console, a control that adjusts the amount of signal sent to an effects device, such as a reverberation or delay unit. Also, the connector in a mixer that you connect to the input of an effects unit. The effects-send or aux-send control adjusts the amount of effects heard on each instrument.

EFFICIENCY   In a loudspeaker, the ratio of acoustic power output to electrical power input.

EIA   Electrical Industries Association.

EIA RATING   A microphone-sensitivity specification that states the microphone output level in dBm into a matched load for a given sound pressure level (SPL). SPL + dB (EIA rating) = dBm output into a matched load.

ELECTRET-CONDENSER MICROPHONE   A condenser microphone in which the electrostatic field of the capacitor is generated by an electret—a material that permanently stores an electrostatic charge.

ELECTROSTATIC FIELD   The force field between two conductors charged with static electricity.

ELECTROSTATIC INTERFERENCE   The unwanted presence of an electrostatic hum field in signal conductors.

547

ENCODED TAPE　An analog tape containing a signal compressed by a noise-reduction unit.

END-ADDRESSED　Referring to a microphone whose main axis of pickup is perpendicular to the front of the microphone. You aim the front of the mic at the sound source. *See* Side-Addressed.

ENVELOPE　The rise and fall in volume of one note. The envelope connects successive peaks of the waves that make up the note. Each harmonic in the note might have a different envelope.

EQUALIZATION (EQ)　The adjustment of frequency response to alter the tonal balance or to attenuate unwanted frequencies.

EQUALIZER　A circuit (usually in each input module of a mixing console, or in a separate unit) that alters the frequency spectrum of a signal passed through it.

ERASE　To remove an audio signal from magnetic tape or disk by applying an ultrasonic varying magnetic field so as to randomize the magnetization of the magnetic particles on the tape or disk.

ERASE HEAD　A head in a tape recorder or hard drive that erases the signal on tape or disk.

EXPANDER　1. A signal processor that increases the dynamic range of a signal passed through it. 2. An amplifier whose gain decreases as its input level decreases. When used as a noise gate, an expander reduces the gain of low-level signals to reduce noise between notes.

FADE-OUT　To gradually reduce the volume of the last several seconds of a recorded song, from full level down to silence, by slowly pulling down the master fader, or by selecting a fade-out process in a digital editing program.

FADER　A linear or sliding potentiometer (volume control), used to adjust signal level.

FEED　1. To send an audio signal to some device or system. 2. An output signal sent to some device or system.

FEEDBACK　1. The return of some portion of an output signal to the system's input. 2. The squealing sound you hear when a PA system microphone picks up its own amplified signal through a loudspeaker.

FEED REEL   The left-side reel on an analog tape recorder that unwinds during recording or playback.

FILTER   1. A circuit that sharply attenuates frequencies above or below a certain frequency. Used to reduce noise and leakage above or below the frequency range of an instrument or voice. 2. A MIDI filter removes selected note parameters.

FIREWIRE   A standard protocol for high-speed transfer of data between digital devices. Also called IEEE 1394.

FLANGING   A special effect in which a signal is combined with its delayed replica, and the delay is varied between 0 and 20 msec. A hollow, swishing, ethereal effect like a variable-length pipe, or like a jet plane passing overhead. A variable comb filter produces the flanging effect.

FLETCHER–MUNSON EFFECT   Named after the two people who discovered it, the psychoacoustical phenomenon in which the subjective frequency response of the ear changes with program level. Because of this effect, a program played at a lower volume than the original level subjectively loses low- and high-frequency response.

FLOAT   To disconnect from ground.

FLUTTER   A rapid periodic variation in tape speed.

FLUTTER ECHOES   A rapid series of echoes that occurs between two parallel walls.

FLUX   Magnetic lines of force.

FLUXIVITY   The measure of the flux density of a magnetic recording tape, per unit of track width.

FLY-IN (LAY-IN)   To copy part of a recorded track onto another recorder, then re-record that copy back onto the original multitrack tape in a different part of the song, in sync with other recorded tracks. For example, copy the vocal track from the first chorus of the song onto an external DAW or sampler. Re-record (fly-in) that copy onto the multitrack tape at the second chorus. Then the first and second choruses have identical vocal performances.

FOLDBACK (FB)   *See* Cue System.

FREQUENCY   The number of cycles per second of a sound wave or an audio signal, measured in hertz (Hz). A low frequency (for example,

100 Hz) has a low pitch; a high frequency (for example, 10,000 Hz) has a high pitch.

FREQUENCY RESPONSE 1. The range of frequencies that an audio device will reproduce at an equal level (within a tolerance, such as ±3 dB). 2. The range of frequencies that a device (mic, human ear, etc.) can detect.

FULL-DUPLEX Describing a sound card that can record and play back simultaneously. The card works on two DMA channels.

FULL TRACK A single tape track recorded across the full width of an analog tape.

FUNDAMENTAL The lowest frequency in a complex wave.

FX Abbreviation for Effects.

GAIN Amplification. The ratio, expressed in decibels, between the output voltage and the input voltage, or between the output power and the input power.

GAP In a tape-recorder head, the thin break in the electromagnet that contacts the tape. In an audio program, the space or silence between songs.

GATE 1. To turn off a signal when its amplitude falls below a preset value. 2. The signal-processing device used for this purpose. *See* also Noise Gate.

GATED REVERB Reverberation with the reverberant "tail" cut off before it fades out.

GENERAL MIDI FILE (GM file) A MIDI file containing a standard set of musical instrument sounds. A General MIDI file produces the same sounds on any MIDI-capable instrument that supports the GM spec.

GENERATION A copy of a tape or a bounce of a track. A copy of the original master recording is a first generation tape. A copy made from the first generation tape is a second generation, and so on.

GENERATION LOSS The degradation of signal quality (the increase in noise and distortion) that occurs with each successive generation of an analog tape recording.

GOBO A moveable partition used to prevent the sound of an instrument from reaching another instrument's microphone. Short for go-between.

GRAPHIC EQUALIZER   An equalizer with a horizontal row of faders; the fader-knob positions indicate graphically the frequency response of the equalizer. Usually used to equalize monitor speakers for the room they are in. Sometimes used for complex EQ of a track.

GROUND   The zero-signal reference point for a system of audio components.

GROUND BUS   A common connection to which equipment is grounded, usually a heavy copper plate.

GROUNDING   Connecting pieces of electronic equipment to ground. Proper grounding ensures that there is no voltage difference between equipment chassis. An electrostatic shield needs to be grounded to be effective.

GROUND LOOP   1. A loop or circuit formed of ground leads. 2. The loop formed when unbalanced components are connected together via two ground paths—the connecting-cable shield and the power ground. Ground loops cause hum and should be avoided.

GROUP   *See* Submix. 1. To select several faders to make them act in unison. For example, select all the faders for the drum tracks to so that you can adjust the overall level of the drums by pushing one fader. 2. To assign the output of several input modules to a single group or bus, whose level is controlled by a single group fader. For example, assign all the input modules of the drum mics to a single "drums" group. 3. A bus or channel in a mixer that contains the signals from several input modules. For example, a drums group is a group of all the signals of the drum-set mics.

GROUP FADER(submaster fader)   In the output section of a mixing console, a potentiometer (fader or volume control) that controls the output level of a bus or group.

GSIF (GigaSampler Interface)   Nemesys' Windows sound card driver that achieves low latency by interfacing directly between the sound card and the audio application software.

GUARD BAND   The spacing between tracks on a multitrack tape or tape head, used to prevent crosstalk.

HALF-TRACK   A tape track recorded across approximately half the width of a tape. A half-track recorder usually records two such tracks simultaneously in the same direction to make a stereo recording.

HARD DISK A random-access storage medium for computer data. A hard-disk drive contains a stack of magnetically coated hard disks that are read by, and written to by, an electromagnetic head.

HARD-DISK RECORDER (hard-drive recorder) A device dedicated to recording digital audio on a hard-disk drive. A hard-disk recorder-mixer includes a built-in mixer.

HARMONIC An overtone whose frequency is a whole-number multiple of the fundamental frequency.

HARMONIZER A signal processor that provides a wide variety of pitch-shifting and delay effects.

HD Abbreviation for hard-disk drive.

HEAD An electromagnet in a tape recorder that either erases the audio signal on tape, records a signal on tape, or plays back a signal that is already on tape. A hard disk drive also has heads with similar functions.

HEAD GAP *See* Gap.

HEADPHONES A head-worn transducer that covers the ears and converts electrical audio signals into sound waves.

HEADROOM The safety margin, measured in decibels, between the signal level and the maximum undistorted signal level. In a tape recorder, the dB difference between standard operating level (corresponding to a 0 VU reading) and the level causing 3% total harmonic distortion. High-frequency headroom increases with analog tape speed.

HERTZ (Hz) Cycles per second, the unit of measurement of frequency.

HIGHPASS FILTER A filter that passes frequencies above a certain frequency and attenuates frequencies below that same frequency. A low-cut filter.

HISS A noise signal containing all frequencies, but with greater energy at higher octaves. Hiss sounds like wind blowing through trees. It is usually caused by random signals generated by microphones, electronics, and magnetic tape.

HOST A DAW recording program that supports plug-ins. *See* Plug-in.

HOT 1. A high recording level causing slight distortion, may be used for special effect. 2. High average level on a CD making it relatively loud, produced by peak limiting and normalization, or by compression and normalization. 3. A condition in which a chassis or circuit has a potentially dangerous voltage on it. 4. Referring to the conductor in a microphone cable that has a positive voltage on it at the instant that sound pressure moves the diaphragm inward.

HUM   An unwanted low-pitched tone (60 Hz and its harmonics) heard in the monitors. The sound of interference generated in audio circuits and cables by AC power wiring. Hum pickup is caused by such things as placing audio cables near power cables or transformers, faulty grounding, poor shielding, and ground loops.

HYPERCARDIOID MICROPHONE   A directional microphone with a polar pattern that has 12 dB attenuation at the sides, 6 dB attenuation at the rear, and two nulls of maximum rejection at 110 degrees off-axis.

IMAGE   An illusory sound source located somewhere around the listener. An image is generated by two or more loudspeakers. In a typical stereo system, images are located between the two stereo speakers.

IMPEDANCE   The opposition of a circuit to the flow of alternating current. Impedance is the complex sum of resistance and reactance. Abbreviated as Z.

INPUT   The connection going into an audio device. In a mixer or mixing console, a connector for a microphone, line-level device, or other signal source.

INPUT ATTENUATOR   *See* Attenuator.

INPUT MODULE   In a mixing console, the set of controls affecting a single input signal. An input module usually includes an attenuator (trim), fader, equalizer, aux sends, and channel-assign buttons.

INPUT/OUTPUT (I/O) CONSOLE (IN-LINE CONSOLE)   A mixing console arranged so that input and output sections are aligned vertically. Each module (other than the monitor section) contains one input channel and one output channel.

INPUT SECTION   The row of input modules in a mixing console.

INSERT JACKS   Two jacks (send and return) in a console input module or output module that allow access to points in the signal path, usually for connecting a compressor. Plugging into the access jacks breaks the signal flow and allows you to insert a signal processor or recorder in series with the signal. In many mixers, a single insert jack has both send and return terminals. Also called Access Jacks.

I/O Referring to Input and Output connectors.

I/O BOX A Breakout Box type of audio interface.

JACK A female or receptacle-type connector for audio signals into which a plug is inserted.

KEYBOARD WORKSTATION Several MIDI components in one chassis—a keyboard, a sample player, a sequencer, and perhaps a synthesizer and disk drive.

KILO A prefix meaning one thousand. Abbreviated k.

LATENCY The signal delay through an A/D, D/A converter, through a software program, or through a computer operating system. Monitoring latency is the delay between the time when a musician plays a note and when she hears the monitored signal of that note. Latency can make you play out-of-sync during overdubs.

LAY-IN *See* Fly-In.

LEADERING The process of splicing leader tape between program selections.

LEADER TAPE Plastic or paper tape without an oxide coating, used for a spacer between takes (for silence between songs) on analog tape.

LEAKAGE The overlap of an instrument's sound into another instrument's microphone. Also called bleed or spill.

LEDE Abbreviation for Live-End/Dead-End, a type of control room acoustic treatment in which the front half of the control room prevents early reflections to the mixing position, while the back half of the control room reflects diffused sound to the mixing position.

LED INDICATOR A recording-level indicator using one or more light-emitting diodes.

LEVEL The degree of intensity of an audio signal—the voltage, power, or sound pressure level. The original definition of level is the power in watts.

LEVEL SETTING In a recording system, the process of adjusting the input-signal level to obtain maximum level on the recording medium without distortion. A VU meter, LED meter, or other indicator shows recording level.

LIGHTPIPE   An Alesis connection protocol that transfers 8 digital audio channels at once over a Toslink fiber-optic cable.

LIMITER   A signal processor whose output is constant above a preset input level. A compressor with a compression ratio of 10:1 or greater, with the threshold set just below the point of distortion of the following device. Used to prevent distortion of attack transients or peaks.

LINE LEVEL   In balanced professional recording equipment, a signal whose level is approximately 1.23 volts (+4 dBm). In unbalanced equipment (most home hi-fi or semipro recording equipment), a signal whose level is approximately 0.316 volt (−10 dBV).

LIVE   1. Having audible reverberation. 2. Occurring in real time, in person.

LIVE RECORDING   A recording made at a concert. Also, a recording made of a musical ensemble playing all at once, rather than overdubbing.

LOCALIZATION   The ability of the human hearing system to tell the direction of a real or illusory sound source.

LOCATE (Autolocate)   A recorder function that makes the tape or disk head go to a specified program address (counter time) at the press of a button.

LOOP   In a sampling program, to play the sustain portion of a sound's envelope repeatedly. Also, a repeated rhythmic or musical pattern.

LOUDSPEAKER   A transducer that converts electrical energy (the signal) into acoustical energy (sound waves).

LOWPASS FILTER   A filter that passes frequencies below a certain frequency and attenuates frequencies above that same frequency. A high-cut filter.

M   Abbreviation for mega, or one million (as in megabytes).

MAGNETIC RECORDING TAPE   A recording medium made of magnetic particles (usually ferric oxide) suspended in a binder and coated on a long strip of thin plastic (usually Mylar).

MAGNETO-OPTICAL DRIVE (MO DRIVE) A drive that stores data (such as audio) on a 3.5-inch rewritable magneto-optical disk. The drive uses a laser and magnetic head to write data, and a laser to read data.

MASK To hide or cover up one sound with another sound. To make a sound inaudible by playing another sound along with it. Masking is used in many data reduction schemes.

MASTER A completed tape or CD used to generate tape copies or compact discs.

MASTER FADER A volume control that affects the level of all program busses simultaneously. It is the last stage of gain adjustment before the 2-track recorder.

MD Abbreviation for MiniDisc.

MDM Abbreviation for Modular Digital Multitrack.

MEMORY A group of integrated circuit chips used to store digital data temporarily or permanently (such as an audio signal in digital format).

MEMORY RECORDER A device that records audio on a memory chip, such as Compact Flash. Usually the recorded audio can be uncompressed wave files or compressed MP3 files.

MEMORY REWIND A tape-recorder function that rewinds the tape to a preset tape-counter position.

METER A device that indicates voltage, resistance, current, or signal level.

MFX The architecture or protocol for MIDI Effects. *See* MIDI Effects.

MIC An abbreviation for microphone.

MIC LEVEL The level or voltage of a signal produced by a microphone, typically 2 millivolts.

MIC PREAMP *See* Preamplifier.

MICROPHONE A transducer or device that converts an acoustical signal (sound) into a corresponding electrical signal.

MICROPHONE TECHNIQUES The selection and placement of microphones to pick up sound sources.

MIDI   Abbreviation for Musical Instrument Digital Interface, a specification for a connection between synthesizers, drum machines, and computers that allows them to communicate with and/or control each other.

MIDI CHANNEL   A route for transmitting and receiving MIDI signals. Each channel controls a separate MIDI musical instrument or synth patch. Up to 16 channels can be sent on a single MIDI cable.

MIDI CONTROLLER   A musical performance device (keyboard, drum pads, breath controller, etc.) that outputs a MIDI signal designating note numbers, note on, note off, and so on.

MIDI/DIGITAL AUDIO SOFTWARE   Software that combines a MIDI sequencer with a multitrack digital audio recorder/editor.

MIDI EFFECTS (MFX)   Non-audio processes applied to MIDI signals, such as an arpeggiator, echo/delay, chord analyzer, quantize, transpose MIDI event filter, or velocity change. They can be used as real-time, non-destructive plug-ins in MIDI tracks.

MIDI IN   A connector in a MIDI device that receives MIDI messages.

MIDI INTERFACE   A circuit that plugs into a computer and converts MIDI data into computer data for storage in memory or on hard disk. The interface also converts computer data into MIDI data.

MIDI OUT   A connector in a MIDI device that transmits MIDI messages.

MIDI THRU   A connector in a MIDI device that duplicates the MIDI information at the MIDI-IN connector. Used to connect another MIDI device in the series.

MID-SIDE   A coincident-pair stereo microphone technique using a forward-facing unidirectional, omnidirectional, or bidirectional mic and a side-facing bidirectional mic. The microphone signals are summed and differenced to produce right- and left-channel signals.

MIKE   To pick up with a microphone.

MILLI   A prefix meaning one thousandth, abbreviated m.

MINIDISC (MD)   A rewritable, magneto-optical storage medium that is read by a laser. It resembles a compact disc in a 2.5-inch square housing. MD recorders use a data compression scheme called ATRAC.

MIX   1. To combine two or more different signals into a common signal. 2. A control on an effects processor that varies the ratio between the dry (unprocessed) signal and the processed signal.

MIXDOWN   The process of playing recorded tracks through a mixing console and mixing them to two stereo channels for recording on a two-track recorder. Also applies to a surround-sound mixdown to 6 or 8 channels.

MIXER   A device that mixes or combines audio signals and controls the relative levels of the signals.

MIXING CONSOLE   A large mixer with additional functions such as equalization or tone control, pan pots, monitoring controls, solo functions, channel assigns, and control of signals sent to external signal processors.

MLP (Meridian Lossless Packing)   Used in DVD-Audio discs, a data-reduction method that compresses six full-range channels of 24-bit, 96 kHz audio without data loss.

MMC (MIDI Machine Control)   A set of MIDI commands by which one device can control another. Some commands include Start, Stop, and Locate. MMC does not include sync information, but MTC does.

MODELER   A device or software that simulates the sound of a microphone, guitar amp, or room.

MO DRIVE   *See* Magneto-Optical Drive.

MODULAR DIGITAL MULTITRACK (MDM)   A multitrack tape recorder that records 8 tracks digitally on a videocassette. Several 8-track modules can be linked together to add more tracks in sync. Two examples of MDMs are the Alesis ADAT and TASCAM DA-88.

MONAURAL   Referring to listening with one ear. Often incorrectly used to mean monophonic.

MONITOR   A loudspeaker in a control room, or headphones, used for judging sound quality. Also, a video display screen used with a computer.

MONITORING   Listening to an audio signal with a monitor.

MONO, MONOPHONIC   1. Referring to a single channel of audio. A monophonic program can be played over one or more loudspeakers, or one or more headphones. 2. Describing a synthesizer that plays only one note at a time (not chords).

MONO-COMPATIBLE   A characteristic of a stereo program, in which the program channels can be combined to a mono program without altering the frequency response or balance. A mono-compatible stereo program has the same frequency response in stereo or mono because there is no delay or phase shift between channels to cause phase interference.

MOVING-COIL MICROPHONE   A dynamic microphone in which the conductor is a coil of wire moving in a fixed magnetic field. The coil is attached to a diaphragm that vibrates when struck with sound waves. Usually called a dynamic microphone.

MP3 (MPEG Level-1 Layer-3)   A data compression format for audio. In an MP3 file (.mp3), the data has been compressed or reduced to one-tenth of its original size or less. Compressed files take up less memory, so they download faster. You download MP3 files to your hard drive, then listen to them. MP3 audio quality at a 128 kbps rate is nearly the same as that of CDs (depending on source material).

MP3PRO   A data-compression format for audio. MP3Pro is an improvement over MP3. Songs encoded at 64 kbps with MP3Pro are said to sound as good as songs encoded at 128 kbps with MP3. MP3Pro offers faster downloads and nearly double the amount of music you can put on a flash-memory player. MP3 and MP3Pro files are compatible with each other's players, but an MP3Pro player is needed to hear MP3Pro's improvement in sound quality.

M-S RECORDING   *See* Mid-Side.

MTC (MIDI Time Code)   A form of time code transmitted over MIDI, used for synchronizing MIDI devices. Unlike SMPTE, MTC is not sample-accurate.

MUDDY   Unclear sounding; having excessive leakage, reverberation, or overhang.

MULTIEFFECTS PROCESSOR   *See* Multiprocessor.

MULTIPLE-D MICROPHONE   A directional microphone that has multiple sound-path lengths between its front and rear sound entries. This type of microphone has minimal proximity effect.

MULTIPROCESSOR   A signal processor that can perform several different signal-processing functions.

MULTITIMBRAL   In a synthesizer, the ability to produce two or more different patches or timbres at the same time.

MULTITRACK   Referring to a recorder with more than two tracks.

MUTE   To turn off an input signal on a mixing console by disconnecting the input-module output from channel assign. During mixdown, the mute function is used to reduce tape noise and leakage during silent portions of tracks, or to turn off unused performances. During recording, mute is used to turn off mic signals.

NEAR COINCIDENT   A stereo microphone technique in which two directional microphones are angled apart symmetrically on either side of center and spaced a few inches apart horizontally.

NEARFIELD MONITORING   A monitor-speaker arrangement in which the speakers are placed very near the listener (usually just behind the mixing console) to reduce the audibility of control-room acoustics.

NOISE   Unwanted sound, such as hiss from electronics or tape. An audio signal with an irregular, non-periodic waveform.

NOISE GATE   A gate used to reduce or eliminate noise between notes.

NOISE-REDUCTION SYSTEM   A Dolby signal processor used to reduce analog tape hiss (and sometimes print-through) caused by the recording process. The object is to compress the high frequencies during recording and expand them in a complementary way during playback.

NOISE SHAPING   Filtering the noise added in dithering in order to make the noise less audible. Usually the filter reduces the level in the upper midrange and increases the level at high frequencies.

NONDESTRUCTIVE EDITING   In a digital audio workstation, editing done by changing pointers (location markers) to information on the hard disk. A nondestructive edit can be undone.

NONLINEAR   1. Referring to a storage medium in which any data point can be accessed or read almost instantly in a random fashion, rather than sequentially. Examples are a hard disk, compact disc, and MiniDisc. *See* Random Access. 2. Referring to an audio device that is distorting the signal.

NORMALIZE   To raise the level of a digital audio signal so that the highest peak in the program is at the highest level allowed by the recording. For example, in a normalized 16-bit recording, the highest peak in the program has all 16 bits on: the highest possible level in a 16-bit recording.

OCTAVE   The interval between any two frequencies where the upper frequency is twice the lower frequency.

OFF-AXIS   Not exactly in front of a microphone or loudspeaker.

OFF-AXIS COLORATION   In a microphone, the deviation from the on-axis frequency response that sometimes occurs at angles off the axis of the microphone. The coloration of sound (alteration of tone quality) for sounds arriving off-axis to the microphone.

OGG   The file extension for Ogg Vorbis, a data-reduction encoding scheme.

OMNIDIRECTIONAL MICROPHONE   A microphone that is equally sensitive to sounds arriving from all directions.

ON-LOCATION RECORDING   A recording made outside the studio, in a room or hall where the music usually is performed or practiced.

OPEN TRACKS   On a multitrack tape recorder, tracks that have not yet been used, or have already been bounced and are available for use.

ORTF   Named after the French broadcasting network (Office de Radio diffusion Television Française), a near-coincident stereo mic technique that uses two cardioid mics angled 110 degrees apart and spaced 17 cm horizontally.

OUTBOARD EQUIPMENT   Signal processors that are external to the mixing console.

OUTPUT   A connector in an audio device from which the signal comes and feeds successive devices.

OUTTAKE   A take, or section of a take, that is to be removed or not used.

OVERDUB   To record a new musical part on an unused track in synchronization with previously recorded tracks.

OVERHANG   The continuation of a signal at the output of a device after the input signal has ceased. Sometimes called ringing.

OVERLOAD   The distortion that occurs when an applied signal exceeds a system's maximum input level.

OVERSAMPLING   Sampling an audio signal at a higher rate than is needed to reproduce the highest frequency in the signal. For example, sampling an audio signal at 8 times 44.1 kHz is called "8x oversampling." This process is followed by a digital low-pass filter and a gentle-slope analog anti-alias filter. The result is less phase shift compared to a steep, "brick-wall" analog filter used alone. See Anti-alias filter. Oversampling also can be applied in D/A conversion.

OVERTONE   In a complex wave, a frequency component that is higher than the fundamental frequency.

PAD   *See* Attenuator.

PAN POT   Abbreviation for panoramic potentiometer. In each input module in a mixing console, a control that divides a signal between two channels in an adjustable ratio. By doing so, a pan pot controls the location of a sonic image between a stereo pair of loudspeakers.

PARAMETRIC EQUALIZER   An equalizer with continuously variable parameters, such as frequency, bandwidth, and amount of boost or cut.

PATCH   1. To connect one piece of audio equipment to another with a cable. 2. A setting of synthesizer parameters to achieve a sound with a certain timbre.

PATCH BAY (PATCH PANEL)   An array of connectors, usually in a rack, to which equipment inputs and outputs are wired. A patch bay makes it easy to interconnect various pieces of equipment in a central, accessible location.

PATCH CORD   A short length of cable with a phone plug on each end, used for signal routing in a patch bay.

PCM   Abbreviation for Pulse Code Modulation, a method of analog-to-digital conversion in which the instantaneous amplitude of an analog waveform is measured or sampled several thousand times a second, and each measurement is assigned a binary value of a certain number of bits (ones and zeroes).

PDM   Abbreviation for Pulse Density Modualtion, a method of of analog-to-digital conversion in which the instantaneous amplitude of an analog waveform is coded as variations in the average number of fixed-width pulses per unit of time.

PEAK   On a graph of a sound wave or signal, the highest point in the waveform. The point of greatest voltage or sound pressure in a cycle.

PEAK AMPLITUDE   *See* Amplitude, Peak.

PEAKING EQUALIZER   An equalizer that provides maximum cut or boost at one frequency, so that the resulting frequency response of a boost resembles a mountain peak.

PEAK PROGRAM METER (PPM)   A meter that responds fast enough to closely follow the peak levels in a program.

PERIOD   The time between the peak of one wave and the peak of the next. The time between corresponding points on successive waves. Period is the inverse of frequency.

PERSONAL STUDIO   A minimal group of recording equipment set up for one's personal use, usually using a 4-track cassette or memory recorder-mixer. Also, a simple 4-track recorder-mixer for one's personal use.

PERSPECTIVE   In the reproduction of a recording, the audible sense of distance to the musical ensemble, the point of view. A close perspective has a high ratio of direct sound to reverberant sound; a distant perspective has a low ratio of direct sound to reverberant sound.

PFL   Abbreviation for pre-fader listen. *See* also Solo.

PHANTOM POWER   A DC voltage (usually 12 to 48 volts) applied to microphone signal conductors to power condenser microphones.

PHASE   The degree of progression in the cycle of a wave, where one complete cycle is 360 degrees.

PHASE CANCELLATION, PHASE INTERFERENCE   The cancellation of certain frequency components of a signal that occurs when the signal is combined with its delayed replica. At certain frequencies, the direct and delayed signals are of equal level and opposite polarity (180 degrees out of phase), and when combined, they cancel out. The result is a comb-filter frequency response having a periodic series of peaks and dips. Phase interference can occur between the signals of two microphones picking up the same source at different distances, or can occur at a microphone picking up both a direct sound and its reflection from a nearby surface.

PHASING   A special effect in which a signal is combined with its phase-shifted replica to produce a variable comb-filter effect. *See* also Flanging.

PHASE SHIFT   The difference in degrees of phase angle between corresponding points on two waves. If one wave is delayed with respect to another, there is a phase shift between them of 2piFT, where pi = 3.14, F = frequency in Hz, and T = delay in seconds.

PHONE PLUG   A cylindrical, coaxial plug (usually 1/4-inch diameter). An unbalanced phone plug has a tip for the hot signal and a sleeve for the shield, which connects to ground. A balanced phone plug has a tip for the signal hot signal, a ring for the return signal, and a sleeve for the shield.

PHONO PLUG   A coaxial plug with a central pin for the hot signal and a ring of pressure-fit tabs for the shield or ground. Also called RCA plug. Phono plugs are used on Tascam modular digital multitrack recorders and on consumer stereo equipment.

PINCH ROLLER   In a tape-recorder transport, the rubber wheel that pinches or traps the tape between itself and the capstan, so that the capstan can move the tape.

PICKUP   A piezoelectric transducer that converts mechanical vibrations to an electrical signal. Used in acoustic guitars, acoustic basses, and fiddles. Also, a magnetic transducer in an electric guitar that converts string vibration to a corresponding electrical signal.

PING-PONGING   *See* Bouncing Tracks.

PINK NOISE   A noise signal containing all frequencies (unless band-limited), with equal energy per octave. Pink noise is a test signal used for equalizing a sound system to the desired frequency response, and for testing loudspeakers.

PITCH   The subjective lowness or highness of a tone. The pitch of a tone usually correlates with the fundamental frequency.

PITCH CONTROL   A control on a tape recorder that varies the tape speed, thereby varying the pitch of the signal on tape. The pitch control can be used to match the pitch of prerecorded instruments with that of an instrument to be overdubbed. It is also used for special effects, such as "chipmunk voices," and to play prerecorded tracks slowly so that fast musical passages can be overdubbed more easily.

PITCH SHIFTER   A signal processor that changes the pitch of an instrument without changing its duration.

PLAYBACK EQUALIZATION   In analog tape-recorder electronics, fixed equalization applied to the signal during recording to compensate for certain losses.

PLAYBACK HEAD   The head in a tape recorder that picks up a prerecorded magnetic signal from the moving tape and converts it to a corresponding electrical signal. The playback head is not the same as the sel-sync or sync head.

PLAYLIST   *See* Edit Decision List.

PLUG   A male connector that inserts into a jack. Also, short for Plug-in.

PLUG-IN   Effects software that you load into your DAW recording program (called the host). The plug-in becomes part of the host program and can be called up from within the host. Some manufacturers make plug-in bundles, which are a variety of effects in a single package.

POLAR PATTERN   The directional pickup pattern of a microphone. A plot of microphone sensitivity plotted versus angle of sound incidence. Examples of polar patterns are omnidirectional, bidirectional, and unidirectional. Subsets of unidirectional are cardioid, supercardioid, and hypercardioid.

POLARITY   Referring to the positive or negative direction of an electrical, acoustical, or magnetic force. Two identical signals in opposite polarity are 180 degrees out-of-phase with each other at all frequencies.

POLYPHONIC   Describing a synthesizer that can play more than one note at a time (chords).

POP   1. A thump or little explosion sound heard in a vocalist's microphone signal. Pop occurs when the user says words with "p," "t," or "b" so that a turbulent puff of air is forced from the mouth and strikes the microphone diaphragm. 2. A noise heard when a mic is plugged into a monitored channel, or when a switch is flipped.

POP FILTER   A screen placed on a microphone grille that attenuates or filters out pop disturbances before they strike the microphone diaphragm. Usually made of open-cell plastic foam or silk, a pop filter reduces pop and wind noise.

PORTABLE STUDIO   A combination recorder and mixer in one portable chassis.

POST-ECHO   A repetition of a sound, following the original sound, caused by print-through.

POWER AMPLIFIER   An electronic device that amplifies or increases the power level fed into it to a level sufficient to drive a loudspeaker.

POWER GROUND (SAFETY GROUND)   A connection to the power company's earth ground through the U-shaped hole in a power outlet. In the power cable of an electronic component with a 3-prong plug, the U-shaped prong is wired to the component's chassis. This wire conducts electricity to power ground if the chassis becomes electrically hot, preventing shocks.

PREAMPLIFIER (PREAMP)   In an audio system, the first stage of amplification that boosts a mic-level signal to line level. A preamp is a standalone device or a circuit in a mixer.

PRE-DELAY   Short for pre-reverberation delay. The delay (about 30 to 150 msec) between the arrival of the direct sound and the onset of reverberation. Usually, the longer the pre-delay, the greater the perceived room size.

PRE-ECHO   A repetition of a sound that occurs before the sound itself, caused by print-through in analog tape.

PRE-FADER/POST-FADER SWITCH   A switch that selects a signal either ahead of the fader (pre-fader) or following the fader (post-fader). The level of a pre-fader signal is independent of the fader position; the level of a post-fader signal follows the fader position.

PREPRODUCTION   Planning in advance what you're going to do at a recording session, in terms of track assignments, overdubbing, studio layout, and microphone selection.

PRESENCE   The audible sense that a reproduced instrument is present in the listening room. Some synonyms are closeness, definition, and punch. Presence is often created by an equalization boost in the midrange or upper midrange, and by a high direct-to-reverb ratio.

PRESSURE ZONE MICROPHONE   A boundary microphone constructed with the microphone diaphragm parallel with, and facing, a reflective surface.

PREVERB   A special effect in which the reverberation of a note precedes it, rather than follows it. It turns a snare-drum hit into a whip sound, like ssSSHHK! Chapter 10, Signal Processors and Effects, describes how to create it. Also see Reverse Echo.

PRINT   To record on tape or disc.

PRINT-THROUGH   The transfer of a magnetic signal from one layer of analog tape to the next on a reel, causing an echo preceding or following the program.

PRODUCTION   1. A recording that is enhanced by effects. 2. The supervision of a recording session to create a satisfactory recording. This involves getting musicians together for the session, making musical suggestions to the musicians to enhance their performance, and making suggestions to the engineer for sound balance and effects.

PROGRAM BUS   A bus or output that feeds an audio program to a recorder track.

PROGRAM MIXER   In a mixing console, a mixer formed of input-module outputs, combining amplifiers, and program busses.

PRO TOOLS   A popular digital audio editing platform for professional use. It offers computer multitrack recording, overdubbing, mixing, editing, and a variety of plug-in effects.

PROXIMITY EFFECT   The bass boost that occurs with a single-D directional microphone when it is placed a few inches from a sound source. The closer the microphone, the greater the low-frequency boost due to proximity effect.

PULSE CODE MODULATION (PCM)   A method of analog-to-digital conversion. The analog signal voltage is measured several thousand times a second, and each measurement is a digital word (a string of 1s and 0s) of a certain word length or bit depth.

PULSE DENSITY MODULATION   See PDM.

PUNCH-IN/OUT   A feature in a multitrack recorder that lets you insert a recording of a corrected musical part into a previously recorded track by going into and out of record mode as the tape or disk is rolling.

PURE WAVEFORM   A waveform of a single frequency; a sine wave. A pure tone is the perceived sound of such a wave.

QUARTER-TRACK A tape track recorded across one-quarter of the width of an analog magnetic tape. A quarter-track recorder usually records two stereo programs (one in each direction).

RACK A 19-inch-wide wooden or metal cabinet used to hold audio equipment.

RADIO-FREQUENCY INTERFERENCE (RFI) Radio-frequency electromagnetic waves induced in audio cables or equipment, causing various noises in the audio signal.

RANDOM ACCESS Referring to a storage medium in which any data point can be accessed or read almost instantly. Examples are a hard disk, compact disc, and MiniDisc.

RAREFACTION The portion of a sound wave in which molecules are spread apart, forming a region with lower-than-normal atmospheric pressure. The opposite of compression.

R-DAT *See* DAT.

REALAUDIO A highly compressed audio file format used for streaming audio. Generally, RealAudio has lower fidelity (less treble) than MP3, but the fidelity depends on modem speed. RealAudio files (.ra or.rm) are often used as short excerpts or previews of songs.

REAL-TIME RECORDING 1. Recording notes into a sequencer in the correct tempo, for later playback at the same tempo as recorded. 2. A recording made direct to lacquer disc or direct to 2-track without any overdubs or mixdown.

RE-AMPING Recording a guitar amp that is fed the signal from a direct-recorded electric guitar track. This technique lets you work on the amp's sound during mixdown, rather than during recording.

RECIRCULATION (REGENERATION) Feeding the output of a delay device back into its input to create multiple echoes. Also, the control on a delay device that affects how much delayed signal is recycled to the input.

RECORD To store an event in permanent form. Usually, to store an audio signal in magnetic form on magnetic tape or disk, or to store an audio signal in optical form on a CD-R or CD-RW. Recording is also possible on magneto-optical disk, on MiniDisc, in RAM, and on memory cards.

RECORD EQUALIZATION   In analog tape-recorder electronics, equalization applied to the signal during recording to compensate for certain losses.

RECORDER-MIXER   A combination multitrack recorder and mixer in one chassis.

RECORD HEAD   The head in a tape recorder or hard drive that puts the audio signal on tape or disk by magnetizing the tape or disk particles in a pattern corresponding to the audio signal.

RECORDING/REPRODUCTION CHAIN   The series of events and equipment that are involved in sound recording and playback.

REFLECTED SOUND   Sound waves that reach the listener after being reflected from one or more surfaces.

REGENERATION   *See* Recirculation.

REGION   In a digital audio editing program, a defined segment of the audio program. Also called clip or zone.

RELEASE   The final portion of a note's envelope in which the note falls from its sustain level back to silence.

RELEASE TIME   In a compressor, the time it takes for the gain to return to normal after the end of a loud passage.

REMIX   To mix again; to do another mixdown with different console settings or different editing.

REMOTE RECORDING   *See* On-Location Recording.

REMOVABLE HARD DRIVE   A hard-disk drive that can be removed and replaced with another, used in a DAWor hard-drive recorder to store a program temporarily.

RENDER   To convert one audio format to another in a DAW or hard-drive recorder. Examples: mix a multitrack recording to a 2-channel wave, aiff, rm, or mp3 file; convert a track with real-time effects to a track with embedded effects; convert MIDI tracks to wave tracks; combine multiple clips into a single track.

RESISTANCE   The opposition of a circuit to a flow of direct current. Resistance is measured in ohms, abbreviated w, and may be calculated by dividing voltage by current.

RESISTOR   An electronic component that opposes current flow.

RETURN-TO-ZERO   *See* Memory Rewind.

REVERBERATION   Natural reverberation in a room is a series of multiple sound reflections that make the original sound persist and gradually die away or decay. These reflections tell the ear that you're listening in a large or hard-surfaced room. For example, reverberation is the sound you hear just after you shout in an empty gymnasium. A reverb effect simulates the sound of a room—a club, auditorium, or concert hall-by generating random multiple echoes that are too numerous and rapid for the ear to resolve. The timing of the echoes is random, and the echoes increase in number with time as they decay. An echo is a discrete repetition of a sound; reverberation is a continuous fade-out of sound.

REVERBERATION TIME (RT60)   The time it takes for reverberation to decay to 60 dB below the original steady-state level.

REVERSE ECHO   A multiple echo that precedes the sound that caused it, building up from silence into the original sound. This special effect is created in a manner similar to preverb.

RFI   *See* Radio Frequency Interference.

RHYTHM TRACKS   The recorded tracks of the rhythm instruments (guitar, bass, drums, and sometimes keyboards).

RIBBON MICROPHONE   A dynamic microphone in which the conductor is a long metallic diaphragm (ribbon) suspended in a magnetic field.

RIDE GAIN   To turn down the volume of a microphone when the source gets louder, and turn up the volume when the source gets quieter, in an attempt to reduce dynamic range.

RINGING   *See* Overhang.

RMF (Rich Music Format)   A format for a MIDI file with General MIDI sounds plus custom sounds. Designed to be played on a Beatnik player.

ROOM MODES   *See* Standing Wave.

RT60   *See* Reverberation Time.

SACD   *See* Super Audio CD.

SAFETY COPY   A copy of the master tape or CD, to be used if the master is lost or damaged.

SAFETY GROUND   *See* Power Ground.

SAMPLE   1. To digitally record a short sound event, such as a single note or a musical phrase, into computer memory. 2. A recording of such an event. 3. A measurement of an analog waveform that is done several thousand times a second during a PCM A/D conversion.

SAMPLING   1. Recording a short sound event into computer memory. The audio signal is converted into digital data representing the signal waveform, and the data is stored in memory chips, tape, or disc for later playback. 2. In PCM digital recording, measuring an analog waveform periodically, several thousand times a second.

SAMPLING RATE   In PCM digital recording, the frequency at which an analog waveform is sampled or measured. The sampling rate of CD-quality audio is 44,100 samples per second. The higher the sampling rate, the higher the high-frequency response of the recording.

SATURATION   Overload of magnetic tape. The point at which a further increase in magnetizing force does not cause an increase in magnetization of the tape oxide particles. Distortion is the result.

SCENE AUTOMATION   A form of console automation in which the console settings are stored in memory. A "snapshot" or reading of many of the settings is taken and stored for later recall. In contrast, dynamic automation continuously follows the fader and knob moves, and the automation data is usually stored as a MIDI file.

SCMS (Serial Copy Management System)   An anti-copy scheme in consumer digital audio devices (those with S/PDIF connectors). SCMS circuits read flags in the data stream that allow users to make only first-generation copies, not copies of copies.

SCRATCH VOCAL   A vocal performance that is done simultaneously with the rhythm instruments so that the musicians can keep their place in the song and get a feel for the song. Because it contains leakage, the scratch-vocal recording is usually erased. Then the singer overdubs the vocal part that is to be used in the final recording.

SCRUB   To manually move an open-reel tape slowly back and forth across a recorder playback head in order to locate an edit point. Some digital editing software has an equivalent scrubbing function.

571

SCSI (Small Computer Systems Interface)   A standard spec for high-speed computer input and output. Commonly used for hard drives, CD-ROM drives, and CD-R recorders. Supports up to seven devices.

SENSITIVITY   1. The output of a microphone in volts for a given input in sound pressure level. 2. The sound pressure level a loudspeaker produces at one meter when driven with 1 watt of pink noise. *See* also Sound Pressure Level.

SEQUENCE   A MIDI data file of musical-performance note parameters, recorded by a sequencer.

SEQUENCER   A device or computer program that records a musical performance done on a MIDI controller (in the form of note numbers, note on, note off, etc.) into computer memory or hard disk for later playback. During playback, the sequencer plays synthesizer sound generators or samples.

SESSION   1. A time period set aside for recording musical instruments, voices, or sound effects. 2. On a CD-R, a lead-in, program area, and lead-out.

SHELVING EQUALIZER   An equalizer that applies a constant boost or cut above or below a certain frequency, so that the shape of the frequency response resembles a shelf.

SHIELD   A conductive enclosure (usually metallic) around one or more signal conductors, used to keep out electrostatic fields that cause hum or buzz. A shield in a mic cable is a cylindrical mesh of fine wires.

SHOCK MOUNT   A suspension system that mechanically isolates a microphone from its stand or boom, preventing the transfer of mechanical vibrations.

SIBILANCE   In a speech recording, excessive frequency components in the 5 to 10 kHz range, which are heard as an overemphasis of "s" and "sh" sounds.

SIDE-ADDRESSED   Referring to a microphone whose main axis of pickup is perpendicular to the side of the microphone. You aim the side of the mic at the sound source. *See* also End-Addressed.

SIGNAL   A varying electrical voltage that represents information, such as a sound.

SIGNAL PATH   The path a signal takes from input to output in a piece of audio equipment.

SIGNAL PROCESSOR   A device that is used to alter a signal in a controlled way.

SIGNAL-TO-NOISE RATIO (S/N)   The ratio in decibels between signal voltage and noise voltage. An audio component with a high S/N has little background noise accompanying the signal; a component with a low S/N is noisy.

SINE WAVE   A wave following the equation $y = \sin x$, where x is degrees and y is voltage or sound pressure level. The waveform of a single frequency. The waveform of a pure tone without harmonics.

SINGLE-ENDED   1. An unbalanced line. 2. A single-ended noise reduction system is one that works only during tape playback (unlike Dolby or dbx, which work both during recording and playback).

SINGLE-D MICROPHONE   A directional microphone having a single distance between its front and rear sound entries. Such a microphone has proximity effect.

SLAP, SLAP BACK   An echo following the original sound by about 50 to 200 msec, sometimes with multiple repetitions.

SLATE   At the beginning of a recording, a recorded announcement of the name of the tune and its take number. The term is derived from the slate used in the motion-picture industry to identify the production and take number being filmed.

SMPTE TIME CODE   A modulated 1200 Hz square-wave signal used to synchronize two or more recorders or MIDI devices. SMPTE uses the format HH:MM:SS:FF (hours, minutes, seconds, frames) to specify locations in the recorded program. SMPTE is an abbreviation for the Society of Motion Picture and Television Engineers, who developed the time code.

SNAKE   A multipair or multichannel mic cable. Also, a multipair mic cable attached to a connector junction box.

SNAPSHOT AUTOMATION   *See* Scene Automation.

SOFT SYNTH   A synthesizer in software form.

573

SOLO   On an input module in a mixing console, a switch that lets you monitor that particular input signal by itself. The switch routes only that input signal to the monitor system.

SOUND   Longitudinal vibrations in a medium (such as air) in the frequency range 20 to 20,000 Hz.

SOUND CARD   A circuit card that plugs into a computer and converts an audio signal into computer data for storage in memory or on hard disk. The sound card also converts computer data into an audio signal. A type of audio interface (*see* audio interface).

SOUND EFFECTS   Recordings of non-musical sounds—such as a door slam, gunfire, thunderstorm, car, or telephone—used in dramatic productions, radio spots and commercials. Not to be confused with Effects.

SOUND MODULE (SOUND GENERATOR)   1. A synthesizer without a keyboard, containing several different timbres or voices. These sounds are triggered or played by MIDI signals from a sequencer program, or by a MIDI controller. A sound module can be a standalone device or a circuit on a sound card. 2. An oscillator.

SOUND PRESSURE LEVEL (SPL)   The acoustic pressure of a sound wave, measured in decibels above the threshold of hearing. The higher the SPL of a sound, the louder it is. $dB\ SPL = 20\log(P/P_{ref})$, where P = the measured acoustic pressure and $P_{ref} = 0.0002\ dyne/cm^2$.

SOUND WAVE   The periodic variations in air pressure radiating from an object vibrating between 20 Hz and 20,000 Hz.

SPACED-PAIR   A stereo microphone technique using two identical microphones spaced several feet apart horizontally, usually aiming straight ahead toward the sound source.

SPATIAL PROCESSOR   A signal processor that allows images to be placed beyond the limits of a stereo pair of speakers   even behind the listener or toward the sides.

S/PDIF (Sony Philips Digital Interface; IEC 958 Type II)   A 2-channel digital signal interface format that uses a 75-ohm coaxial cable with RCA connectors, or an optical cable with TOSLINK connectors. *See* also AES/EBU.

SPEAKER   *See* Loudspeaker.

SPECTRUM   The output level versus frequency of a sound source, including the fundamental frequency and overtones.

SPL   *See* Sound Pressure Level.

SPLICE   To join the ends of two lengths of magnetic tape or leader tape with tape. Also, a splice is the taped joint between two lengths of magnetic tape or leader tape.

SPLICING BLOCK   *See* Editing Block.

SPLIT CONSOLE   A console with a separate monitor-mixer section. *See* also Input/Output (I/O) Console.

SPLITTER   A transformer or circuit used to divide a microphone signal into two or more identical signals to feed different sound systems.

SPOT MICROPHONE   In classical music recording, a close-placed microphone that is mixed with more distant microphones to add presence or to improve the balance.

STANDING WAVE   An apparently stationary waveform created by multiple reflections between opposite room surfaces. At certain points along the standing wave, the direct and reflected waves cancel, and at other points the waves add together or reinforce each other.

STEM   A submix in ProTools lingo-a drums submix, keys submix, left-front submix, etc.

STEP-TIME RECORDING   Recording notes into a sequencer one at a time without regard to tempo, for later playback at a normal tempo.

STEREO, STEREOPHONIC   An audio recording and reproduction system with correlated information between two channels (usually discrete channels), meant to be heard over two or more loudspeakers to give the illusion of sound-source localization and depth.

STEREO BAR, STEREO MICROPHONE ADAPTER   A microphone stand adapter that mounts two microphones on a single stand for convenient stereo miking.

STEREO IMAGING   The ability of a stereo recording or reproduction system to form clearly defined audio images at various locations between a stereo pair of loudspeakers.

575

STEREO MICROPHONE   A microphone containing two mic capsules in a single housing for convenient stereo recording. The capsules usually are coincident.

STREAMING AUDIO   Audio sent over the Internet in real time. A streaming file plays as soon as you click on its title. A downloaded file doesn't play until you copy the entire file to your hard disk. Streaming audio is heard almost instantly, but usually sounds muffled and can be interrupted by Net congestion. With downloaded audio, you must wait up to several minutes to hear the song, but the sound is high-fidelity and continuous.

STUDIO   A room used or designed for sound recording.

SUBMASTER   1. A master volume control for an output bus. 2. A recorded tape that is used to form a master tape.

SUBMIX   A small preset mix within a larger mix, such as a drum mix, keyboard mix, vocal mix, etc. Also a cue mix, monitor mix, or effects mix.

SUBMIXER   A smaller mixer within a mixing console (or standalone) that is used to set up a submix, a cue mix, an effects mix, or a monitor mix.

SUPER AUDIO CD (SACD)   An alternative to DVD-Audio, Super Audio CD was developed by Sony and Philips. It is the size of a CD but offers higher sound quality and 5.1 surround sound. SACD uses the Direct Stream Digital (DSD) process, which encodes a digital signal in a 1-bit (bit-stream) format at a 2.8224 MHz sampling rate. This system offers a frequency response from DC to 100 kHz with 120 dB dynamic range.

SUPERCARDIOID MICROPHONE   A unidirectional microphone that attenuates side-arriving sounds by 8.7 dB, attenuates rear-arriving sounds by 11.4 dB, and has two nulls of maximum sound rejection at 125 degrees off-axis.

SUPPLY REEL   *See* Feed Reel.

SURROUND SOUND   A multichannel recording and reproduction system that plays sound all around the listener. The 5.1 surround system uses the following speakers-front-left, center, front-right, left-surround, right-surround, and subwoofer.

SUSTAIN   The portion of the envelope of a note in which the level is constant. Also, the ability of a note to continue without noticeably decaying, often aided by compression.

SWEETENING   The addition of strings, brass, chorus, etc., to a previously recorded tape of the basic rhythm tracks.

SYNC, SYNCHRONIZATION   Aligning two separate audio programs in time, and maintaining that alignment as the programs play.

SYNC, SYNCHRONOUS RECORDING   Using a tape record head temporarily as a playback head during an overdub session, to keep the overdubbed parts in synchronization with the recorded tracks.

SYNC TONE   *See* Tape Sync.

SYNC TRACK   A track of a multitrack recorder that is reserved for recording an FSK sync tone or SMPTE time code. This allows audio tracks to synchronize with virtual tracks recorded with a sequencer. A sync track also can synchronize two audio tape machines or an audio recorder and a video recorder, and can be used for console automation.

SYNTHESIZER   A musical instrument (usually with a piano-style keyboard) that creates sounds electronically and allows control of the sound parameters to simulate a variety of conventional or unique instruments.

TAIL   The end of a reverberation signal where the reverb fades down to silence.

TAIL-OUT   Referring to a reel of tape wound with the end of the program toward the outside of the reel. Analog tape stored tail-out is less likely to have audible print-through.

TAKE   A recorded performance of a song. Usually, several takes are done of the same song, and the best one   or the best parts of several— becomes the final product.

TAKE SHEET   A list of take numbers for each song, plus comments on each take.

TAKE-UP REEL   The right-side reel on a tape recorder that winds up the tape as it is playing or recording.

TALKBACK   An intercom in the mixing console for the engineer and producer to talk to the musicians in the studio.

TAPE   *See* Magnetic Recording Tape.

TAPE LOOP   An endless loop formed from a length of recording tape spliced end-to-end, used for continuous repetition of several seconds of recorded signal.

TAPE RECORDER   A device that converts an electrical audio signal into a magnetic audio signal on magnetic tape, and vice versa. A tape recorder includes electronics, heads, and a transport to move the tape.

TAPE SYNC   A frequency-modulated signal recorded on a tape track, used to synchronize a tape recorder to a sequencer. Tape sync also permits the synchronized transfer of sequences to tape. *See* also Sync Track.

TDIF (TASCAM DIGITAL INTERFACE)   An interface protocol for connecting digital audio signals to and from TASCAM DTRS multitrack recorders, such as the DA-78 and DA-88.

3-PIN CONNECTOR   A 3-pin professional audio connector used for balanced signals. Pin 1 is soldered to the cable shield, pin 2 is soldered to the signal hot or in-polarity lead, and pin 3 is soldered to the signal cold or opposite-polarity lead. *See* also XLR-Type Connector.

THREE-TO-ONE RULE (3:1 RULE)   A rule in microphone applications. When multiple mics are mixed to the same channel, the distance between mics should be at least three times the distance from each mic to its sound source. This prevents audible phase interference.

THRESHOLD   In a compressor or limiter, the input level above which compression or limiting takes place. In an expander, the input level below which expansion takes place.

TIE   To connect electrically, for example, by soldering a wire between two points in a circuit.

TIGHT   1. Having very little leakage or room reflections in the sound pickup. 2. Referring to well-synchronized playing of musical instruments. 3. Having a well-damped, rapid decay.

TIMBRE   The subjective impression of spectrum and envelope. The quality of a sound that allows us to differentiate it from other sounds. For example, if you hear a trumpet, a piano, and a drum, each has a different timbre or tone quality that identifies it as a particular instrument.

TIME CODE   A modulated 1200-Hz square-wave signal used to synchronize two or more tape or disc transports. *See* also Tape Sync, Sync Track, SMPTE Time Code.

TONAL BALANCE   The balance or volume relationships among different regions of the frequency spectrum, such as bass, mid-bass, midrange, upper midrange, and highs.

TOSLINK (Toshiba Link)   A fiber-optic cable connection for S/PDIF data transfers.

TRACK   A path on magnetic tape containing a single channel of audio. A group of bytes in a digital signal (on tape, on hard disk, on compact disc, or in a data stream) that represents a single channel of audio or MIDI. Usually one track contains a performance by one musical instrument.

TRANSDUCER   A device that converts energy from one form to another, such as a microphone or loudspeaker.

TRANSFORMER   An electronic component made of two magnetically coupled coils of wire. The input signal is transferred magnetically to the output, without a direct connection between input and output.

TRANSIENT   A short signal with a rapid attack and decay, such as a drum stroke, cymbal hit, or acoustic-guitar pluck.

TRANSIENT RESPONSE   The ability of an audio component (usually a microphone or loudspeaker) to follow a transient accurately.

TRANSPORT   The mechanical system in a reproduction device that moves the medium past the read/write heads. In a tape recorder, the transport controls tape motion during recording, playback, fast forward, and rewind.

TRIM   1. In a mixing console, a control for fine adjustment of level, as in a Bus Trim control. 2. In a mixing console, a control that adjusts the gain of a mic preamp to accommodate various signal levels.

TRS (Tip Ring Sleeve)   A phone-plug connector used for aux send/return, unbalanced stereo, or balanced mono connections.

TUBE   A vacuum tube, an amplifying component made of electrodes in an evacuated glass tube. Tube sound is characterized as being "warmer" than solid-state or transistor sound.

TWEETER   A high-frequency loudspeaker.

UNBALANCED LINE   An audio cable having one conductor surrounded by a shield that carries the return signal. The shield is at ground voltage.

UNIDIRECTIONAL MICROPHONE   A microphone that is most sensitive to sounds arriving from one direction—in front of the microphone. Examples are cardioid, supercardioid, and hypercardioid.

UNITY GAIN   A condition in which the input and output levels of a device are equal.

USB (UNIVERSAL SERIAL BUS)   A Mac/PC computer serial interface for connecting external devices such as MIDI interfaces and audio interfaces to a computer. Faster than a standard serial port.

VALVE   British term for vacuum tube. *See* Tube.

VBR(Variable Bit Rate MP3 encoding)   The bit rate varies with the audio signal.

VIRTUAL CONTROLS   Audio-equipment controls that are simulated on a computer monitor screen. You adjust them with a mouse or with a controller surface.

VIRTUAL TRACK   A recording of a single take of one instrument or vocal on a random-access medium. The best parts of several virtual tracks can be bounced a single track, forming a complete, perfect performance. This process is called comping tracks.

VST (VIRTUAL STUDIO TECHNOLOGY)   Steinberg's virtual instrument integration standard format for plug-ins.

VU METER   A voltmeter with a specified transient response, calibrated in VU or volume units, used to show the relative volume of various audio signals, and to set recording levels in analog tape recorders.

WAV (.WAV)   A computer audio file format for Windows. It encodes sound without any data reduction, using pulse code modulation. Its audio resolution is 16-bit, 44.1 kHz, or higher.

WAVEFORM   A graph of a signal's sound pressure or voltage versus time. The waveform of a pure tone is a sine wave.

WAVELENGTH   The physical length between corresponding points of successive waves. Low frequencies have long wavelengths; high frequencies have short wavelengths.

WEBER   A unit of magnetic flux.

WEIGHTED   Referring to a measurement made through a filter with a specified frequency response. An A-weighted measurement is taken through a filter that simulates the frequency response of the human ear.

WINDOWS MEDIA   *See* WMA.

WINDSCREEN   *See* Pop Filter.

WMA (WINDOWS MEDIA AUDIO)   A popular compressed audio file format for streaming audio and for downloads. Windows Media 8 promises performance similar to that of MP3Pro-near-CD quality at 48 kbps and CD quality at 64 kbps.

WOOFER   A low-frequency loudspeaker.

WORD   A binary number that is the value of each sample in an analog-to-digital conversion. A sample is a measurement of an analog waveform that is done several thousand times a second during a PCM A/D conversion.

WORD LENGTH   *See* Bit Depth.

WORKSTATION   A system of MIDI- or computer-related equipment that works together to help you compose and record music. Usually, this system is small enough to fit on a desktop or equipment stand. *See* also Keyboard Workstation and Digital Audio Workstation.

WOW   A slow periodic variation in tape speed.

WRAPPER   Software that converts a plug-in from an unsupported format to a supported format. For example, a DAW recording program (host) that cannot use VST plug-ins directly might include a wrapper that converts VST plug-ins to Direct-X plug-ins.

XLR-TYPE CONNECTOR   An ITT Cannon part number that has become the popular definition for a 3-pin professional audio connector. *See* alsoThree-Pin (3-Pin) Connector.

X-Y   *See* Coincident-Pair.

Y-ADAPTER   A cable that divides into two cables in parallel to feed one signal to two destinations.

Z   Abbreviation for impedance.

ZONE   *See* Region.

INDEX

CD LINER NOTES

Welcome to *Practical Recording Techniques*. This CD demonstrates various topics in the book.

All tracks were engineered by Bruce Bartlett, except for the electric-guitar samples (Track 23) recorded by Steve Mills of S.E.M. Labs. All music on this CD is copyrighted by the original artists and may not be reproduced without permission. Thanks to Joe Probst, Station One, One Word, Ryan Vice, Josh Hilliker, and Grooveside for the use of their recordings.

This CD was produced with Cakewalk® Sonar Producer digital recording software. To provide the best sound quality, this CD was not processed to maximize its level.

Track
1   Introduction
2   Five Ways to Record

Sound, Signals, and Studio Acoustics
3   Amplitude
4   Frequency
5   Harmonics
6   Envelope
7   Frequency Response
8   Noise and Signal-to-Noise Ratio
9   Distortion: Digital distortion and analog-cassette-tape distortion
10   Echoes and Reverberation
11   Leakage
12   Hum and Buzz

Microphones
13   Transducer Type: Condenser, dynamic, ribbon
14   Polar Pattern: Cardioid mic on and off axis; cardioid versus omni at 6 feet
15   Microphone Frequency Response: Flat, contoured
16   Breath Pops: Effect of pop filter, effect of mic placement

Mic-Technique Basics
17   Miking Distance
18   Effects of Mic Placement on Tonal Balance
19   Stereo Miking the Acoustic Guitar—Close and Distant
20   Boundary Mics
21   The Three-To-One Rule
22   Stereo Mic Techniques: Coincident, near-coincident, spaced

Specific Mic Techniques
23   Electric Guitar
24   Drum Set
25   Grand Piano (the PZM recording is a mix of instruments)
26   Lead Vocal

Signal Processors and Effects
27   EQ
28   Compression
29   No Effects on Voice
30   Echo
31   Reverberation
32   Preverb
33   Chorus
34   Flanging
35   Pitch Shift
36   Tube Modeling
37   Analog Tape Modeling
38   Guitar-amp Modeling

Mixdown
39   A sample of the finished mix, individual tracks before and after processing, panning, level balancing, the mix before and after processing and level balancing

Mastering
40   Normalization, peak limiting and normalization, multiband compression

Data Compression
41   Uncompressed audio file, MP3 data compression, WMA data compression

Decibels
42   0 dB, -10 dB, -6 dB, -3 dB, -1 dB

Outro
43   Thanks for listening!

Total running time 43:58.